INFERENCE AND LINEAR MODELS

McGRAW-HILL INTERNATIONAL BOOK COMPANY

New York
St. Louis
San Francisco
Auckland
Beirut
Bogotá
Düsseldorf
Johannesburg
Lisbon
London
Lucerne
Madrid
Mexico
Montreal
New Delhi
Panama
Paris
San Juan
São Paulo
Singapore
Sydney
Tokyo
Toronto

D. A. S. FRASER
University of Toronto
Department of Statistics

Inference and Linear Models

This book was set in Times Roman 327

British Library Cataloging in Publication Data

Fraser, Donald Alexander Stuart
   Inference and linear models.
   1. Mathematical statistics 2. Probabilities
   I. Title
   519.5′4   QA276   77-30563

ISBN 0-07-021910-9

INFERENCE AND LINEAR MODELS

Copyright © 1979 McGraw-Hill Inc. All rights reserved.
Printed in Great Britain. No part of this publication may be
reproduced, stored in a retrieval system, or transmitted in
any form or by any means, electronic, mechanical, photocopying,
recording or otherwise, without the prior permission of the publisher.

1 2 3 4 5 WJM 8 0 7 9

Printed and bound in Great Britain

CONTENTS

Preface		ix
Acknowledgments		xii
1	**Model and Data : The Inference Base**	**1**
	1-1 The System and the Model	2
	1-2 The Model and the Location-Scale Example	5
	References and Bibliography	15
2	**Location-Scale Analysis**	**16**
	2-1 Core Methods of Analysis	17
	2-2 Terminal Methods of Analysis	22
	2-3 Analysis of an Inference Base	26
	2-4 Lifetesting and the Weibull	30
	2-5 Robustness and Resistance	37
	References and Bibliography	47
3	**Necessary Methods**	**49**
	3-1 On the Parameter Space	49
	3-2 On the Sample Space	54
	3-3 Factorization	59
	3-4 By Reexpression	61
	3-5 By Reexpression; a Parameter Component	65
	References and Bibliography	68

4 Density Allocation Methods — 69

- 4-1 Sufficiency Reduction — 70
- 4-2 Ancillarity Reduction — 75
- 4-3 Sufficiency-Ancillarity Reduction — 80
- 4-4 Weak Sufficiency and Ancillarity — 84
- References and Bibliography — 86

5 Terminal Methods of Inference — 89

- 5-1 Tests of Significance — 89
- 5-2 Confidence Intervals — 91
- 5-3 Likelihood — 98
- 5-4 Inference and Decisions — 103
- References and Bibliography — 108

6 The Regression Model — 109

- 6-1 Core Methods of Analysis — 109
- 6-2 Terminal Methods of Analysis — 116
- 6-3 Regression with Serial Correlation — 120
- 6-4 Regression with Nonnormal Variation — 125
- References and Bibliography — 132

7 Coherent Models — 133

- 7-1 The Structural Model — 133
- 7-2 Change of Variable — 139
- 7-3 Inference, Tests and Confidence Regions — 149
- 7-4 Multiple Tests and Confidence Regions — 158
- References and Bibliography — 165

8 Some Multivariate Models — 167

- 8-1 Location-Scale Multivariate Model — 167
- 8-2 Multivariate Model: Progression — 174
- 8-3 Multivariate Model: Normal Progression — 183
- 8-4 Multivariate Model: Linear — 191
- 8-5 Multivariate Model: Normal Linear — 197
- References and Bibliography — 205

9 Distributions on the Circle and Sphere — 206

- 9-1 The Circle — 206
- 9-2 The Sphere — 212
- 9-3 Generalized Distribution Form — 219
- References and Bibliography — 232

10	**Bioassay and Dilution Series**	**233**
	10-1 The Model	234
	10-2 The Analysis: Theory and Examples	240
	References and Bibliography	250
11	**Extended Likelihood Methods**	**251**
	11-1 Some Likelihood Components	252
	11-2 Extended Likelihood	253
	11-3 Group-Based Likelihood and the Transformed Regression Model	260
	References and Bibliography	266
12	**Multivariate Regression Models**	**267**
	12-1 Multivariate Regression Model with Progressive Variation	267
	12-2 Normal Multivariate Regression Model with Progressive Variation	276
	12-3 Multivariate Regression Model with Linear Variation	281
	12-4 Normal Multivariate Regression Model with Linear Variation	287
	References and Bibliography	293
Index		**295**

PREFACE

THE SUBJECT

Statistical *inference* is the process by which conclusions about unknown characteristics and properties of a real world system are reached from background information and current data from an investigation of the system. The starting point for this process is the formal mathematical presentation of the background information together with the data—the *model* with the *data*, called the *inference base*.

Linear Models are models in which some or all of the parameters—representing unknowns of the real world system—have a linear effect on primary variables of the model. The term linear model commonly refers to the standard regression model, perhaps with a parameter for distribution form; here the term covers the same linearity but in more general contexts.

This book examines statistical inference and gives some special attentions to the more fruitful areas in which models have linearity.

THE BACKGROUND

The common statistical model is really just a class of density functions, or even just a class of probability measures. Inevitably, the real world system supports more, even such minimum properties as continuity and relevant metrics. The common model seems a deplorable minimum omitting ingredients that would be viewed essential in component areas of science. Do these ingredients matter in statistics?

The starting point for inference is often taken to be a pair, (E, x), an experiment and an observed response. The term experiment of course, does not cover all investigations. And an investigation as such does not include the background information concerning it. Perhaps the formal process of inference needs a formal starting point.

Many methods exist and are proposed in statistical applications. An appealing and seemingly successful method can invite explanation and plausible grounds. Bayesian priors and procedural principles can then receive reverse support. Are there solid and independent bases for some statistical methods?

Several principles are commonly available for the derivation or support of inference methods. The most prominent is that of sufficiency providing support for many methods associated with normal distributions. Sufficiency, and some related principles, involve allocating sample points to sample-space contours to achieve a type of density-function balance, call this a density-allocation method. The support for some of these methods and the related principles is often derived largely from a few attractive examples. Perhaps the examples are attractive because of some more fundamental properties?

The normal distribution has attractive mathematical and stochastic properties. Part of this is connected with multidimensional rotational symmetry. And as part of this symmetry much of statistical theory centers around the normal distribution. Should not the ubiquitous presence of nonnormal distributions in applications suggest a more broadly based theory covering inference methods?

THE CONTENTS

This book examines statistical inference. Various current reduction methods of inference are surveyed in Chapter 3, and various terminal methods are explored in detail in Chapter 5. Five new necessary reduction methods are developed and discussed in Chapter 4; these allow a broad extension of inference methods to nonnormal statistical models. General methods for building on linearity and extending from linearity are explored in Chapters 7 and 11.

An important part of clarifying the process of statistical inference is being clear about the starting point. Chapter 1 examines the statistical model for a real world system and involves requirements for the *model for an investigation*. This leads to the concept of the *inference base* as the starting point for the process of statistical inference.

In Chapter 2 the model and the inference base are examined in detail for this very basic system, the location-scale system. Inference methods and computer programs are discussed for handling the location-scale case with say a Student family, or a Weibull model, or a numerically presented family—for the basic distribution form. Robustness and resistance properties are discussed in a concluding section.

After the initial survey and development of inference method in Chapters 3, 4, 5, the regression model is used in Chapter 6 as a major illustration. An example is given of the computer analysis of the regression model with Student distribution form and nonnormal data.

The introducing Chapter 1 develops the concept of the inference base as the formal starting point for the process of inference. This clarification of the starting point allows the separation of the given from the arbitrary and leads to the necessary reduction methods in Chapter 4. The advantages of this clarification become apparent with the necessary extensions of current methods to handle the nonnormal cases for location-scale (Chapter 2) and regression (Chapter 6). In these chapters we find an unequivocal separation of variables to handle key component parameters.

A central notion in Chapters 2 and 6 is the objective nature of distribution form in the location-scale and the regression contexts. This objectivity of distribution form is examined in detail in Chapter 7 and various statistical and mathematical methods are developed for models with objective distribution form. The models are structural models.

The general methods of Chapter 7 are then applied to the multivariate model with nonnormal error (Chapter 8), distributions on the circle and sphere (Chapter 9), dilution series and bioassay (Chapter 10), and multivariate regression (Chapter 12). Computer methods and data are examined for a three parameter model on the circle, and for dilution series and bioassay models leading to significant extensions beyond the usual maximum likelihood approach.

Chapter 11 explores some likelihood and extended likelihood methods for estimating parameters of distribution form or response reexpression. Power transformation of the regression model provide an illustration.

In the text we see that the clarifications connected with the definition of the model and of the inference base lead unequivocally to certain conditional methods of analysis and to parameter variable separations. This is illustrated throughout the book with the computer analyses for location-scale, Weibull, extreme value, bioassay, dilution-series, circle and sphere, regression and multivariate problems. A range of new directions are opened.

ACKNOWLEDGMENTS

The material in this book has developed in very close interaction with many colleagues, staff and students at the University of Toronto. From the development of the concepts and ideas through the organization, the write-ups, and the proof reading, my deep appreciation goes to David Brenner, Malcolm Cairns, Douglas Chan, Nick Cheng, Alok Dobriyal, Michael Evans, Gordon Fick, Mike Green, Anna Lubiw, Louis Mao, Hélène Masson, Allen McIntosh, Georges Monette, Rick Moorhouse, Kai W. Ng, Daryl Pregibon, Mark Reimers, Hari Shanmugadhasan, Kathy Sykora, Maureen Trudeau, Martin Wong. My special thanks go to Frances Mitchell for very careful and accurate typing and for proof reading.

CHAPTER
ONE

MODEL AND DATA: THE INFERENCE BASE

In this book we examine statistical inference.

Statistical inference arises in the context of an *investigation* of a real world *system*. The system can be physical, biological, social, or any other system that is controllable to the extent that the system responses are random. The investigation involves a prescription that enunciates what variables are to be examined, what performances are to be made, and what unknowns are the real purpose of the investigation.

There are two initial ingredients for statistical inference: the *specification* of the system and the *data* from the investigation. The specification gives the background information concerning the system; it establishes what *are* and *are not* possible performance patterns for the system—to some reasonable approximation. The data are the response results or values from the current performances of the system, the performances as detailed by the investigation.

Statistical inference, then, is concerned with *what the specification and data say concerning the unknowns of the system*.

This book is concerned with the basic and unifying ideas in statistical inference, with the common threads and purposes. Accordingly, the emphasis is on the wide range of problems involving continuous response variables. This is not to suggest that the discrete problems such as with the binomial, Poisson, multinomial, and hypergeometric are not important, but rather that the continuous problems represent a wide range of possibilities for examining ideas and purposes of inference, and the discrete problems do have some very special simplicities.

1-1 THE SYSTEM AND THE MODEL

We consider further some real world system that is under investigation, a system that is controllable to the extent that the response variable is random.

1-1-1 The System

We find it convenient at times to view the system as a box with input variables to the box and response variables from the box. How the system performs can then be viewed in terms of the response pattern of the system that corresponds to each chosen array of input values to the system. A system is called *random* or *stable* or *under statistical control* if each relevant input pattern gives response performances that can be described by probability distributions; for a recent discussion, see Fraser (1976, Chaps. 1 and 4).

In many texts the term *experiment* is used for what we have called a random system. In scientific contexts, however, the term experiment is used for investigations where controllable variables are changed explicitly and purposefully to gain information on cause-effect relationships. This usage of the term experiment has fundamental importance within science and it is prudent not to dilute its meaning when discussing statistical inference. Accordingly we prefer the term random system and reserve the term experiment for those very specific investigations of a system in which input variables are changed in a designed goal-directed manner.

Typically, a system has many input variables that may affect the response performance of the system. Hopefully these variables will be *controllable* in their input values to the system. The specification then establishes what are and what are not possible performance patterns for the system in relation to values for the controllable input variables. For designated response variables and designated ranges for the controllable input variables a *model for the system* is a mathematical-probabilistic presentation of the specification in a form appropriate to the designated variables and ranges.

For an investigation of the system, the controllable variables can be of two different kinds: the *design* variables that are changed in a planned pattern in different performances of the system; and the remaining controllable variables that are kept constant (*controlled*). The investigation becomes an experiment if the design variables are changed purposefully to gain information on cause-effect relationships—from cause in the design variables to effect in the response variables.

There may, however, be input variables that are not directly controllable. A basic experimental procedure, then, is to use external *randomization* to assign the planned input pattern to the different performances of the system; the effect of the uncontrolled input variables then appears as a random effect on the response.

The external randomization of the input pattern to the different performances of the investigation provides an effective compensation for a certain lack of control of particular input variables. Also, as a component of the investigation, the external randomization is a random system separate from and additional to the basic random system under investigation.

1-1-2 The Model

The specification gives the background information concerning the random system being investigated. The details of the investigation enunciate what variables are to be examined, what performances are to be made, what randomization is to be applied, and what unknowns are the purpose of the investigation.

A specification may provide background information on response variables additional to those under investigation, or on sequences of performances different from that detailed by the investigation. Indeed, the specification may provide information on a wide variety of aspects of the system beyond those detailed by the investigation. Besides, a specification is typically not recorded in a formal mathematical-probabilistic form.

In short, the specification as it stands is not formal and specific to the details of the investigation—with the particular group of variables and performances, with the particular input conditions and randomization pattern, and with the particular unknowns of interest.

Accordingly, the term *model for the investigation* is used for the set of mathematical-probabilistic descriptions needed for the particular variables, performances, randomization, and conditions as detailed by the investigation. The model is not an arbitrary construct; rather it is the specification made formal and specific to the investigation. There are four components to the definition of the model for the investigation:

(1) The model is *descriptive*. The components and variables of the model correspond to objective components and variables for the performances as detailed by the investigation. Thus the components and variables are real, not arbitrary.

(2) The model is *exhaustive*. There is a component or variable in the model that corresponds to each objective component or variable for the performances as detailed by the investigation. Thus the description is full, not partial.

(3) The model is *probabilistic*. Without probability, inference is trivial and not statistical. By probabilistic we mean that the use of the model and its components conforms to the requirements of probability theory. In a sense this is already covered by the term descriptive. However, two requirements of probability theory are often overlooked and accordingly should be made explicit. Both involve conditional probability.

The *first* requirement is concerned with the observed value of an objective variable with known objective probabilities: a value on a component probability space. The requirement then is that all probability descriptions be conditional probabilities given the observed value.

The *second* requirement is concerned with information about an objective variable, information that takes the form of an observed value of a function of the variable. The requirement then is that the marginal probability description be used for the observable function and the conditional probability description be used for the variable itself given the observed value of the function.

(4) The detailed investigation of a random system is concerned with unknown characteristics of the system—unknown characteristics of the distributions or unknown characteristics of the relations among the variables. Accordingly the

model is the *set* of possible descriptions for the investigation thus embracing the range of possibilities for the unknown characteristics—and such additional descriptions as are needed for an internally *consistent* model.

For the presentation of the model in notation, it is important to have an index set for the set of available descriptions. Often we will use $\Omega = \{\theta\}$ to designate this index set and refer to a *free* variable θ on Ω as the parameter of the model. For an application, the inference problem is to determine what can be said concerning the *true value* of θ, that value of θ that designates the particular true description for the system—of course to some reasonable approximation.

We will frequently use the letter $\mathcal{M}$ to designate a model, adding subscripts or superscripts as needed. The model then has the form

$$\mathcal{M} = (\Omega; \ldots) \qquad (1\text{-}1)$$

where Ω is the index set and the remainder is the set of descriptions indexed by θ in Ω.

Note that if we were to use some transformed response in place of the initial response then, typically, each of the descriptions would change accordingly. *The index and index set would*, however, *remain the same*. For this recall that a particular one of the values in Ω is the *true value* of θ, the value that designates the true description for the response.

In conclusion we note that for our purposes here it is convenient and appropriate to restrict attention to models in which probability is given in terms of density functions—either with respect to a volume measure on real spaces, or with respect to a counting measure, or with respect to a mixture of the preceding; this gives a wide range of practical generality. Also, in conformity with the requirement that a model be descriptive, we assume that the density functions are density functions in the obvious limiting sense.

1-1-3 The Inference Base

We have obtained the model $\mathcal{M}$ by formalizing the relevant material from the specification and making it specific to the investigation. The model, then, is the set of descriptions needed for the particular investigation of the system: descriptive, exhaustive, probabilistic and consistent.

The investigation produces the outcomes for the variables on the particular performances examined. The *data* for the investigation are the values for the observable variables as designated by the model. We will use the letter $\mathcal{D}$ for the data.

We have now formalized the ingredients for statistical inference: the model $\mathcal{M}$ and the data $\mathcal{D}$. The combination of the model and the data is called the *inference base* and is designated

$$\mathcal{I} = (\mathcal{M}, \mathcal{D}) \qquad (1\text{-}2)$$

The model and data as initially assembled to form an inference base may contain arbitrary elements, elements beyond those specified by the definitions we have been discussing. In referring to the combination as the inference base we will be referring to the substance and ignoring the arbitrary. One of our first concerns will be the elimination of arbitrary elements from the inference base as presented; we consider this in detail in Chap. 3.

For the remainder of this chapter we investigate in detail the formation of a model. For a central illustration we consider a real-valued response with unknown location and unknown scaling and yet with a distribution for variation that is known or known up to a shape parameter, say λ.

In Chap. 2 we then explore key methods of inference for this particular location-scale illustration.

1-1-4 Personal Preferences

The specification and the model have established what *are* and what *are not* possible performance patterns for the system under investigation.

An investigator, however, may have personal feelings and preferences concerning the various possibilities allowed by the specification and the model. For himself—at a given place and time—he may want to combine these personal preferences with the results obtained from an inference base. In such a case he would be resorting to a terminal procedure, a decision procedure for a specific person at a specific place and time. We examine this briefly in Sec. 5-4 on terminal decision methods.

The available statistical methods at any given time seem almost always to fall short of the expectations of theoretical and applied statisticians. In certain areas this can be due in part to the lack of developed methods or even the absence of any method. Such shortcomings, however, can be a powerful stimulus to search for new methods.

A statistician should feel free to explore any plausible method. Certainly just about anything can be proffered as a method. A method, however, is only as good as its validity and reliability, and these qualities depend on properties deduced from the model or determined pragmatically. Accordingly, we find no place for a formal input of personal preferences to an inference base. The inference base as presented thus remains the basis for the objective assessment of possibilities in relation to data.

The preceding formally excludes Bayesian methods from the discussion of statistical inference in this book. However, we do not exclude an area that is often called *empirical Bayes*. This latter has an objective distribution for an inaccessible variable. In our view this is not connected in any substantive way with Bayesian methods proper, and the title empirical Bayes is inaccurate. We view this special area as involving standard probability and statistical modeling subject to the requirements we have been discussing.

1-2 THE MODEL AND THE LOCATION-SCALE EXAMPLE

In the remainder of this chapter we discuss the formation of a statistical model for a random system. To do this we focus our discussion on an important and

6 INFERENCE AND LINEAR MODELS

common type of system, a system with a real-valued response. As a special case of this we then consider a response with unknown location, unknown scaling, and unknown distribution form, say within some parametric family.

In Chap. 2 we examine in detail statistical inference for the location-scale case. We then return in Chap. 7 to discuss in a more general way the formation of a statistical model for a random system.

1-2-1 The Model : Response Based

For our central example we consider the formation of a statistical model for a random system with a real-valued response. We picture ourselves in the typical situation where density functions with respect to length measure provide a reasonable approximation for the distributions. In particular we examine the case with unknown location, unknown scaling, and unknown distribution form within some parametric family.

For our example, the random system has a real-valued response; let y designate a particular expression of this response. The specification of the random system allows a spectrum of distributions for this response y; let θ in Ω be a parameter indexing these possible distributions for y and let $f(y|\theta)$ be the density function for y corresponding to the index value θ. The range Ω for the parameter θ would be such as to give all the possible distributions for the response y, to some reasonable approximation, of course.

The parameter θ indexing the spectrum of possible distributions for y could be real or vector valued. For the special case with unknown location scaling and form, we might reasonably use an index θ that had separate coordinates for these different characteristics; for example, we could use $\theta = (\mu, \sigma, \lambda)$ where μ referred to location in some way, σ to scaling in some way, and λ to distribution form.

We now form a model based on the set of possible density functions for the response. As a minimum *response-based model* $\mathcal{M}_R$ we then have

$$\mathcal{M}_R = (\Omega; \mathbb{R}, \mathcal{B}, \mathcal{F}) \tag{1-3}$$

where Ω is the parameter space for the index θ, $\mathcal{B}$ is the class of Borel sets, and

$$\mathcal{F} = \{f(y|\theta) : \theta \in \Omega\} \tag{1-4}$$

is the class of density functions for the particular expression of the response variable. Informally we can write the model as

$$f(y|\theta)\, dy \tag{1-5}$$

For the special case with unknown location scaling and form, the class of density function could be written

$$\mathcal{F} = \{f(y|\mu, \sigma, \lambda) : \mu \in \mathbb{R}, \sigma \in \mathbb{R}^+, \lambda \in \Lambda\} \tag{1-6}$$

where, as mentioned earlier, μ refers to location in some way, σ to scaling in some way, and λ to distribution form; the full parameter space is $\Omega = \mathbb{R} \times \mathbb{R}^+ \times \Lambda$.

The use of density functions implicitly assumes some differential continuity of the line ℝ as the sample space. It also assumes some length or support measure for the densities. For the model (1-3) with $\mathscr{F}$ as given by (1-4), the particular choice of length measure could be arbitrary, not just the ordinary and familiar length measure suggested by (1-5). Alternatively a particular choice of length measure could be explicitly included in the model. For this recall the definition of the model in Sec. 1-1-2.

In general the statistical model includes all the relevant available properties concerning the system. Any such properties would be appended to the minimum model (1-3) giving

$$\mathscr{M}_R = (\Omega; \mathbb{R}, \mathscr{B}, \mathscr{F}, \ldots) \tag{1-7}$$

This allows the formalization of properties that are often implicitly used in analyses though rarely stated explicitly. For example, the Euclidean distance measure, in certain cases, could be explicitly included.

Note that we are following the pattern in Sec. 1-1-2 and have separately recorded the index Ω in the model $\mathscr{M}$ in (1-3) or (1-7). Also, for the class $\mathscr{F}$ of possible density functions, we assumed that each density function retains its particular θ value as an index. One possible way of seeing the significance of this is in terms of response reexpressions or transformations. Suppose the investigator chooses to think in terms of some alternative presentation

$$\tilde{y} = h(y) \tag{1-8}$$

for the response where h is, say, a monotone increasing continuously differentiable function. The density function for $\tilde{y}$ corresponding to the index θ is

$$g(\tilde{y} \mid \theta) = f(h^{-1}(\tilde{y}) \mid \theta) \left| \frac{dh^{-1}(\tilde{y})}{d\tilde{y}} \right| \tag{1-9}$$

As the minimum model relative to the alternate response expression we then have

$$\tilde{\mathscr{M}}_R = (\Omega; \mathbb{R}, \mathscr{B}, \tilde{\mathscr{F}}) \tag{1-10}$$

where Ω is the same indexing set as before and

$$\tilde{\mathscr{F}} = \{g(\tilde{y} \mid \theta) : \theta \in \Omega\} \tag{1-11}$$

is the class of densities for the alternative response expression. For some particular investigation there is of course a true value for θ; this value of θ indexes the true density in $\mathscr{F}$ for the given response expression and correspondingly the true density in $\tilde{\mathscr{F}}$ for the alternate response expression. To emphasize this point we note explicitly that θ and Ω do not change under a transformation of the response.

In a typical investigation we would of course be concerned with multiple performances of the system. The minimum *response-based model* $\mathscr{M}_R^n$ for the compound response $\mathbf{y} = (y_1, \ldots, y_n)'$ from n independent performances is obtained by direct compounding:

$$\mathscr{M}_R^n = (\Omega; \mathbb{R}^n, \mathscr{B}^n, \mathscr{F}^n) \tag{1-12}$$

where $\mathscr{B}^n$ is the Borel class on $\mathbb{R}^n$ and $\mathscr{F}^n$ is the class

$$\mathscr{F}^n = \{\Pi_1^n f(y_i|\theta): \theta \in \Omega\} \quad (1\text{-}13)$$

Again we, of course, assume that each density in the class retains its θ value as a label.

Now suppose that for the n performances of the system we obtain the following data: $\mathbf{y}^0 = (y_1^0, \ldots, y_n^0)'$. We then have the model $\mathscr{M}_R^n$ giving what is known about the compound system and the data $\mathscr{D} = \mathbf{y}^0$ giving what was obtained from the performances: this gives the inference base

$$\mathscr{I} = (\mathscr{M}_R^n, \mathbf{y}^0) \quad (1\text{-}14)$$

for making inferences concerning the true value of the parameter θ in Ω.

1-2-2 Distribution Form

Now consider further the special case of a system with a real-valued response y with unknown location, unknown scaling, and either known or unknown distribution form. First suppose the distribution form is known.

Let f be the density function for the response expressed in some suitable relocated and rescaled manner. The various possible one–one equivalent presentations for the response then give the class

$$\mathscr{C} = \{c^{-1}f(c^{-1}(y-a)): a\in\mathbb{R}, c\in\mathbb{R}^+\} \quad (1\text{-}15)$$

with arbitrary relocation a in $\mathbb{R}$ and rescaling c in $\mathbb{R}^+ = (0, \infty)$. There is just one true distribution for the response, f is its density in some suitable standard presentation, and $\mathscr{C}$ is the class for the one–one equivalent presentations.

Now let σ be a parameter designating the scaling for the response distribution and μ be a parameter for the location. If we take some density f in the class $\mathscr{C}$ as a standard, then the density for location μ and scaling σ is

$$\sigma^{-1}f(\sigma^{-1}(y-\mu)) \quad (1\text{-}16)$$

The meaning or interpretation that we give to μ and σ will clearly depend on what density f in $\mathscr{C}$ we choose as the standard. Of course, in general, to work with an equivalence class it is useful and convenient to have a *representative* from the class—often a representative standardized in some simple and easily understood manner.

One possibility that first appears is to standardize with respect to mean and standard deviation: thus, we would use the representative f having

$$\int_{-\infty}^{\infty} zf(z)\,dz = 0 \qquad \int_{-\infty}^{\infty} z^2 f(z)\,dz = 1 \quad (1\text{-}17)$$

For the response distribution (μ, σ) given by (1-16) we would then interpret μ as the mean and σ as the standard deviation—for the particular response expression. This standardization would not, of course, work for long-tailed distributions such as the Cauchy.

Another possibility that has particular intuitive appeal for symmetric distributions is to standardize with respect to the median and standard error: that is, we would use the representative f having

$$\int_{-\infty}^{0} f(z)\,dz = 0.5 \qquad \int_{-1}^{+1} f(z)\,dz = 0.6826 \qquad (1\text{-}18)$$

For the normal this agrees with the first standardization (1-17). For the response distribution (μ, σ) given by (1-16) we would then interpret μ as the median and σ as the standard error.

Another possibility that seems appealing for asymmetric distributions and agrees with the preceding for symmetric distributions is to standardize so that $(-1, +1)$ is a central 68.26 percent interval: i.e., we would use the representative f having

$$\int_{-\infty}^{-1} f(z)\,dz = 0.1587 = \int_{1}^{\infty} f(z)\,dz \qquad (1\text{-}19)$$

For the response distribution (μ, σ) we would then interpret σ as the standard error and μ as the midpoint of the central 68.26 percent interval. Other standardizations may be suitable or appropriate depending on the particular distribution form.

We have considered several ways of choosing a standardized representative from the equivalence class $\mathscr{C}$ in (1-15). Note, however, that an element in the class has the form $c^{-1}f(c^{-1}(y - a))$ for some a, c; and yet we have casually referred to the representative as f and implicitly assumed that this f when used in the expression (1-15) would generate the equivalence class as initially given. This turns out to be true, for we can see that a location-scale transformation of a location-scale transformation *is* a location-scale transformation. This closure property is important from several viewpoints and we now make it more explicit.

To do this we find it useful to introduce specialized notation for location-scale transformations. Let $[a, c]$ designate the transformation of $\mathbb{R}$ to $\mathbb{R}$ given by

$$[a, c]y = a + cy \qquad (1\text{-}20)$$

or of $\mathbb{R}^n$ to $\mathbb{R}^n$ given by applying the preceding, coordinate by coordinate: that is,

$$[a, c]\mathbf{y} = [a, c]\begin{pmatrix} y_1 \\ \vdots \\ y_n \end{pmatrix} = \begin{pmatrix} a + cy_1 \\ \vdots \\ a + cy_n \end{pmatrix} = a\mathbf{1} + c\mathbf{y}$$

where $\mathbf{1} = (1, \ldots, 1)'$ is the appropriate one-vector. For two transformations applied in succession the composition rule is easily verified as

$$[A, C]\,[a, c] = [A + Ca, Cc] \qquad (1\text{-}21)$$

and the inverse transformation is easily seen to be

$$[a, c]^{-1} = [-c^{-1}a, c^{-1}] \qquad (1\text{-}22)$$

We can then note immediately that the class

$$G = \{[a, c] : a \in \mathbb{R}, c \in \mathbb{R}^+\} \tag{1-23}$$

is *closed* under the formation of products and inverses: the product of two elements of G is in G, and the inverse of an element of G is in G. This formalizes the concluding remarks in the preceding paragraph, and gives justification for referring to (1-15) as an equivalence class of similarly shaped densities.

In the more general case in which we have a shape or form parameter λ we would have an equivalence class $\mathscr{C}_\lambda$ for each λ and we would choose a representative or standardized density f_λ for each distribution form λ. The parameter λ could be a real parameter allowing thicker tails as, for example, with the degrees of freedom λ for the Student family. Or it could be a two-dimensional parameter allowing different tail thicknesses for the two tails of the distribution. For notation we will use f_λ for the *standardized* representative from the class (1-15) for the particular distribution form.

1-2-3 Observing Distribution Form

We consider further the special case of a system with a real-valued response y with unknown location μ, with unknown scaling σ, and with distribution form that is known or known up to a shape parameter λ—in a space, say Λ. Also we continue with the notation in the preceding section and let f_λ be the density for the response in some suitable standard presentation, and $\mathscr{C}$ be the equivalence class

$$\mathscr{C}_\lambda = \{c^{-1} f_\lambda(c^{-1}(y - a)) : a \in \mathbb{R}, c \in \mathbb{R}^+\} \tag{1-24}$$

for the one–one equivalent presentations given λ. The full class of possible densities for the response variable is of course

$$\mathscr{F} = \{f(y \mid \mu, \sigma, \lambda) : \mu \in \mathbb{R}, \sigma \in \mathbb{R}^+, \lambda \in \Lambda\}$$
$$= \{\sigma^{-1} f_\lambda(\sigma^{-1}(y - \mu)) : \mu \in \mathbb{R}, \sigma \in \mathbb{R}^+, \lambda \in \Lambda\} \tag{1-25}$$

For any particular application there is just *one* true distribution. To what extent can this distribution be observed or identified directly from the response variable? More particularly we consider the question whether the distribution form can be observed or identified apart from its location and scaling? In other words is the distribution form something objective in itself?

Consider a response sample $(y_1, \ldots, y_n)$ from an application. Let μ and σ designate the true values for the location and scaling and λ be the value for the distribution form. Then

$$([\mu, \sigma]^{-1} y_1, \ldots, [\mu, \sigma]^{-1} y_n) = \left(\frac{y_1 - \mu}{\sigma}, \ldots, \frac{y_n - \mu}{\sigma} \right) \tag{1-26}$$

is the corresponding sample from the standard density f_λ. The corresponding equivalence class under location-scale change is

$$\{([a, c] [\mu, \sigma]^{-1} y_1, \ldots, [a, c] [\mu, \sigma]^{-1} y_n) : a \in \mathbb{R}, c \in \mathbb{R}^+\}$$

which can, of course, be rewritten as

$$\{([A, C]y_1, \ldots, [A, C]y_n): A \in \mathbb{R}, C \in \mathbb{R}^+\} \tag{1-27}$$

In this second form, we see that the equivalence class for samples does *not* involve the $[\mu, \sigma]$; this is a direct consequence of the closure property noted with formula (1-23). Thus (1-27) provides direct estimation of (1-24) and thereby also for the density f_λ.

From one point of view we can think of the location-scale transformations as providing a movable platform that allows us to observe and identify the distribution form apart from its location and scaling. A simple analogy might be a car somewhere on a highway. This could be conceptualized in terms of a car at each of the possible locations. The location transformations, however, used in the manner we have been discussing provide a moving platform that allows us to observe and identify the car apart from its location.

We have seen that the distribution form f_λ is directly observable and identifiable apart from response location and scaling. We will refer to this distribution form f_λ as the *distribution for the variation* and we refer to a corresponding variable, say z, as the *variation*. The variation is an objective characteristic of a location-scale system.

We return in Chap. 7 to a more formal and detailed study of this identifiability of distribution form. We will determine the precise criteria needed for this identifiability and we will see that the closure properties—group properties—are central.

1-2-4 The Model : Variation Based

Consider the formation of a statistical model for a system with a real-valued response y; we examine the special case with unknown location μ, unknown scaling σ, and distribution form either known or known up to a shape parameter λ. For this we freely use the notation and results from Secs. 1-2-2 and 1-2-3.

From Sec. 1-2-3 we have that the distribution form, the distribution f_λ for variation, is an objective characteristic of the system. Then, in accord with Sec. 1-1-2, the variation must be an explicit component of the statistical model. Accordingly we let z be a variable for the variation and f_λ be the distribution for the variation. The response variable y is then a location-scale presentation $[\mu, \sigma]$ of the variation z:

$$y = [\mu, \sigma]z = \mu + \sigma z \tag{1-28}$$

In order to write the model formally we let

$$\mathcal{T} = \{[\mu, \sigma] : \mu \in \mathbb{R}, \sigma \in \mathbb{R}^+\} \tag{1-29}$$

be the class of location-scale transformations (or presentations) $[\mu, \sigma]$ with μ in $\mathbb{R}$ and σ in $\mathbb{R}^+$. We also let

$$\mathcal{V} = \{f_\lambda(z) : \lambda \in \Lambda\} \tag{1-30}$$

be the class of density functions for the variation z. We then have the following minimum *variation-based model*:

$$\mathcal{M}_V = (\Omega; \mathbb{R}, \mathcal{B}, \mathcal{V}, \mathcal{T}) \tag{1-31}$$

where the parameter space Ω is $\{(\mu, \sigma, \lambda)\} = \mathbb{R} \times \mathbb{R}^+ \times \Lambda$ and $\mathcal{B}$ is the Borel class on the line $\mathbb{R}$. Note that we have, of course, separately recorded the parameter space Ω; we also assume that each density in $\mathcal{V}$ and each transformation in $\mathcal{T}$ retains its λ and (μ, σ) value as a label. Informally we can write the model as

$$y = \mu + \sigma z \qquad f_\lambda(z) \tag{1-32}$$

Now consider the model for multiple performances of the system. Let $\mathbf{y} = (y_1, \ldots, y_n)'$ designate the compound response variable and $\mathbf{z} = (z_1, \ldots, z_n)'$ designate the corresponding compound variation. We then have the following minimum *variation-based model*:

$$\mathcal{M}_V^n = (\Omega; \mathbb{R}^n, \mathcal{B}^n, \mathcal{V}^n, \mathcal{T}) \tag{1-33}$$

where, for example,

$$\mathcal{V}^n = \{\Pi f_\lambda(z_i): \lambda \in \Lambda\} \tag{1-34}$$

is the class of densities for the compound variation and $\mathcal{T}$ is as before but applies coordinate by coordinate to $\mathbf{z}$ and thus presents the response $\mathbf{y}$. Informally we can write this as

$$\mathbf{y} = \mu\mathbf{1} + \sigma\mathbf{z} \qquad \Pi f_\lambda(z_i)\, d\mathbf{z} \tag{1-35}$$

but with the interpretations as given with the formal model.

Now suppose we have n performances with the data $\mathbf{y}^0 = (y_1^0, \ldots, y_n^0)'$ corresponding to the variation $\mathbf{z}^0 = (z_1^0, \ldots, z_n^0)'$. The model and data then give the inference base

$$\mathcal{I} = (\mathcal{M}_V^n, \mathbf{y}^0) \tag{1-36}$$

In Sec. 1-2-3 we saw how the variation was an objective characteristic of the system. In this section we have followed the criteria in Sec. 1-1-2 and have introduced the variation as an explicit component of the model. Interestingly enough this gives us a model that is simpler by a whole order of magnitude, for with a specified value λ_0 we have a model containing a *single* distribution that describes the variation. By contrast the corresponding response-based model contains a doubly infinite class of different distributions $f(y|\mu, \sigma, \lambda_0)$. We have obtained a much simpler model by properly modeling components of the location-scale system.

1-2-5 The Model: Pivotal Types

Consider further our system with a real-valued response, with unknown location μ and scaling σ and with distribution form that is known up to a shape parameter λ. In this section we survey some other models that have been proposed for the location-scale system. Then in Sec. 1-2-6 we make some comparisons among these models.

As a *pivotal-type location-scale model* consider the following:

$$\mathcal{M}_P = (\Omega; \mathbb{R}, \mathcal{B}, p, \mathcal{E}) \tag{1-37}$$

where $\Omega = \{(\mu, \sigma): \mu \in \mathbb{R}, \sigma \in \mathbb{R}^+\}$ is the location-scale parameter space, $\mathbb{R}$ is the real line with Borel class $\mathcal{B}$ for the response variable y, p is a function

$$p(y, \mu, \sigma) = \frac{y - \mu}{\sigma} \tag{1-38}$$

called a *pivotal function* that maps $\mathbb{R} \times \Omega$ into the *pivotal space* $\mathbb{R}$, and $\mathcal{E}$ is a class of distributions given, say, by density functions on the pivotal space $\mathbb{R}$.

The preceding differs in form from the basic pattern specified in Sec. 1-1: the "parameter" space Ω includes only the location-scale parameters μ and σ and omits an index λ for the distributions recorded in the class $\mathcal{E}$. This is a difference in organization coupled with the notational detail of omitting an index set Λ for the distributions in $\mathcal{E}$.

Some close connections can be established between the variation-based model $\mathcal{M}_V$ in (1-31) and the pivotal-type model $\mathcal{M}_P$ in (1-37). For example, the pivotal function (1-38) provides the inverses of the transformations in the class $\mathcal{T}$ in (1-29); the class $\mathcal{E}$ of pivotal distributions corresponds to the class $\{f_\lambda : \lambda \in \Lambda\}$ of distributions describing variation.

Dempster (1966) has proposed a model that is closer to the variation-based model $\mathcal{M}_V^n$ than to the compound version $\mathcal{M}_P^n$ of the present pivotal model. In the pattern of Dempster's model, values would occur for $y_1, \ldots, y_n, \mu, \sigma$ such that

$$\left(\frac{y_1 - \mu}{\sigma}, \ldots, \frac{y_n - \mu}{\sigma} \right) \tag{1-39}$$

would be a sample from a fixed distribution; for the case examined, the pivotal class $\mathcal{E}$ would consist of a single distribution. There is no requirement of a distribution for the response y; just that from $(y_1, \ldots, y_n, \mu, \sigma)$ there should be a "marginal" distribution for the pivotal expression (1-39).

Dempster (1966), in fact, examined the location version of the model as just presented; the obvious ingredients for the location case are $(y_1, \ldots, y_n, \mu)$ and the location relation $y_i - \mu$. The preceding material records the location-scale version of the model.

An earlier version (Dempster, 1963) assumed a joint distribution for $(y_1, \ldots, y_n, \mu, \sigma)$ but specified only the particular marginal of the pivotal expression (1-39).

Beran (1971, 1972) examined in considerable detail a model that blended components of the variation-based and Dempster models. The location-scale version of the Beran model has the following form: values of z are obtained from a fixed distribution which could be a single distribution forming the class $\mathcal{E}$ in (1-37); the observable response values for y are obtained from z by a mapping

$$y = \mu + \sigma z \tag{1-40}$$

involving parameters μ, σ. Beran tended to emphasize a distribution for the response y, and there seemed to be less freedom for having a distribution for (μ, σ)

than with Dempster. The Beran analysis covered the continuous case of interest here and provided more detail than the Dempster analysis.

Barnard (1974) proposed a model of the pivotal type (1-37). Various definitions were introduced to relate the model to various inference methods commonly used with the response-based model in Sec. 1-2-1. The notion of a pivotal function being maximal in relation to other pivotal functions was introduced.

Pivotal functions were considered by Fisher (1930, 1933) in his early papers concerned with the construction and inversion of tests of significance. Their use typically appeared at the terminal stages of inference and not at the model presentation stage.

1-2-6 Some Comparisons

We have developed the response-based model in Sec. 1-2-1 and the variation-based model in Sec. 1-2-4, all with attention to the criteria in Sec. 1-1-2. In particular we have emphasized that the components of the model be objective—observable from the particular real world system being examined.

The response-based model is a common and familiar model used in statistics. The variation-based model in Sec. 1-2-4 is a structural model (Fraser, 1965a, 1965b, 1966, 1967, 1968a, 1968b) but presented in a more general context of applications. The identifiability of the distribution for variation is closely tied to the identifiability of events for variation by means of a response sample; some discussion on this related topic may be found in Fraser (1971, 1976, p. 162) and Brenner and Fraser (1977). We return to this question in Chap. 7.

For the pivotal-type models in Sec. 1-2-5 we presented a brief survey but did not discuss any questions connected with the objective nature of the components of the models. Indeed the various papers proposing these models have largely avoided such descriptive characteristics and have presented a model as a possible model for a system, as a construct, without attention to the requirements we have assembled in Sec. 1-1-2. In particular the papers have not discussed the origins for the pivotal functions.

For a simple illustration consider a location-scale system involving normal distributions: the response y is normal (μ, σ). The function

$$\frac{y - \mu}{\sigma}$$

is a pivotal function with standard normal distribution. But also the function

$$p(y, \mu, \sigma) = \begin{cases} \dfrac{y - \mu}{\sigma} & \text{if } |y - \mu| < \sigma \\ \dfrac{\mu - y}{\sigma} & \text{if } |y - \mu| \geq \sigma \end{cases}$$

is a pivotal function with standard normal distribution. Indeed a wealth of possibilities exist. Of course some seem nicer than others.

The essential question, however, is not whether a component of a construct or a model is nice but whether the component is objective.

Accordingly, we refer back to the requirements for a model as assembled in Sec. 1-1-2. For the location-scale system these give us the variation-based model in Sec. 1-2-4.

REFERENCES AND BIBLIOGRAPHY

Barnard, G. A.: Conditionality, Pivotals and Robust Estimation (with Discussion), in O. Barndorff-Nielsen, P. Blaesild, and G. Schou (eds), "Proc. of the Conference on Foundation Questions in Statistical Inference". Department of Theoretical Statistics. Aarhus University. pp. 61–80, 1974.

Beran, R. J.: On Distribution-Free Statistical Inference with Upper and Lower Probabilities, *Ann. Math. Stat.*, vol. 42, pp. 157–168, 1971.

————: Upper and Lower Risks and Minimax Procedures. *Proc. Sixth Berkeley Symposium*, vol. 1, pp. 1–16, 1972.

Brenner, D., and D. A. S. Fraser: "When is a Class of Functions a Function?" Department of Statistics, University of Toronto, 1977.

Dempster, A. P.: On Direct Probabilities, *J. Roy. Stat. Soc.*, ser. B, vol. 25, pp. 102–107, 1963.

————: New Methods for Reasoning Toward Posterior Distributions Based on Sample Data, *Ann. Math. Stat.*, vol. 37, pp. 355–374, 1966.

————: Upper and Lower Probabilities Induced by a Multivalued Mapping, *Ann. Math. Stat.*, vol. 38, pp. 325–339, 1967.

————: Upper and Lower Probability Inferences Based on a Sample from a Finite Univariate Population, *Biometrika*, vol. 54, pp. 515–528, 1967.

Fisher, R. A.: Inverse Probability, *Proc. Camb. Phil. Soc.*, ser. 2, vol. 26, pp. 528–535, 1930.

————: The Concepts of Inverse Probability and Fiducial Probability Referring to Unknown Parameters, *Proc. Roy. Stat. Soc.*, ser. A, vol. 139, pp. 343–348, 1933.

Fraser, D. A. S.: On Information in Statistics, *Ann. Math. Stat.*, vol. 36, pp. 890–896, 1965a.

————: "Lecture Notes on Statistical Inference," University of Toronto, Toronto, 1965b.

————: Structural Probability and a Generalization, *Biometrika*, vol. 53, pp. 1–9, 1966.

————: Data Transformations and the Linear Model, *Ann. Math. Stat.*, vol. 38, pp. 1456–1465, 1967.

————: A Black Box or a Comprehensive Model, *Technometrics*, vol. 10, pp. 219–229, 1968a.

————: "The Structure of Inference," Krieger Publishing Company, Huntington, N.Y., 1968b.

————: Events, Information Processing, and the Structural Model, in V. P. Godambe and D. A. Sprott (eds.), "Proc. Symposium on the Foundations of Statistical Inference," Holt, Rinehart and Winston of Canada, Toronto and Montreal, pp. 32–55, 1971.

————: "Probability and Statistics, Theory and Applications," Duxbury Press, North Scituate, Mass., 1976.

CHAPTER
TWO

LOCATION-SCALE ANALYSIS

In this chapter we consider the analysis of data from a real-valued response with unknown location and scaling but with a distribution pattern for the variation that is known or known up to a shape parameter λ. Our starting point, then, as formalized in Chap. 1, is an inference base $\mathscr{I} = (\mathscr{M}, \mathscr{D})$ consisting of a model $\mathscr{M}$ as in Sec. 1-2 and data $\mathscr{D}$.

We examine two sets of empirical data: the Darwin data recorded in Fisher (1960, p. 37) on the difference in heights of cross- and self-fertilized plants, and some industrial lifetime data usually analyzed with a Weibull model. We also examine some computer generated data.

The common analysis of location-scale data is based on the assumption of a normal, or approximately normal, model. For the normal model a sample of n leads almost unequivocally to the Student $(n-1)$ distribution for the t-statistic and the chi $(n-1)$ distribution for the residual length; these provide the basis for inference for the location parameter μ and the scale parameter σ. The normal, of course, has some very special symmetries and simplicities that lead very easily to the preceding Student and chi analyses.

For the situation without the normality assumption, various methods have been proposed. For example, the sign-test procedure leads to tests of significance and confidence intervals for the location parameter μ; the procedure is valid for Λ as large as all continuous distributions. In general, however, such methods do not provide for estimates or inference for the shape parameter λ in Λ.

The sign-test and its confidence procedure are nonparametric procedures. There are many such nonparametric procedures both for the location parameter μ and for the scale parameter σ. Selections among these can sometimes be made on the basis of apparent better performance under certain distribution forms of interest.

Various estimators of location and scale parameters have been developed as part of the study of robustness. These methods, however, are targeted primarily on estimation and not on general tests and confidence intervals. From the viewpoint developed in this book, estimation is largely a terminal decision procedure and is not in itself central to inference. This viewpoint is discussed further in Sec. 5-4.

In this chapter we first examine some core methods of inference for the location-scale inference base. We then examine some terminal methods of inference that build on the core methods. We follow this with an examination of two sets of empirical data. The chapter concludes with a discussion of the resistance and robustness properties of the procedures; these are examined by means of computer simulations.

2-1 CORE METHODS OF ANALYSIS

Consider the inference base $\mathscr{I} = (\mathscr{M}, \mathscr{D})$ where $\mathscr{M}$ is a location-scale model, one or a blend of the models $\mathscr{M}_R$, $\mathscr{M}_V$, $\mathscr{M}_P$ discussed in Sec. 1-2, and $\mathscr{D}$ is an observed-response vector $\mathbf{y}^0 = (y_1^0, \ldots, y_n^0)'$.

2-1-1 Preliminaries

For each of the models we have that

$$[\mu, \sigma]^{-1}\mathbf{y} = \sigma^{-1}(\mathbf{y} - \mu\mathbf{1}) = \mathbf{z} \tag{2-1}$$

has the distribution of a sample from a distribution in

$$\mathscr{V} = \{f_\lambda(z) : \lambda \in \Lambda\} \tag{2-2}$$

Under the model $\mathscr{M}_V$ the vector $\mathbf{z}$ describes the *objective* variation in the system being investigated; *under the model* $\mathscr{M}_R$ the vector $\mathbf{z}$ provides a standardized way of describing the distribution form for the response; and *under the models* $\mathscr{M}_R$ and $\mathscr{M}_P$ the vector $\mathbf{z}$ is a pivotal quantity that has some preferential properties in relation to other pivotal quantities.

The inference base provides an observed value $\mathbf{y}^0$; let $\mathbf{z}^0$ be the corresponding realized value for $\mathbf{z}$ given by (2-1). We ask: how much of $\mathbf{z}^0$ is identifiable from $\mathbf{y}^0$ without information as to values of μ, σ? The equation (2-1) with realized values inserted gives

$$\mathbf{z}^0 = [\mu, \sigma]^{-1}\mathbf{y}^0 = -\sigma^{-1}\mu\mathbf{1} + \sigma^{-1}\mathbf{y}^0$$

This identifies $\mathbf{z}^0$ as a point on the half two-space given by

$$\mathscr{L}^+(\mathbf{1}; \mathbf{y}^0) = \{a\mathbf{1} + c\mathbf{y}^0 : a \in \mathbb{R}, c \in \mathbb{R}^+\}$$

$$= \{[a, c]\mathbf{y}^0 : a \in \mathbb{R}, c \in \mathbb{R}^+\} \tag{2-3}$$

which is a half plane subtended by the line $\mathscr{L}(\mathbf{1})$ and passing through the observed $\mathbf{y}^0$. Thus the *possible* values for $\mathbf{z}^0$ are not n dimensional but are essentially two dimensional. We can present this more formally by saying that we have observed the value of the function $\mathscr{L}^+(\mathbf{1};\mathbf{z})$ of $\mathbf{z}$:

$$\mathscr{L}^+(\mathbf{1};\mathbf{z}) = \mathscr{L}^+(\mathbf{1};\mathbf{z}^0) = \mathscr{L}^+(\mathbf{1};\mathbf{y}^0) \tag{2-4}$$

but do not have further information concerning the value of $\mathbf{z}^0$ on the identified half plane.

Under the model $\mathscr{M}_V$ we have observed all but two dimensions of the variation $\mathbf{z}$ for a performance of the system; under the models $\mathscr{M}_R$ and $\mathscr{M}_P$ we have observed all but two dimensions of a constructed pivotal function (2-1). This leads us to consider the marginal model for what has been observed and the conditional model for the unobserved. Under the model $\mathscr{M}_V$ this separation into marginal and conditional models follows from the necessary method in Sec. 3-3. Under the models $\mathscr{M}_R$ and $\mathscr{M}_P$ the introduction of the weak sufficiency principle in Sec. 4-4 can be used to give the marginal model for inference concerning λ; and given λ the introduction of an ancillarity principle in Sec. 4-2 can be used to give the conditional model for inference concerning (μ, σ).

2-1-2 Suitable Coordinates

In the preceding section we have seen that an n-dimensional vector $\mathbf{z}$ should be examined in terms of where $\mathbf{z}$ lies in a two-dimensional region $\mathscr{L}^+(\mathbf{1};\mathbf{z})$ and in terms of which two-dimensional region contains $\mathbf{z}$. For this it is convenient to have simple familiar coordinates remembering, of course, that there is nothing absolute in a *choice* of coordinates, a choice just provides a means of saying where $\mathbf{z}$ is in $\mathbb{R}^n$.

For the half plane $\mathscr{L}^+(\mathbf{1};\mathbf{z})$ the vector $\mathbf{1}$ is a natural choice as a basis vector. Let $\mathbf{d}(\mathbf{z})$ be a vector orthogonal to $\mathbf{1}$, of unit length and lying in $\mathscr{L}^+(\mathbf{1};\mathbf{z})$; we use $\mathbf{d}(\mathbf{z})$ as the second basis vector. We have

$$\mathbf{d}(\mathbf{z}) = s^{-1}(\mathbf{z})(\mathbf{z} - \bar{z}\mathbf{1}) = [\bar{z}, s(\mathbf{z})]^{-1}\mathbf{z} \tag{2-5}$$

where

$$s^2(\mathbf{z}) = |\mathbf{z} - \bar{z}\mathbf{1}|^2 = \sum(z_i - \bar{z})^2$$

Thus

$$\mathbf{z} = \bar{z}\mathbf{1} + s(\mathbf{z})\mathbf{d}(\mathbf{z}) \tag{2-6}$$

From this we see that the vector $\mathbf{z}$ has coordinates $[\bar{z}, s(\mathbf{z})]$ with respect to the basis $(\mathbf{1}, \mathbf{d}(\mathbf{z}))$ on the half plane and that $\mathbf{d}(\mathbf{z})$ determines the half plane. Note that $\mathbf{d}(\mathbf{z})$ generates the unit sphere in the orthogonal complement $\mathscr{L}^\perp(\mathbf{1})$ of $\mathscr{L}(\mathbf{1})$.

Again we find it appropriate to emphasize that there is nothing absolute in any particular choice of coordinates—a choice just provides a means of saying where $\mathbf{z}$ is in $\mathbb{R}^n$. We could, for example, use the median $\tilde{z}$ and the range $R(\mathbf{z})$, or the first order statistic $z_{(1)}$ and the first interval $z_{(2)} - z_{(1)}$, or any other coordinates even without the pleasant location-scale properties of the preceding. Such arbitrary choices have no effect on the inference.

Now consider the equation

$$\sigma^{-1}(\mathbf{y} - \mu \mathbf{1}) = \mathbf{z} \qquad (2\text{-}7)$$

in terms of these coordinates. The points $\mathbf{y}$ and $\mathbf{z}$ differ by a location-scale transformation and thus, of course, lie on the same half plane

$$\mathscr{L}^+(\mathbf{1}; \mathbf{z}) = \mathscr{L}^+(\mathbf{1}; \mathbf{y})$$

and have, accordingly, the same identifying basis vector

$$\mathbf{d}(\mathbf{z}) = \mathbf{d}(\mathbf{y}) \qquad (2\text{-}8)$$

We use the decomposition (2-6) for both $\mathbf{z}$ and $\mathbf{y}$ and substitute in (2-7):

$$\sigma^{-1}[\bar{y}\mathbf{1} + s(\mathbf{y})\mathbf{d}(\mathbf{y}) - \mu \mathbf{1}] = \bar{z}\mathbf{1} + s(\mathbf{z})\mathbf{d}(\mathbf{z})$$

This gives

$$\sigma^{-1}(\bar{y} - \mu) = \bar{z}$$
$$\sigma^{-1}s(\mathbf{y}) = s(\mathbf{z}) \qquad (2\text{-}9)$$

for the coefficients for $\mathbf{1}$ and $\mathbf{d}(\mathbf{z}) = \mathbf{d}(\mathbf{y})$.

We now have convenient coordinates for the half plane as given by (2-8) and for points on the half plane as given by (2-9).

2-1-3 Marginal and Conditional Distributions

The preliminaries in Sec. 2-1-1 lead us to consider the *marginal distribution* for the observed half plane as given by $\mathbf{d}(\mathbf{z}) = \mathbf{d}(\mathbf{y})$ and the *conditional distribution* for points on the half plane as given by $[\bar{y}, s(\mathbf{y})]$ and $[\bar{z}, s(\mathbf{z})]$ with (2-9) as the connecting relation.

The initial distribution describing $\mathbf{y}$ and $\mathbf{z}$ is given by

$$\sigma^{-n}\Pi f_\lambda(\sigma^{-1}(y_i - \mu)) \Pi dy_i = \Pi f_\lambda(z_i) \Pi dz_i \qquad (2\text{-}10)$$

To get the marginal and conditional distributions we first need to make the change of variable:

$$\mathbf{y} \leftrightarrow (\bar{y}, s(\mathbf{y}), \mathbf{d}(\mathbf{y}))$$
$$\mathbf{z} \leftrightarrow (\bar{z}, s(\mathbf{z}), \mathbf{d}(\mathbf{z})) \qquad (2\text{-}11)$$

The substitution for the density function is straightforward. The substitution for the differential can be obtained easily by noting that $\bar{z}$, $s(\mathbf{z})$, $\mathbf{d}(\mathbf{z})$ provide locally orthogonal coordinates at the point $\mathbf{z} = \bar{z}\mathbf{1} + s(\mathbf{z})\mathbf{d}(\mathbf{z})$; for $\bar{z}$ we have Euclidean length $\sqrt{n}\, d\bar{z}$; for $s = s(\mathbf{z})$ we have Euclidean length ds; and for $\mathbf{d}(\mathbf{z})$ we have Euclidean volume $s^{n-2}\, da$ where da is surface volume on the unit sphere for $\mathbf{d}(\mathbf{z})$ in $\mathscr{L}^\perp(\mathbf{1})$ and thus $s^{n-2}\, da$ is surface volume on the sphere for $s\,\mathbf{d}(\mathbf{z})$ with radius *s*. This gives

$$\Pi dz_i = \sqrt{n}\, d\bar{z}\, ds\, s^{n-2}\, da \tag{2-12}$$

for the differential. By substitution in (2-10) we then obtain

$$\sigma^{-n}\Pi f_\lambda(\sigma^{-1}(\bar{y} - \mu + sd_i))s^{n-2}\sqrt{n}\, d\bar{y}\, ds\, da \tag{2-13}$$

for $\bar{y}$, $s(\mathbf{y}) = s$, $d_i(\mathbf{y}) = d_i$, and correspondingly

$$\Pi f_\lambda(\bar{z} + sd_i)s^{n-2}\sqrt{n}\, d\bar{z}\, ds\, da \tag{2-14}$$

for $\bar{z}$, $s(\mathbf{z}) = s$, $d_i(\mathbf{z}) = d_i$.

The *marginal distribution* for $\mathbf{d}(\mathbf{z}) = \mathbf{d}(\mathbf{y}) = \mathbf{d}$ is obtained by integration from either of (2-13) or (2-14):

$$h_\lambda(\mathbf{d})\, da = \int_{-\infty}^{\infty} \int_0^{\infty} \Pi f_\lambda(\bar{z} + sd_i)s^{n-2}\sqrt{n}\, d\bar{z}\, ds \cdot da \tag{2-15}$$

Typically this integration cannot be completed in closed form but is readily available by computer integration.

The *conditional distribution* for $\bar{y}$, $s(\mathbf{y}) = s$, given $\mathbf{d}$, is obtained by division:

$$h_\lambda^{-1}(\mathbf{d})\sigma^{-n}\Pi f_\lambda(\sigma^{-1}(\bar{y} - \mu + sd_i))s^{n-2}\sqrt{n}\, d\bar{y}\, ds \tag{2-16}$$

The corresponding *conditional distribution* for $\bar{z}$, $s(\mathbf{z}) = s$ given $\mathbf{d}$ is

$$h_\lambda^{-1}(\mathbf{d})\Pi f_\lambda(\bar{z} + sd_i)s^{n-2}\sqrt{n}\, d\bar{z}\, ds \tag{2-17}$$

These conditional distributions are distributions on the two-dimensional half plane $\mathscr{L}^+(\mathbf{1};\mathbf{y}) = \mathscr{L}^+(\mathbf{1};\mathbf{z})$. They are recorded here in terms of the choice of coordinates $[\bar{y}, s(\mathbf{y})]$ and $[\bar{z}, s(\mathbf{z})]$, but could equally have been recorded in terms of any other choice of coordinates.

The preceding marginal and conditional distributions are the particular distributions discussed in Sec. 2-1-1.

2-1-4 Parameter Components

For the location-scale model we have the equation

$$[\mu, \sigma]^{-1}\mathbf{y} = \sigma^{-1}(\mathbf{y} - \mu\mathbf{1}) = \mathbf{z}$$

For the model $\mathscr{M}_V$ the equation presents the objective variation $\mathbf{z}$. For the models $\mathscr{M}_R$ and $\mathscr{M}_P$ the equation presents a pivotal function. For both cases we have, as noted in Secs. 2-1-1 and 2-1-2, that the equation separates into

$$\mathbf{d}(\mathbf{y}) = \mathbf{d}(\mathbf{z}) \tag{2-18}$$

for the observable part of $\mathbf{z}$ and

$$\begin{aligned} \sigma^{-1}(\bar{y} - \mu) &= \bar{z} \\ \sigma^{-1}s(\mathbf{y}) &= s(\mathbf{z}) \end{aligned} \tag{2-19}$$

for the unobservable. The relevant distributions are recorded in Sec. 2-1-3.

Now consider separately the two parameter components μ and σ. The equations (2-19) can be rearranged so that μ and σ are separated:

$$s^{-1}(\mathbf{y})(\bar{y} - \mu) = s^{-1}(\mathbf{z})\bar{z}$$
$$\sigma^{-1}s(\mathbf{y}) = s(\mathbf{z}) \qquad (2\text{-}20)$$

This separation from (2-19) to (2-20) is *unique*. Some constants can be used to rewrite (2-20) into a more familiar form:

$$\frac{\sqrt{n}(\bar{y} - \mu)}{s_y} = \frac{\sqrt{n}\bar{z}}{s_z} = t(\mathbf{z}) \qquad (2\text{-}21)$$

$$\frac{s_y}{\sigma} = s_z \qquad (2\text{-}22)$$

where $t(\mathbf{z})$ is the common t variable and s_y and s_z are the standard deviations for $\mathbf{y}$ and $\mathbf{z}$. That is, $s_y = s(\mathbf{y})/\sqrt{n-1}$, and $s_z = s(\mathbf{z})/\sqrt{n-1}$.

Again, a few remarks about the choice of coordinates. The equations (2-20) are essentially relations concerning positions on the two-dimensional half plane $\mathscr{L}^+(\mathbf{1};\mathbf{y}) = \mathscr{L}^+(\mathbf{1};\mathbf{z})$. The use of other coordinates in place of $[\bar{y}, s(\mathbf{y})]$ and $[\bar{z}, s(\mathbf{z})]$ would produce relations (2-20) that would be identical as relations on the half plane $\mathscr{L}^+(\mathbf{1};\mathbf{y}) = \mathscr{L}^+(\mathbf{1};\mathbf{z})$; that is, each (μ, σ) would yield the same mapping of points on the half plane.

The discussions in Sec. 2-1-1 lead us to consider the equations (2-19) in relation to the distributions (2-16) and (2-17) describing $[\bar{y}, s(\mathbf{y})]$ and $[\bar{z}, s(\mathbf{z})]$—in each case conditional on the value of $\mathbf{d}(\mathbf{y}) = \mathbf{d}(\mathbf{z}) = \mathbf{d}(\mathbf{y}^0)$. We now have a unique separation of equations (2-19) into equation (2-21) concerning μ and equation (2-22) concerning σ.

Consider the parameter component μ. For the model $\mathscr{M}_V$ we need the distribution of the function of the variation (2-21),

$$\frac{\sqrt{n}\bar{z}}{s_z} = t$$

conditional on the value of $\mathbf{d} = \mathbf{d}(\mathbf{y}^0)$ as cited. And for the models $\mathscr{M}_R$ and $\mathscr{M}_P$ we need the distribution of the pivotal function (2-21),

$$\frac{\sqrt{n}(\bar{y} - \mu)}{s_y} = t$$

also conditional on the value of $\mathbf{d} = \mathbf{d}(\mathbf{y}^0)$ as just cited. These distributions are, of course, the same and the common distribution is easily derived from (2-16) or (2-17) by expressing, say, $\bar{z}$ in terms of t and s and then integrating out the s:

$$g_\lambda(t:\mathbf{d})\,dt = h_\lambda^{-1}(\mathbf{d}) \int_0^\infty \Pi f_\lambda\!\left(s\!\left(\frac{t}{\sqrt{n^2 - n}} + d_i\right)\right) s^{n-1}\,ds \cdot (n-1)^{-1/2}\,dt \qquad (2\text{-}23)$$

Consider the parameter component σ. For the model $\mathscr{M}_V$ we need the distribution of the function of the variation (2-22),

$$s_z$$

conditional on the value of $\mathbf{d} = \mathbf{d}(\mathbf{y}^o)$. And for the models $\mathcal{M}_R$ and $\mathcal{M}_P$ we need the distribution of the pivotal function (2-22),

$$\frac{s_y}{\sigma} = s_z$$

also conditional on the value of $\mathbf{d} = \mathbf{d}(\mathbf{y}^o)$. These are the same and the distribution is easily derived from (2-16) or (2-17) by integrating out $\bar{y}$ or $\bar{z}$:

$$g_\lambda(s_z : \mathbf{d}) \, ds_z = h_\lambda^{-1}(\mathbf{d}) \int_{-\infty}^{\infty} \Pi f_\lambda(\bar{z} + \sqrt{n-1}\, s_z d_i) \sqrt{n} \, d\bar{z}\,(n-1)^{(n-1)/2} s_z^{n-2} \, ds_z \tag{2-24}$$

In conclusion for parameter components, we note that the discussion in Sec. 2-1-1 leads us to consider the distribution (2-23) in relation to the parameter μ and the distribution (2-24) in relation to the parameter σ.

2-2 TERMINAL METHODS OF ANALYSIS

In the preceding section we have examined some core methods for the reduction of the location-scale inference base $\mathcal{I} = (\mathcal{M}, \mathbf{y}^o)$. In this section we examine some available terminal inference methods that follow after these core reduction methods.

From Secs. 2-1-1 and 2-1-3 we have a separation of the distribution into a marginal distribution for an identified component and a conditional distribution for an unidentified component. *For the model* $\mathcal{M}_V$ this is a necessary reduction (Sec. 3-3). And *for the models* $\mathcal{M}_R$ *and* $\mathcal{M}_P$ this can be supported by the introduction of a weak sufficiency principle (Sec. 4-4) and an ancillarity principle (Secs. 4-2, 4-4).

The marginal distribution for $\mathbf{d}(\mathbf{y}) = \mathbf{d}(\mathbf{z})$ is recorded in formula (2-15). The observed value of the function is $\mathbf{d}(\mathbf{y}) = \mathbf{d}(\mathbf{z}) = \mathbf{d}(\mathbf{y}^o)$ obtained from the data $\mathbf{y}^o$ in the inference base. Let $\mathbf{d}^o$ designate $\mathbf{d}(\mathbf{y}^o)$.

The distributions of $[\bar{y}, s(\mathbf{y})]$ and $[\bar{z}, s(\mathbf{z})]$ both conditional on $\mathbf{d}^o$ are given by formulas (2-16) and (2-17) respectively. For $\bar{y}$ and $s(\mathbf{y})$ the observed values are $\bar{y}^o$ and $s(\mathbf{y}^o)$. For $\bar{z}$ and $s(\mathbf{z})$ the realized values are the only unobservable characteristics of $\mathbf{z}$; indeed, this is the reason why it is *necessary* to use the conditional distribution for $[\bar{z}, s(\mathbf{z})]$.

With this separation of the distribution as a starting point we now consider terminal methods of inference concerning λ, μ, and σ.

2-2-1 Inference: Shape λ

The marginal distribution for $\mathbf{d}(\mathbf{y}) = \mathbf{d}(\mathbf{z})$ depends on the shape parameter λ. The observed value is $\mathbf{d}(\mathbf{y}) = \mathbf{d}(\mathbf{z}) = \mathbf{d}^o$. We now consider inference for the parameter λ, and make frequent forward references to inference concepts that will be discussed in detail later.

Under the model $\mathcal{M}_V$ there is a distribution $f_\lambda(\mathbf{z})$ for the objective variation and

$\mathbf{d}(\mathbf{z}) = \mathbf{d}^o$ is the only observable value available from the inference base. This is a necessary reduction based on Sec. 3-3.

Under the models $\mathcal{M}_R$ *and* $\mathcal{M}_P$ the function $\mathbf{d}(\mathbf{y}) = \mathbf{d}(\mathbf{z})$ is weakly sufficient (see Sec. 4-4). The reduction to the function $\mathbf{d}(\mathbf{y}) = \mathbf{d}(\mathbf{z})$ can also be viewed as a necessary reduction if we permit attention to be focused on λ as described in Sec. 3-5.

The distribution for $\mathbf{d}(\mathbf{y}) = \mathbf{d}(\mathbf{z})$ is a distribution on the unit sphere generated by $\mathbf{d}$ in $\mathcal{L}^\perp(\mathbf{1})$; this sphere is an $(n-2)$-dimensional manifold.

For any given $\mathbf{d}$ and λ the density function can be calculated by straightforward computer integration. Thus for the observed $\mathbf{d} = \mathbf{d}^o$ we can readily obtain the likelihood function

$$L(\mathbf{d}^o; \lambda) = ch_\lambda(\mathbf{d}^o) \qquad (2\text{-}25)$$

For one-, two-, and three-dimensional parameters λ various computer graphing techniques allow the display and assessment of this likelihood function.

The likelihood function (2-25) is a marginal likelihood function as introduced in Fraser (1965, 1967, 1968); it has been adapted to the classical model by Sprott and Kalbfleisch (1969). It is also an invariant likelihood; see Hájek (1971) The likelihood function (2-25) can also be obtained by Bayesian procedures, provided a certain one of the many arbitrary possibilities for a prior distribution is chosen at the beginning of the analysis so as to target on the particular likelihood function (2-25) examined here; see Box and Tiao (1973), but also note that the requisite flat prior is currently unacceptable to some Bayesians (Lindley, 1973).

The likelihood function $L(\mathbf{d}^o; \lambda)$ for λ based on the observable $\mathbf{d}(\mathbf{y}) = \mathbf{d}(\mathbf{z}) = \mathbf{d}^o$ allows a direct assessment of the various λ values in relation to the inference base. Certainly we would want to go beyond this and form tests and confidence intervals for λ. Unfortunately the distribution (2-15) and the space for that distribution are complicated and seemingly intractable for the typical parametric model $f_\lambda(z)$ for the variation.

Some understanding of the model in relation to an observed likelihood is available from experience—in particular from computer simulations in which realized likelihood functions are obtained for computer samples from various distributions in a model.

2-2-2 Inference: Location μ

We now consider inference concerning the location parameter μ *given a value for* λ.

From Secs. 2-1-1 and 2-1-3 we see that the initial distribution gives rise to a conditional distribution (2-16) for $[\bar{y}, s(\mathbf{y})]$ and (2-17) for $[\bar{z}, s(\mathbf{z})]$; the conditional distribution has $\mathbf{d} = \mathbf{d}^o$. Then from Sec. 2-1-4 we can separate out the parameter μ obtaining the equation

$$\frac{\sqrt{n}(\bar{y} - \mu)}{s_y} = \frac{\sqrt{n}\,\bar{z}}{s_z} = t(\mathbf{z}) \qquad (2\text{-}26)$$

The conditional distribution for $t = t(\mathbf{z})$ with $\mathbf{d} = \mathbf{d}^o$ is given by (2-23).

For the model $\mathcal{M}_V$ the *necessary* description for the unobserved location $\bar{z}$ and scale $s(\mathbf{z})$ is given by the conditional probability distribution in formula (2-17). A value for μ then identifies a contour for $t(\mathbf{z}) = \sqrt{n}\bar{z}/s_z$, and the marginal distribution for $t(\mathbf{z})$ in formula (2-23) is then the necessary basis for tests and confidence intervals for μ (see Sec. 3-5).

For the models $\mathcal{M}_R$ and $\mathcal{M}_P$ the restriction to the conditional distribution (2-16) can be supported by the introduction of a weak ancillarity principle related to that in Secs. 4-2 and 4-4. The *further* reduction to the distribution (2-23) relating to μ alone has some practical appeal, and for hypothesis testing it can be supported by a specialized extension of Sec. 3-4 involving the use of invariance.

Consider testing the hypothesis: $\mu = \mu_0$. On the assumption that $\mu = \mu_0$ we find that the value of $t(\mathbf{z}) = \sqrt{n}\bar{z}/s_z$ is observable:

$$t = \frac{\sqrt{n}\bar{z}}{s_z} = \frac{\sqrt{n}(\bar{y}^0 - \mu_0)}{s_y^0} \tag{2-27}$$

This observed value can be compared with the distribution (2-23), with $\mathbf{d} = \mathbf{d}^0$, to see whether it is a reasonable high-density value, or a 'marginal' value, or an almost impossible value far out on the tails where the density is essentially zero. The hypothesis can then be assessed accordingly.

Now consider forming a confidence region or confidence interval for the parameter μ. We note that the observed response $\mathbf{y}^0$ gives various values for

$$t = \frac{\sqrt{n}\bar{z}}{s_z} = \frac{\sqrt{n}(\bar{y}^0 - \mu)}{s_y^0} \tag{2-28}$$

depending on the value being considered for μ. Let (t_1, t_2) be a $(1 - \alpha)$ interval, say a central interval, for the distribution (2-23) for t:

$$\int_{t_1}^{t_2} g_\lambda(t:\mathbf{d}^0)\, dt = 1 - \alpha \tag{2-29}$$

Then we obtain the observed $(1 - \alpha)$ confidence interval for μ:

$$\left(\bar{y}^0 - t_2 \frac{s_y^0}{\sqrt{n}}, \bar{y}^0 - t_1 \frac{s_y^0}{\sqrt{n}}\right) \tag{2-30}$$

The preceding is, of course, an ordinary $(1 - \alpha)$ confidence interval based on (2-26) and the conditional distribution (2-23). The random interval has the form

$$\left[\bar{y} - t_2(\mathbf{d}) \frac{s_y}{\sqrt{n}}, \bar{y} - t_1(\mathbf{d}) \frac{s_y}{\sqrt{n}}\right] \tag{2-31}$$

This has conditional confidence $1 - \alpha$ given $\mathbf{d}(\mathbf{y}) = \mathbf{d}$, and thus, of course, has marginal confidence $1 - \alpha$.

By shrinking such *central* confidence intervals down we obtain a median-type estimate for the location μ.

2-2-3 Inference: Scale σ

We now consider inference for the scale parameter. Again as in Sec. 2-2-2 we see that the basic distribution gives rise to a conditional distribution for $[\bar{y}, s(\mathbf{y})]$ in (2-16) and for $[\bar{z}, s(\mathbf{z})]$ in (2-17), in each case with $\mathbf{d} = \mathbf{d}^0$. Then from Sec. 2-1-4 we obtain σ alone in the equation

$$\frac{s_y}{\sigma} = s_z \tag{2-32}$$

with distribution for s_z as recorded in (2-24) with $\mathbf{d} = \mathbf{d}^0$.

The justifications under model $\mathcal{M}_V$ and under models $\mathcal{M}_R$ and $\mathcal{M}_P$ parallel those recorded in Sec. 2-2-2.

Consider testing the hypothesis: $\sigma = \sigma_0$. On the assumption that $\sigma = \sigma_0$ the value of s_z is observable:

$$s_z = \frac{s_y^0}{\sigma_0} \tag{2-33}$$

This observed value can be compared with the distribution (2-24), with $\mathbf{d} = \mathbf{d}^0$, to see whether it is a reasonably high-density value, or a 'marginal' value, or an almost impossible value far out on the tails where the density is essentially zero. The hypothesis can then be assessed accordingly.

Now consider forming a confidence interval for the parameter σ. We note that the observed $\mathbf{y}^0$ gives various values for

$$s_z = \frac{s_y^0}{\sigma} \tag{2-34}$$

depending on the value being considered for σ. Let (s_1, s_2) be a $(1-\alpha)$ interval, say a central interval, for the distribution (2-24) for s_z:

$$\int_{s_1}^{s_2} g_\lambda(s_z : \mathbf{d}^0)\, ds_z = 1 - \alpha \tag{2-35}$$

We then obtain the observed $(1-\alpha)$ confidence interval for σ:

$$\left(\frac{s_y^0}{s_2}, \frac{s_y^0}{s_1} \right) \tag{2-36}$$

The preceding is, of course, a regular $(1-\alpha)$ confidence interval based on (2-32) and the conditional distribution (2-24). The random interval has the form

$$\left[\frac{s_y}{s_2(\mathbf{d})}, \frac{s_y}{s_1(\mathbf{d})} \right] \tag{2-37}$$

This has conditional confidence $1 - \alpha$ given $\mathbf{d}(\mathbf{y}) = \mathbf{d}$ and thus, of course, has marginal confidence $1 - \alpha$.

By shrinking such central confidence intervals down we obtain a median-type estimate for the scale σ.

2-2-4 Further Remarks

The confidence intervals and tests for μ and σ are essentially unique. Of course, for confidence intervals, there is the choice of how much probability should be assigned to each tail in (2-29) for μ and in (2-35) for σ. Other than this, however, the intervals and tests are unique.

A different choice of initial coordinates on the half plane $\mathscr{L}^+(1;\mathbf{y}) = \mathscr{L}^+(1;\mathbf{z})$ would not affect the results. For example, if coordinates different from $(\bar{y}, s_y)$ were used, then a compensating displacement would occur with the distribution for $t(\mathbf{z})$ with the result that the confidence interval would remain unchanged.

The preceding uniqueness is easily overlooked in casual comparison with ordinary unconditional analysis. With such analysis, a change from, say, $\bar{y}$ to the median $\tilde{y}$ or the first-order statistic $y_{(1)}$ would make a considerable difference. Conditionally, all choices of coordinates are equivalent—just ways of attaching labels to points on a well-specified half plane.

Simultaneous confidence methods are readily available. The joint distribution of (t, s) referred to just before (2-23) can be used to calculate the probability for the rectangle $(t_1, t_2) \times (s_1, s_2)$; this would give the joint confidence for the interval (2-31) for μ and the interval (2-37) for σ.

2-3 ANALYSIS OF AN INFERENCE BASE

As an illustration of the methods discussed in Secs. 2-1 and 2-2 we now examine the Darwin data as recorded in Fisher (1960, p. 37).

The data came from an experiment to compare the heights of cross- and self-fertilized plants. The design involved 15 pairs of plants, each pair consisting of a cross- and a self-fertilized plant grown under the same conditions in the same pot. The data available are the 15 differences in height, cross- minus self-fertilized:

$$
\begin{array}{ccccc}
49 & 23 & 24 & -67 & 28 \\
75 & 8 & 41 & 60 & 16 \\
14 & -48 & 6 & 56 & 29
\end{array}
\qquad (2\text{-}38)
$$

It will be of interest later to recall the two extreme values on the negative end of the sample.

Fisher (1960, p. 37) initially analyzed the data, assuming normality. He then reanalyzed the data, assuming a symmetric distribution form (p. 46); he tested the hypothesis of a zero mean difference in height and used input randomization as support for the symmetric distribution form.

Box and Tiao (1973) used the power exponential

$$f_\lambda(z) = a(\lambda) \exp\left[-c^{-1}(\lambda)|z|^\lambda\right] \qquad (2\text{-}39)$$

as a distribution form for variation and presented a Bayesian analysis of the data.

The preceding distribution form has an unnatural cusp at the origin and could only be viewed as a very rough approximation for a reasonable distribution form for variation.

Fraser and Fick (1975) analyzed the data using the Student family

$$f_\lambda(z) = \frac{\Gamma\left(\frac{\lambda+1}{2}\right)}{\Gamma\left(\frac{\lambda}{2}\right)\Gamma\left(\frac{1}{2}\right)}\left[1 + \frac{z^2}{c^2(\lambda)}\right]^{-(\lambda+1)/2} c^{-1}(\lambda) \qquad (2\text{-}40)$$

as the distribution form for the variation. The preceding family provides a continuous spectrum of symmetric density forms ranging from the normal ($\lambda \to \infty$) down to the Cauchy ($\lambda = 1$) and to even thicker tailed sub-Cauchy distributions ($0 < \lambda < 1$). The Student distribution with λ in the 5 to 7 range has been cited as a reasonable distribution form for many actual response variables, providing realistic tail thickness and tail length.

For our analyses in this book we refer extensively to this Student(λ) family as a basic and appropriate family for the variation in many responses—with symmetry and variable tail thickness.

We now analyze the Darwin data using the Student family (2-40) as the distribution form $f_\lambda(z)$ for the variation. For the analysis we use the computer program (Fick, 1976) developed to implement the analyses in Secs. 2-1 and 2-2 and indeed used for the original Student-family analysis of the data in Fraser and Fick (1975). Some standardization values for the Student family are recorded with discussion in Fraser (1976, p. 467).

For the present analysis we start with an inference base $(\mathcal{M}, \mathbf{y}^o)$: the model $\mathcal{M}$ can be one of the models $\mathcal{M}_R$, $\mathcal{M}_V$, $\mathcal{M}_P$ in Sec. 1-2, using the Student family (2-40) for the variation; the data vector $\mathbf{y}^o$ is recorded in (2-38).

2-3-1 Likelihood Analysis for λ

The observed value of the unit residual vector $\mathbf{d}(\mathbf{y}^o) = \mathbf{d}(\mathbf{z}^o) = \mathbf{d}^o$ is recorded in the following array:

$$\begin{array}{ccccc} 0.1987 & 0.0146 & 0.0217 & -0.6226 & 0.0500 \\ 0.3828 & -0.0916 & 0.1421 & 0.2766 & -0.0349 \\ -0.0491 & -0.4881 & -0.1057 & 0.2483 & 0.0571 \end{array} \qquad (2\text{-}41)$$

Note the two somewhat large negative values of -0.6226 and -0.4881; the largest positive value is 0.3828.

The likelihood function $L(\mathbf{d}^o; \lambda)$ in (2-25) contains the customary arbitrary multiplicative constant c. The constant can be avoided by using likelihood ratios, say the likelihood at λ as a proportion of the likelihood at some reference value λ_0. The normal distribution form ($\lambda = \infty$) provides a convenient reference distribution. Accordingly, we examine the likelihood ratio,

$$L^*(\mathbf{d}^0; \lambda) = \frac{h_\lambda(\mathbf{d}^0)}{h_\infty(\mathbf{d}^0)} = A_{n-1} h_\lambda(\mathbf{d}^0)$$

In this we have used the normal reference value $\lambda_0 = \infty$ and have substituted using the density function

$$h_\infty(\mathbf{d}) = \frac{1}{A_{n-1}}$$

for the normal case. For this, recall that the rotational symmetry of the normal gives a *uniform* marginal distribution for $\mathbf{d}$ on the unit sphere in $\mathscr{L}^\perp(\mathbf{1})$ and that the surface area of a unit sphere in $\mathbb{R}^f$ is

$$A_f = \frac{2\pi^{f/2}}{\Gamma(f/2)}$$

The computer program produces the likelihood function $L^*(\mathbf{d}^0; \lambda)$ in both tabulated and graphic form. The likelihood function has a mode of approximately 2.4 at $\lambda = 2.3$. It is greater than 1.5 from $\lambda = 1.2$ to $\lambda = 8.5$ and has, of course, the limiting value 1 as $\lambda \to \infty$. For the remainder of the analysis we examine λ values in the range from 1 to 9 indicated by the likelihood function, and also for comparison we examine the $\lambda = \infty$ value corresponding to the usual normal analysis.

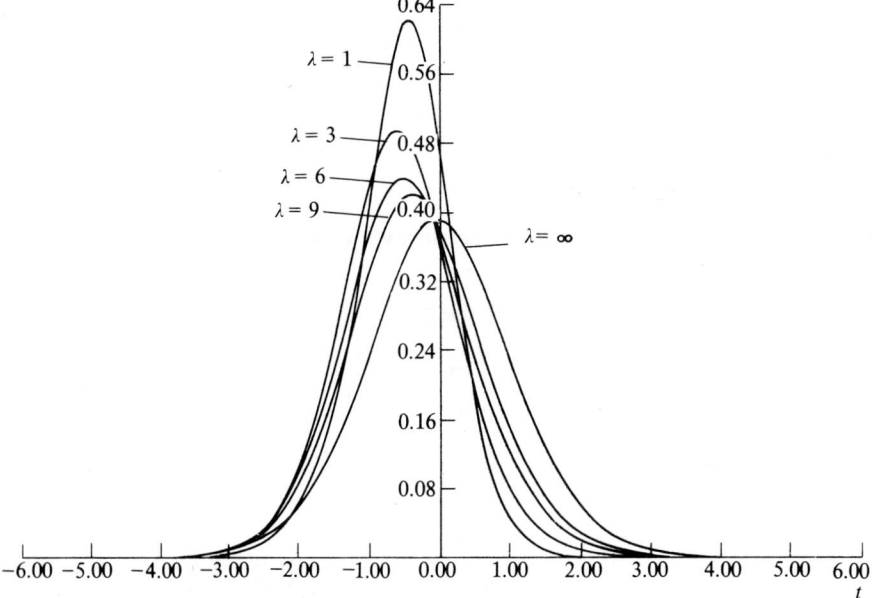

Figure 2-1 Density function for the t statistic for Darwin data; $\lambda = 1, 3, 6, 9, \infty$.

2-3-2 Inference for μ and σ

For the location parameter μ we need the distribution (2-23) for the t function (2-21) for various λ values; as suggested in Sec. 2-3-1 we use the values $\lambda = 1, 3, 6, 9, \infty$.

The computer program produces the distribution (2-23) in graphical form and in a suitably integrated form to give the percentage points and confidence intervals.

The density functions are plotted in Fig. 2-1. Note that they are all somewhat asymmetric except for the normal case with $\lambda = \infty$. Of course, in the normal case, the density of $t(\mathbf{z})$ is the Student(14) distribution appropriate to a 'normal' sample of 15. The mode of the distribution seems to shift to the left as λ decreases and the distribution becomes more concentrated about the mode. On reflection this is a reasonable phenomenon: the use of a Student distribution for variation provides a tolerance for the two extreme values on the left tail of the sample and as λ decreases the analysis gradually compensates for the 'biases' coming from these extreme values.

Consider the hypothesis that $\mu = 0$—that there is a zero median difference between cross- and self-fertilized plants.

The observed value of the t statistic under the hypothesis is

$$t = \frac{\bar{y} - 0}{s_y/\sqrt{n}} = 2.148$$

For $\lambda = \infty$ this value is just beyond the $2\frac{1}{2}$ percent point on the right tail of the Student(14) distribution.

For $\lambda = 1, 3, 6, 9$ and indeed for $\lambda = \infty$ the probability of exceeding the observed t value 2.148 is available by computer integration:

λ	1	3	6	9	∞
$P(t > 2.148)$	0.000397	0.002989	0.007638	0.011126	0.024835

Approximations to these values are available by a rough visual assessment of Fig. 2-1.

The analysis based on smaller values of λ is an analysis involving a distribution form that tolerates extreme values. Accordingly we find from the preceding tabulation a much stronger case against the hypothesis when we use the λ values indicated in the range from 1 to 9.

Now consider the formation of confidence intervals for the location parameter μ. From Sec. 2-2-2 the $(1 - \alpha)$ confidence interval has the form

$$(\hat{\mu}_1, \hat{\mu}_2) = \left(\bar{y} - t_2 \frac{s_y}{\sqrt{n}}, \bar{y} - t_1 \frac{s_y}{\sqrt{n}}\right)$$

where (t_1, t_2) is a $(1 - \alpha)$ interval for the t-statistic distribution (2-23).

The computer calculates these intervals for any chosen λ values and confidence levels. For example, the 95 percent central intervals are as follows:

λ	t_1	t_2	$\hat{\mu}_1$	$\hat{\mu}_2$
1	-2.25	0.529	14.8	42.9
3	-2.28	1.17	9.6	43.1
6	-2.28	1.55	5.8	43.2
9	-2.27	1.73	4.1	43.0
∞	-2.14	2.14	0.03	41.8

The value $\lambda = 3$ is close to the point maximizing the likelihood for λ. For this value we have the 95 percent interval (9.6, 43.1) with the median interval value 26.7. This, of course, contrasts sharply with the normal theory interval (0.03, 41.8) centered at the sample average 20.9.

For the scale parameter σ the analysis proceeds similarly but with details as recorded in Sec. 2-2-3.

Graphs of the s_z distribution display properties similar to those in Fig. 2-1, but now shifting to the right and expanding as we go to smaller λ values.

A hypothesis that σ has some specified value $\sigma = \sigma_0$ could be assessed by comparing an s_z observed under the hypothesis with the distribution just described.

Confidence intervals are readily available on the computer printout from the analysis. For example, for the $\lambda = 3$ value near the point-maximizing likelihood and for the normal $\lambda = \infty$ value we have the following 95 percent central intervals:

λ	s_1	s_2	$\hat{\sigma}_1$	$\hat{\sigma}_2$
3	0.698	2.072	18.2	54.1
∞	0.634	1.366	27.6	59.5

2-4 LIFETESTING AND THE WEIBULL†

Lifetesting is an important and yet somewhat specialized area of statistics. The results from the preceding sections provide the basic and incisive method for lifetesting analysis.

An object placed on lifetest until failure will give a response, *lifetime*, that is, of course, nonnegative. A sample of objects placed on lifetest until all have failed will give a sample of lifetimes. Two variations on this provide protection against prolonged testing. With Type I censoring, a sample of objects is tested until a predetermined time T; this will give the lifetimes that are smaller than T and the

† With Alok Dobriyal.

number of lifetimes greater than or equal to T. With Type II censoring, a sample of objects is tested until a predetermined portion of them has failed; this will give the smaller values, the particular proportion of smaller values, for a sample from lifetimes. The methods in the preceding sections apply directly to the full sample case and to the Type II censoring case.

In this section we illustrate the methods by analyzing a frequently cited data set from Lieblein and Zelen (1956, p. 286). The Weibull is a common distribution for lifetesting and has been the usual distribution for various analyses of the preceding data. We use the location-scale methods from Secs. 2-1 and 2-2 to obtain an exact analysis of the data with the Weibull distribution. For comparison purposes we also give the analysis with the log-normal distribution.

An important advantage of the location-scale analysis is the availability of the likelihood for the shape parameter. Thus as a routine part of the analysis we obtain the likelihood for the log-normal versus the Weibull. Interestingly the data give the log-normal an almost 2 to 1 likelihood preference over the commonly used Weibull.

2-4-1 The Model and the Data

The Weibull is a common distribution for lifetesting and the usual distribution for analyzing the Lieblein and Zelen data.

The Weibull distribution for lifetime t has the density function

$$f(t) = \frac{\beta}{\alpha}\left(\frac{t-\gamma}{\alpha}\right)^{\beta-1} \exp\left[-\left(\frac{t-\gamma}{\alpha}\right)^{\beta}\right] \qquad t > \gamma \qquad (2\text{-}42)$$

$$= 0 \qquad t \leq \gamma$$

with parameters α, β, γ. For most applications the initial point γ of the distribution is known—usually zero—and t is measured from that known value. This gives the special Weibull with density

$$f(t) = \frac{\beta}{\alpha}\left(\frac{t}{\alpha}\right)^{\beta-1} \exp\left[-\left(\frac{t}{\alpha}\right)^{\beta}\right] \qquad t > 0 \qquad (2\text{-}43)$$

$$= 0 \qquad t \leq 0$$

with $\beta > 0$ and $\alpha > 0$.

Note that the power transform $(t/\alpha)^{\beta}$ of the standardized t/α variable from (2-43) has the standard exponential distribution. Or, in a reverse way, note that the various positive power transforms of the scaled exponential generate the special Weibull family.

For the special Weibull family (2-43) the parameter α gives the general "location" of the distribution on the positive axis and the parameter β determines the tightness or scale of the distribution about the "location."

The distribution and parameters of the special Weibull are more easily examined by taking a logarithmic transform of the lifetime variable t, that is, by putting lifetime into multiplicative units. Specifically, we consider the new variable

$$y = \ln t \tag{2-44}$$

The distribution for y obtained from the special Weibull (2-43) is easily verified to have density

$$f(y) = \frac{1}{\sigma} \exp\left(\frac{y-\mu}{\sigma}\right) \exp\left(-\exp\frac{y-\mu}{\sigma}\right) \tag{2-45}$$

This distribution has location parameter μ and scale parameter σ where

$$\mu = \ln \alpha \qquad \sigma = \beta^{-1} \tag{2-46}$$

The distribution form can be represented by the variable

$$z = \frac{y-\mu}{\sigma} \tag{2-47}$$

It has the standardized density function

$$f(z) = \exp(z) \exp(-\exp z) \tag{2-48}$$

The variable $-z$ has the standard extreme value distribution as discussed for example in Fraser (1976, p. 64). The distribution of z is the Type I asymptotic distribution form for the smallest value in a sample.

The Weibull as a distribution for lifetime has been discussed from an empirical viewpoint in Weibull (1951) and Kao (1959) and from a theoretical viewpoint in Gumbel (1958), Mann et al (1974). The representation of the Weibull form as the limiting distribution for the smallest in a sample (failure when the first component ingredient fails) makes it a somewhat natural choice for lifetesting.

The log-normal distribution is also used for lifetesting data. The log-normal distribution for t has the density function

$$\begin{aligned} f(t) &= \frac{1}{\sqrt{2\pi}\,\sigma} \exp\left[-\frac{1}{2}\left(\frac{\ln t - \mu}{\sigma}\right)^2\right] \frac{1}{t} \qquad & t > 0 \\ &= 0 & t \leq 0 \end{aligned} \tag{2-49}$$

with parameters μ, σ.

The distribution is more easily examined by taking a logarithmic transform of the lifetime variable t. The distribution for

$$y = \ln t \tag{2-50}$$

has the following normal density:

$$f(y) = \frac{1}{\sqrt{2\pi}\,\sigma} \exp\left[-\frac{1}{2}\left(\frac{y-\mu}{\sigma}\right)^2\right] \tag{2-51}$$

with location parameter μ and scale parameter σ. The distribution form can be represented by $z = (y - \mu)/\sigma$ with the standard normal density

$$f(z) = \frac{1}{\sqrt{2\pi}} \exp\left(-\frac{z^2}{2}\right) \tag{2-52}$$

We now record the data given in Lieblein and Zelen (1956, p. 286). The data are the lifetimes until failure of 23 deep-groove ball bearings in millions of revolutions:

17.88, 28.92, 33.00, 41.52, 42.12, 45.60,
48.48, 51.84, 51.96, 54.12, 55.58, 67.80,
68.64, 68.64, 68.68, 84.12, 93.12, 98.64,
105.12, 105.84, 127.92, 128.04, 173.40.

2-4-2 The Analysis

We now consider the Weibull and log-normal analyses of the Lieblein and Zelen data as just recorded.

Specifically we consider the analysis of an inference base $\mathscr{I} = (\mathscr{M}_V, \mathscr{D})$. The data $\mathscr{D}$ for use with the location-scale program are the natural logarithms of the lifetimes recorded at the end of Sec. 2-4-1. The model $\mathscr{M}_V$ is the variation-based location-scale model as presented in Sec. 1-2-4. For distribution form we allow, of course, the appropriate Weibull expression for the logarithmic variable, which is, in fact, the extreme value distribution recorded in (2-48). As an alternative for comparison purposes we allow the appropriate log-normal expression for the logarithmic variable, which is, of course, the standard normal (2-52). The basic location-scale program uses the standardization in (1-19): for the normal this is trivial but for the extreme value it introduces location and scale factors. As the Weibull and extreme value parameters are usually given direct interpretation we choose to deactivate this option and to use the extreme value distribution as it stands in (2-48). This does have the effect of making μ and σ somewhat different parameters for the Weibull as opposed to the normal, and accordingly would make the distributions for $t(\mathbf{z})$ and s_z somewhat different in nature for the Weibull as opposed to the normal.

Now consider the results of the location-scale computer analysis of the inference base $(\mathscr{M}_V, \mathscr{D})$.

Under the Weibull (extreme value) model we obtain the following central confidence intervals for the location μ:

Level	t_1	t_2	$\hat{\mu}_1$	$\hat{\mu}_2$
90%	−3.951	−0.636	4.2211	4.5899
95%	−4.281	−0.279	4.1815	4.6265
99%	−4.951	−0.462	4.0991	4.7011

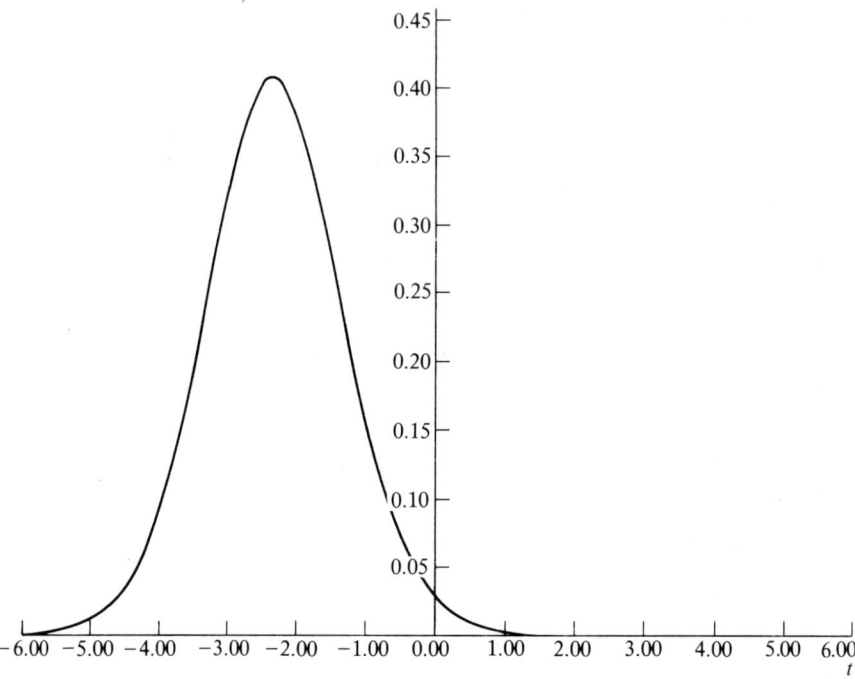

Figure 2-2 The distribution of t for the Weibull analysis of the data.

The median estimate for the parameter μ is 4.4096. The appropriate t distribution is recorded in Fig. 2-2.

The preceding intervals for μ can be transformed to give the intervals for $\alpha = \exp\{\mu\}$:

Level	$\hat{\alpha}_1$	$\hat{\alpha}_2$
90%	68.1084	98.4846
95%	65.4640	102.1559
99%	60.2860	110.0682

The median estimate for α is 82.237.

Under the Weibull we obtain the following central confidence intervals for the scale σ:

Level	s_1	s_2	$\hat{\sigma}_1$	$\hat{\sigma}_2$
90%	0.810	1.382	0.38600	0.65875
95%	0.762	1.444	0.36936	0.69959
99%	0.674	1.569	0.34001	0.79166

The median estimate for the parameter σ is 0.49458.

The preceding intervals for σ can be transformed to give the intervals for $\beta = 1/\sigma$.

Level	$\hat{\beta}_1$	$\hat{\beta}_2$
90%	1.5180	2.5907
95%	1.4294	2.7074
99%	1.2632	2.9411

The median estimate for the parameter β is 2.0219.

We recall in passing that tests of significance are readily available from the location-scale computer program.

Under the log-normal analysis we obtain the following confidence intervals for μ and σ:

Level	$\hat{\mu}_1$	$\hat{\mu}_2$	$\hat{\sigma}_1$	$\hat{\sigma}_2$
90%	3.959	4.341	0.429	0.712
95%	3.920	4.381	0.412	0.755
99%	3.837	4.464	0.382	0.852

The median estimate for μ is 4.15 and for σ is 0.541. The appropriate t distribution is recorded in Fig. 2-3; it is Student (22). Note that these estimates and corre-

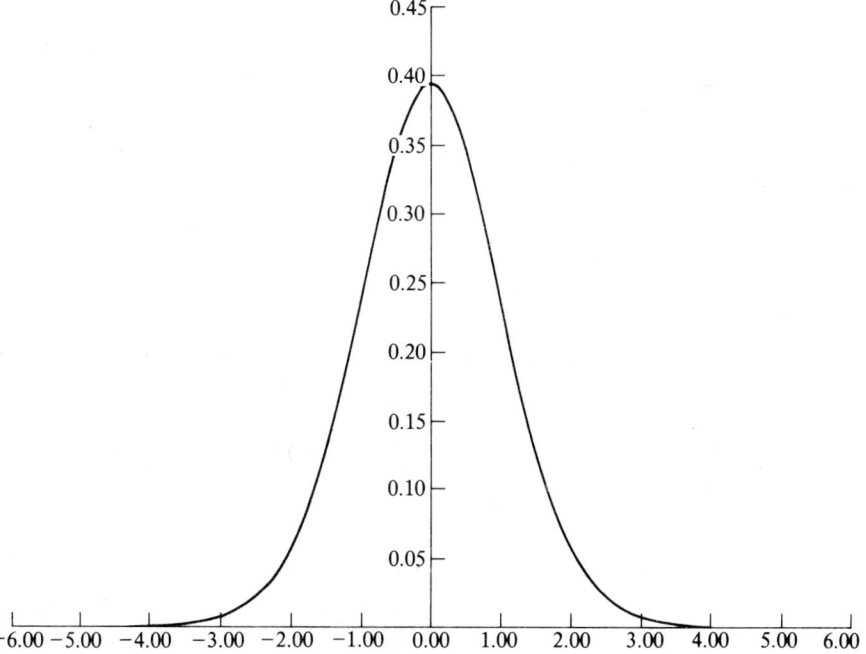

Figure 2-3 The distribution of t for the log-normal analysis.

sponding parameters cannot be immediately related to those earlier, as we did not standardize the extreme value distribution.

The computer analysis gives the likelihood 0.5520 for the Weibull in comparison with the log-normal, or, more casually, the data prefer the log-normal to the Weibull. This suggests that more extensive data be assembled to pin down more carefully the appropriate distribution form. Indeed it suggests that a more general model be widely used—a model that might allow extreme value, normal, logistic, or Student for the log-lifetime distribution form. A compilation of likelihood values from computer analyses could then give solid evidence for appropriate form for particular contexts.

2-4-3 Other Analyses

The original analysis of ball-bearing data by Lieblein and Zelen (1956) was based on the use of linear orderly estimates (generalized Gauss-Markov on the order statistics). They obtained the estimate $\hat{\beta} = 2.23$ for β.

Thoman, Bain, and Antle (1969) obtained the maximum likelihood estimates

$$\hat{\alpha} = 81.99 \qquad \hat{\beta} = 2.102$$

for the parameters α and β. In addition, they used simulations to estimate percentage points of the functions

$$\hat{\beta} \ln(\hat{\alpha}/\alpha) \qquad \hat{\beta}/\beta$$

The resulting 90 percent confidence intervals (unconditional, approximate) for α and β are

$\hat{\alpha}_1$	$\hat{\alpha}_2$	$\hat{\beta}_1$	$\hat{\beta}_2$
68.04	98.75	1.50	2.62

Lawless (1972, 1973, 1974) examined the conditional distribution of

$$\left(\frac{\hat{\mu} - \mu}{\hat{\sigma}}, \frac{\hat{\sigma}}{\sigma}\right)$$

conditional on the standardized sample deviations

$$\left(\frac{y_1 - \hat{\mu}}{\hat{\sigma}}, \ldots, \frac{y_n - \hat{\mu}}{\hat{\sigma}}\right)$$

He obtained a 95 percent interval for α as $(68.05, \infty)$ and a 90 percent interval for β as $(1.52, 2.59)$. This conditional analysis corresponds to the computer location-scale analysis but without the likelihood for comparing possible shapes. The justification given for this conditional analysis was, however, that of an ancillarity principle in Sec. 4-2 or as originally in Fisher (1934). By contrast the necessity here of the conditional analysis and of the comparative likelihoods follows from the discussions earlier in this chapter and the more detailed discussions in Chap. 3.

2-5 ROBUSTNESS AND RESISTANCE

For the typical real-valued response variable we have noted that the normal distribution form is extremely short-tailed and is thus rather unrealistic as a sole distribution form for variation. In Sec. 2-3 we have indicated that the Student(λ) family provides a reasonable spectrum of distribution forms to cover many common response variables, ranging from short-tailed distributions to thick-tailed distributions and embracing those with λ in the 5 to 7 range that are often cited as being particularly appropriate for applications. An additional λ coordinate could easily be added allowing for skewness—for different thicknesses in the two tails.

In this section we examine some statistical properties of the analysis based on the Student family for the variation; see Fick and Fraser (1976). In particular, we examine the robustness of the analysis based on the Student family in comparison with the analysis based on the normal. We also examine how the Student family analysis for small λ values is resistant to spurious observations in the data.

Our approach involved extensive computer simulations using the Marsaglia generator. Some theory was used initially to indicate directions. The conclusions, however, are essentially pragmatic, not distribution theoretic.

2-5-1 Robustness

We compare the normal distribution theory analysis with a Student analysis based on a small value of λ—specifically the value $\lambda = 3$. For this we used computer-generated samples from the normal and from the Student(3) distributions; we used a moderate sample size $n = 30$. We then analyzed each sample by the computer program using the Student family for the variation. In particular we examined how the $\lambda = 3$ and $\lambda = \infty$ analyses compared.

The normal samples were surprisingly consistent in the pattern of the computer analyses. The following is a typical sample from the normal(10, 1) distribution:

10.8189	9.7212	8.6366	9.4186	9.5955
9.1757	10.0207	8.1110	11.1069	11.4919
9.6644	8.1520	8.3642	9.7043	11.0955
10.2122	9.7021	10.2434	9.8919	10.2624
8.6833	11.9849	8.8890	10.6948	10.4537
11.5243	8.9173	8.2356	9.9143	9.0277

$$\bar{y} = 9.7905 \qquad s_y = 1.0572$$

The likelihood function for λ is plotted in Fig. 2-4; it is very flat and non-discriminating among λ values. The distributions for the t statistic for $\lambda = 3$ and for $\lambda = \infty$ are plotted in Fig. 2-5 and for the s_z statistic in Fig. 2-6.

We see that the two t-statistic distributions are very similar and that the two s-statistic distributions are also quite similar. Indeed, the estimates and confidence

38 INFERENCE AND LINEAR MODELS

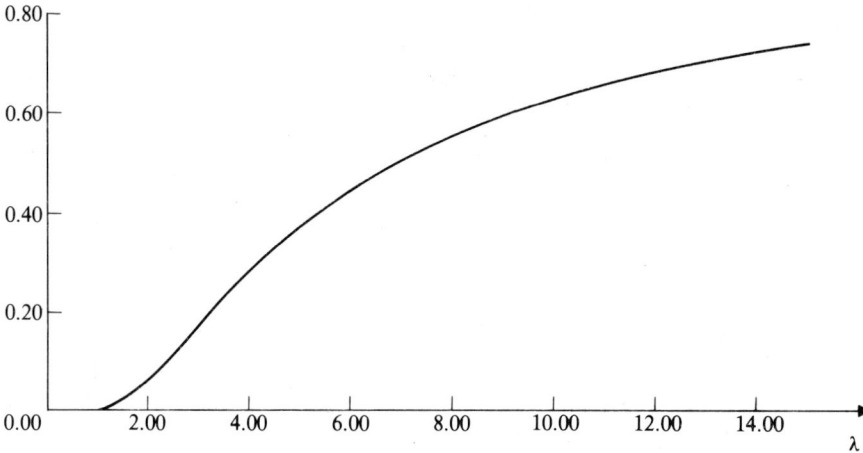

Figure 2-4 The likelihood function for λ for a sample from the $N(10, 1)$ distribution.

intervals are very close. For this sample from the normal it does not seem to matter whether a normal analysis or a Student(3) analysis is used.

Similar results were obtained with other normal samples: sometimes they were somewhat closer, sometimes somewhat farther apart, but basically very similar results were obtained for the two analyses on a given sample. In particular, the larger discrepancies seemed to correlate with the presence of extreme values in the samples.

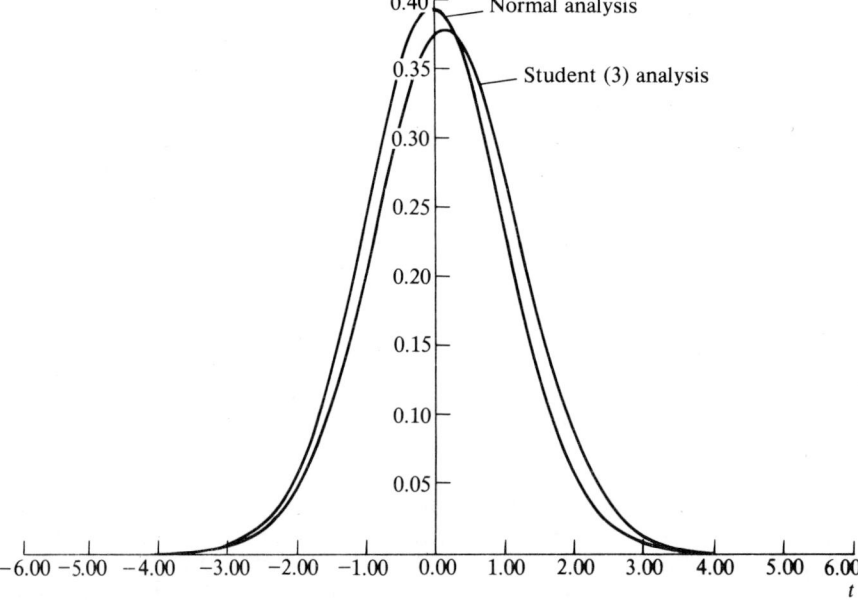

Figure 2-5 The distribution of $t(\mathbf{z})$ for $\lambda = 3, \infty$; the normal sample.

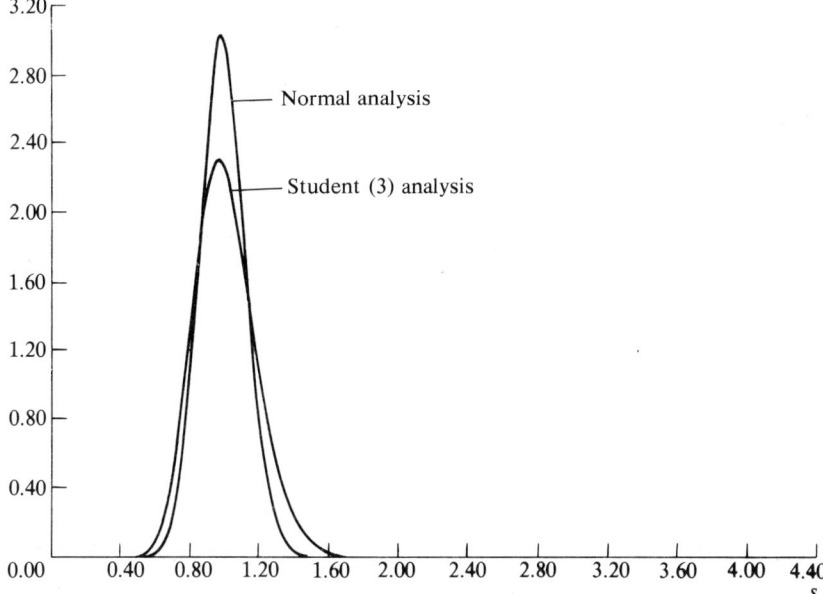

Figure 2-6 The distribution of s_z for $\lambda = 3, \infty$; the normal sample.

The Student(3) samples were also reasonably consistent in the pattern of computer analysis—consistent, however, in a rather interesting and different way. The following is a typical sample from the Student(3) distribution but relocated to the value $\mu = 10$; this distribution has $\sigma = 1.1966$ [Student(3) relative to *standardized* Student(3)]:

11.4426	6.9131	10.0922	11.3484	5.0395
10.6864	9.0702	10.1503	9.0740	10.0486
10.4382	8.6284	9.0450	8.1336	9.7684
20.5017	8.7002	10.6349	9.1181	7.0280
13.5396	10.2305	9.5360	10.2183	8.7759
9.7609	10.2147	9.9614	9.4326	6.9113

The likelihood function for λ is plotted in Fig. 2-7; it is very sharply discriminating, emphasizing rather heavily the smaller values of λ from 1 to 5. The distributions for the t statistic for $\lambda = 3$ and $\lambda = \infty$ are plotted in Fig. 2-8 and for the s_z statistic in Fig. 2-9.

We see that the two t-statistic distributions are very different. If we suppose that the Student(3) origins of the sample are known, then the *correct* t-statistic distribution is the distribution labeled $\lambda = 3$. Compare this with the "incorrect" t-statistic distribution obtained from the usual normal analysis and labeled $\lambda = \infty$; it is shifted to the left and is more diffuse.

The two s_z distributions are also very different. Again the correct distribution is the one labeled $\lambda = 3$. By comparison the "incorrect" distribution is the normal

40 INFERENCE AND LINEAR MODELS

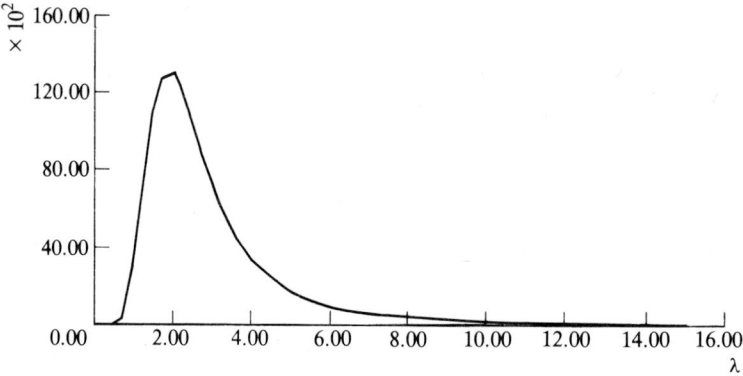

Figure 2-7 The likelihood function for λ for a sample from the distribution with $\mu = 10$, $\sigma = 1.1966$, and Student (3) variation.

analysis distribution labeled $\lambda = \infty$; it is shifted to the left and is more concentrated, so that in its effect on a confidence interval for σ it is shifted to the right and is more diffuse.

Thus, to the degree that our examples here represent the many cases obtained by computer simulation, we can note that for a sample from a normal response, there appears to be little practical difference between a normal analysis and a

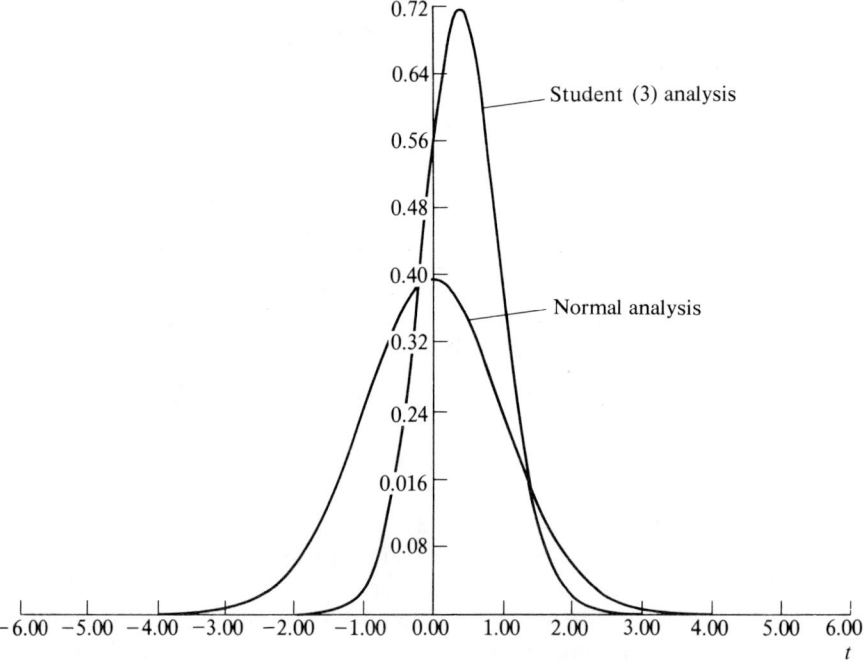

Figure 2-8 The distribution of $t(z)$ for $\lambda = 3, \infty$; the Student sample.

LOCATION-SCALE ANALYSIS **41**

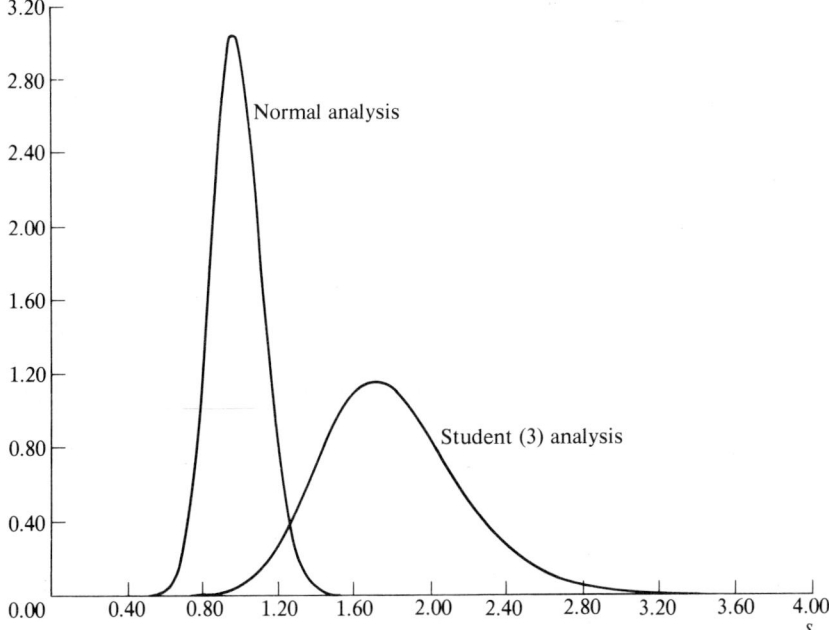

Figure 2-9 The distribution of s_z for $\lambda = 3, \infty$; the Student sample.

Student(3) analysis, whereas for a sample from a Student(3) response the normal analysis can be very far from correct.

The preceding suggests that the Student(3) analysis works reasonably well both for normal responses and Student(3) responses, but that the same does not hold for the commonly used normal analysis.

2-5-2 Resistance

We now consider further the Student family analysis based on small λ values and investigate the resistance of the analysis to spurious observations in the data.

In Sec. 2-5-1 we noted how the Student(3) analysis seems to "tolerate" extreme values in the tails of a sample. We can investigate this tolerance further by deliberately producing extreme values, say by taking a value in an otherwise reasonable sample and moving the value far away from the center of the sample. For this we obtained a sample of size 30 from the normal(10, 1) distribution:

10.8189	9.7212	8.6366	9.4186	9.5955
9.1757	10.0207	8.1110	11.1069	11.4919
9.6644	8.1520	8.3642	9.7043	11.0955
10.2122	9.7021	10.2434	9.8919	10.2624
8.6833	11.9849	8.8890	10.6948	10.4537
11.5243	8.9173	8.2356	9.9143	9.0277

$$\bar{y} = 9.7905 \qquad s_y = 1.0572$$

The value 9.7043 is a central observation in the sample. We take this observation and move it by each of the following amounts:

$$0 \quad -2s_y \quad -4s_y \quad -6s_y \quad -8s_y \qquad (2\text{-}53)$$

We then see what the effects of this spurious observation are on the normal analysis methods and on the Student(3) analysis methods.

For inference concerning μ our methods give a confidence interval

$$\left(\bar{y} - t_2 \frac{s_y}{\sqrt{n}}, \bar{y} - t_1 \frac{s_y}{\sqrt{n}}\right) \qquad (2\text{-}54)$$

where (t_1, t_2) is the $(1 - \alpha)$ central interval for the t-statistic distribution. When we alter the sample by making one observation spurious, then we change the sample average and standard deviation, both of which appear explicitly in the expression (2-54). We also alter the standardized residual and thus alter the distribution used to calculate the t-statistic interval. Specifically for an altered sample we would have

$$\left(\bar{y}^* - T_2 \frac{s_y^*}{\sqrt{n}}, \bar{y}^* - T_1 \frac{s_y^*}{\sqrt{n}}\right) \qquad (2\text{-}55)$$

where $\bar{y}^*$ and s_y^* are the values for the altered sample and (T_1, T_2) is the $(1 - \alpha)$ interval based on the altered standardized residual.

We are interested in the effects of the spurious observation on the inferences concerning μ. We could directly examine confidence intervals as given by (2-54) and (2-55) with varying extremity in the spurious observation, or we could examine the underlying t-statistic distribution provided we compensate for the fact that the reference sample average and standard deviation in (2-55) vary with the different extremity in the spurious observation. We adopt the second route as pictures of distributions can often be more revealing and informative.

Let (T_1, T_2) be the interval calculated with respect to the actual average and standard deviation of the altered sample and let (t_1, t_2) be the corresponding interval as reexpressed with respect to the original sample average and standard deviation:

$$\bar{y} - t_i \frac{s_y}{\sqrt{n}} = \bar{y}^* - T_i \frac{s_y^*}{\sqrt{n}}$$

$$t = \frac{s_y^*}{s_y} T - \frac{(\bar{y}^* - \bar{y})}{s_y/\sqrt{n}} \qquad (2\text{-}56)$$

Thus we need the location-scale transformation (2-56). This transformation adjusts the calculated t-statistic distribution and makes it relevant to the original sample average and standard deviation.

For inference concerning σ our methods give a confidence interval

$$\left(\frac{s_y}{s_2}, \frac{s_y}{s_1}\right) \qquad (2\text{-}57)$$

where (s_1, s_2) is the $(1 - \alpha)$ central interval for the s_z distribution. For an altered sample we would have

$$\left(\frac{s_y^*}{S_2}, \frac{s_y^*}{S_1}\right) \tag{2-58}$$

where (S_1, S_2) is the $(1 - \alpha)$ interval based on the altered standardized residual. As in the preceding paragraph we are then led to the transformation

$$s = \frac{s_y}{s_y^*} S \tag{2-59}$$

to adjust the s_z distribution and make it relevant to the original sample average and standard deviation.

Now consider the t-statistic and s_z distributions as made relevant to the original average $\bar{y}$ and standard deviation s_y.

First consider the normal analyses. We record in Fig. 2-10 the t-statistic distributions for the sample altered as indicated by (2-53). The distributions are, of course, just relocated and rescaled Student(29) distributions. The shift in the distribution as the spurious observation moves out shows clearly the large effect of that observation on inferences concerning μ.

Second consider the Student(3) analyses. We record in Fig. 2-11 the t-statistic distributions for the sample altered as mentioned above. An initial effect occurs as the observation is just made spurious; then the distribution form is reasonably

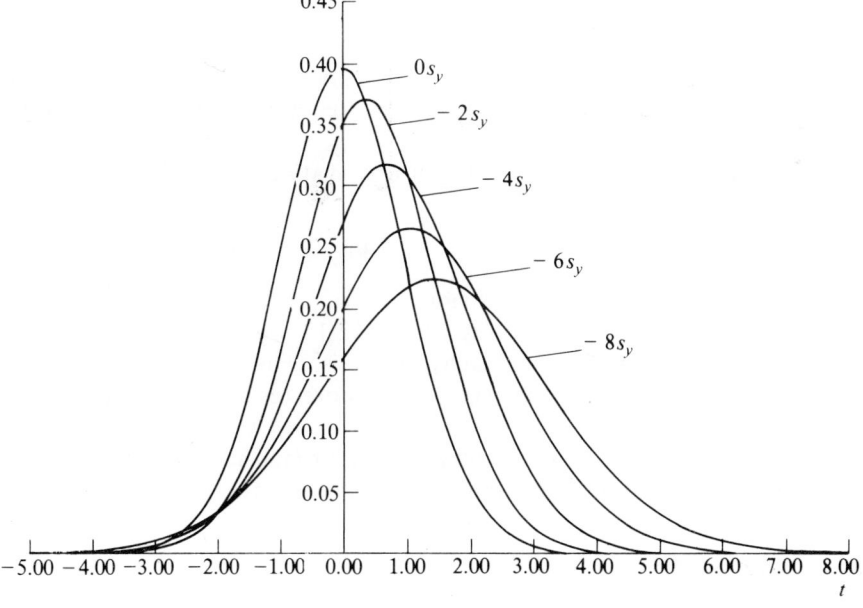

Figure 2-10 One observation displaced 0, $-2s_y$, $-4s_y$, $-6s_y$, $-8s_y$ for a normal sample of 30: the normal analysis t-statistic distribution as made relevant to the unaltered $\bar{y}$, s_y.

44 INFERENCE AND LINEAR MODELS

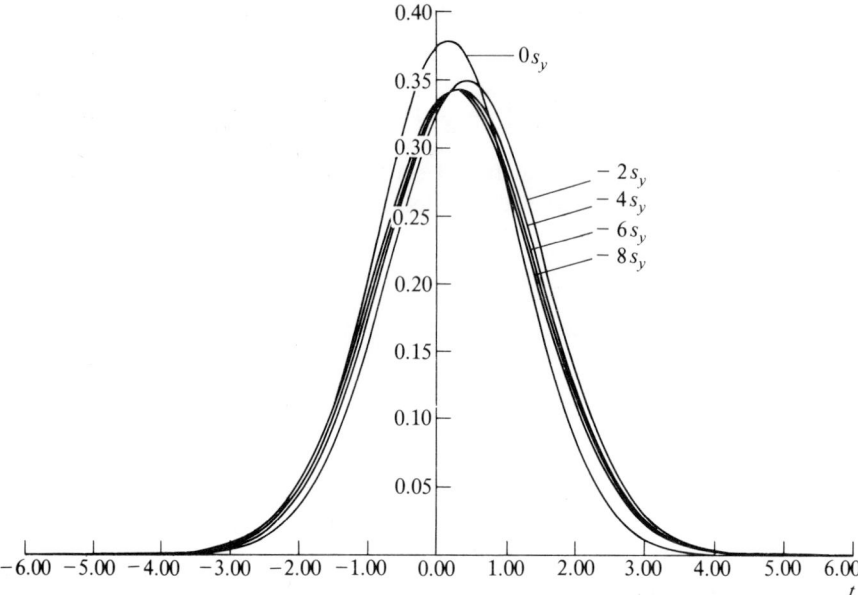

Figure 2-11 One observation displaced 0, $-2s_y$, $-4s_y$, $-6s_y$, $-8s_y$ for a normal sample of 30: the Student(3) analysis t-statistic distribution as made relevant to the unaltered $\bar{y}$, s_y.

stable. The Student(3) analysis has surprising resistance—an initial effect of the spurious observation but little additional effect as the observation moves far away from the sample cluster.

Similar results are obtained with the s_z distribution: a steady drift of the distribution under normal analysis and surprising resistance and stability for the Student(3) analysis.

The preceding results invite us to try an observation moved by each of the following amounts:

$$0 \quad -10s_y \quad -20s_y \quad -30s_y \quad -40s_y \quad (2\text{-}60)$$

In Fig. 2-12 we record the t-statistic distributions for a Student(3) analysis. Some drifting of the distribution does occur. In Fig. 2-13 we record the t-statistic distributions for a Student(1) or Cauchy analysis. The distribution form is reasonably stable. A Student($\frac{1}{2}$) analysis is even more resistant to such extreme outliers.

The parallel results for the s_z distribution are in fact more favourable than the preceding—more resistant to extreme spurious observations.

2-5-3 An Overview

The Darwin example in Sec. 2-3 indicated the flexibility of the Student(λ) analysis. We noted how the Student(λ) model was more realistic in its allowance for the

LOCATION-SCALE ANALYSIS 45

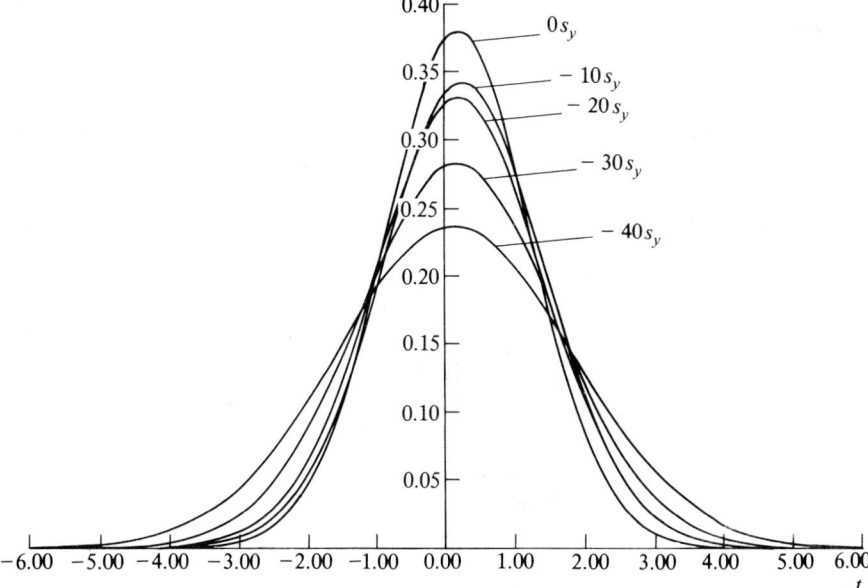

Figure 2-12 One observation displaced 0, $-10s_y$, $-20s_y$, $-30s_y$, $-40s_y$ for a normal sample of 30: the Student (3) analysis t-statistic distribution as made relevant to the unaltered $\bar{y}$, s_y.

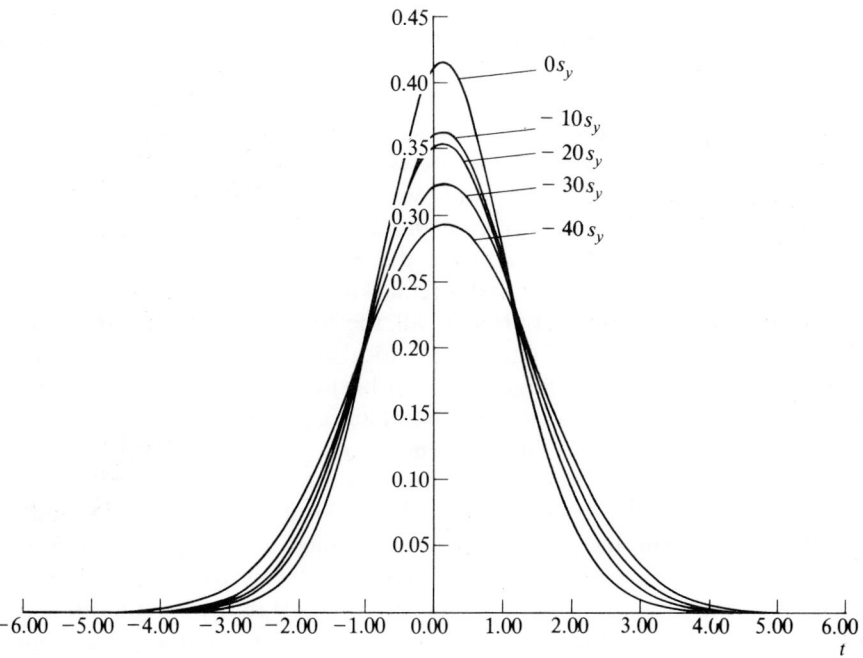

Figure 2-13 One observation displaced 0, $-10s_y$, $-20s_y$, $-30s_y$, $-40s_y$ for a normal sample of 30: the Student (1) analysis t-statistic distribution as made relevant to the unaltered $\bar{y}$, s_y.

thicker-tailed distributions we expect to find in many applications. We also saw how the lower λ analyses tolerated the two extreme values in the Darwin data. This resistance to extreme values was examined in Sec. 2-5-2 by artificially introducing very large spurious observations: the smaller the λ value, the greater the resistance.

The example in Sec. 2-3 also gave some indication that a Student(λ) analysis with λ too small may give results with somewhat overstated precision—confidence limits that are tighter than they should be. By choosing to use analyses with very small λ values we are facing a trade-off: the possibility of overstating the precision as opposed to having very high resistance to extreme spurious values. Extreme spurious values should be pretty obvious in an ordinary sample. However, with more complex models such as the regression model a spurious observation may be far from obvious and we might choose to have very high resistance at the risk of slightly overstating the precision.

The approach in Sec. 2-3 could almost be interpreted as an adaptive method. From the sample we estimate the λ value; thus adapted to the sample we proceed to inference methods for μ and σ. Such an "adaptive" approach could be examined for the case of the spurious observation as discussed in Sec. 2-5-2: as the observation is moved out the likelihood function would indicate a smaller and smaller λ value which would then be increasingly tolerant of the extreme spurious observation. This seems to have the advantage found with the medium λ values when there are moderate spurious observations and the advantage found with very small λ values when there are extreme spurious observations.

The pragmatic assessment in this section emphasizes the flexibilities associated with the location-scale analysis in this chapter. We can accommodate nonnormal shape, uncertainty as to shape, and spurious observations; and, in addition, we obtain all the needed distributions for tests and confidence intervals.

The Princeton robustness study (Andrews et al., 1972) examined numerous statistics for estimating location. The study was concerned with properties of the *marginal* distributions of these *location statistics*. Our viewpoint here has overwhelmingly emphasized the *conditional* distribution of these statistics. And, indeed, from results in Sec. 2-1-2 we find a remarkable property for these location statistics under the conditional approach: all the location statistics are in fact *equivalent*. Thus the problem of choosing a location statistic vanishes. Indeed there is much more. The robustness approach basically gives an estimate; here we have available the full range of test and confidence procedures. The robustness approach depends strongly on the influence function which often does not correspond to a density function for variation; here we obtain results comparable to those with extreme influence functions and, in addition, have a corresponding distribution for variation and have the just mentioned distributional properties that make the significance tests and confidence intervals available. Thus the arbitrariness for the estimation statistic and the unrealizability for the influence function is replaced by uniqueness for the estimation statistic and the full flexibility for the choice of distribution form for variation.

REFERENCES AND BIBLIOGRAPHY

Andrews, D. A., Bickel, Hampel, Huber, Rogers, and Tukey: "Robust Estimates of Location," Princeton University Press, Princeton, N.J., 1972.

Barnard, G. A.: The Use of the Likelihood Function in Statistical Practice, *Proc. Fifth Berkeley Symposium*, vol. 1, pp. 27–40, 1967.

Box, G. E. P., and G. C. Tiao: A Bayesian Approach to Some Outlier Problems, *Biometrika*, vol. 55, 1, pp. 119–129, 1968.

———: "Bayesian Inference in Statistical Analysis," Addison-Wesley, Reading, Mass., 1973.

Buehler, R. J.: Some Validity Criteria for Statistical Inferences, *Ann. Math. Stat.*, vol. 30, pp. 845–863, 1959.

Cox, D. R.: Some Problems Connected with Statistical Inference, *Ann. Math. Stat.*, vol. 29, pp. 357–372, 1958.

Edwards, A. W. F.: "Likelihood," Cambridge Press, Cambridge, 1972.

Fick, G. H.: "Computer Implementation Program for Location-Scale Analysis," Department of Mathematics, University of Toronto, 1976.

———, and D. A. S. Fraser: Robustness with Structural Methods, in "Proc. of Statistics Days," Technical Report, Department of Mathematical Sciences, Ball State University, pp. 75–93, 1976.

———: Two New Properties of Mathematical Likelihood, *Proc. Roy. Soc.*, ser. A, vol. 144, pp. 285–304, 1934.

Fisher, R. A.: "The Design of Experiments," 7th ed., Hafner, New York, 1960.

Fraser, D. A. S.: "Lecture Notes on Statistical Inference," University of Toronto, 1965.

———: Data Transformations and the Linear Model, *Ann. Math. Stat.*, vol. 38, pp. 1456–1465, 1967.

———: "The Structure of Inference," John Wiley and Sons, New York, and Krieger Publishing Company, Huntington, N.Y., 1968.

———: "Probability and Statistics, Theory and Applications," Duxbury Press, North Scituate, Mass., 1976.

———: Necessary Analysis and Adaptive Inference, *Jour. Amer. Stat. Assoc.*, vol. 71, pp. 99–113, 1976.

———, and G. H. Fick: Necessary Analysis and Its Implementation, in A. K. Md. Ehsanes Saleh (ed.), "Proc. Symposium on Statistics and Related Topics," Carleton University, Ottawa, pp. 5.01–5.30, 1975.

Gumbel, E. J.: "Statistics of Extremes," Columbia University Press, New York, 1958.

Hájek, J.: Limiting Properties of Likelihoods and Inference, in V. P. Godambe and D. A. Sprott (eds.), "Proc. Symposium on the Foundations of Statistical Inference," Holt, Rinehart and Winston of Canada, Toronto and Montreal, pp. 142–162, 1971.

Johnson, N. L., and Kotz, S.: "Continuous Univariate Distributions—I," Houghton-Mifflin, Boston, 1970.

Kalbfleisch, J. D.: Likelihood Methods of Prediction, in V. P. Godambe and D. A. Sprott (eds.), "Proc. Symposium on the Foundations of Statistical Inference"; Holt, Rinehart and Winston of Canada, Toronto and Montreal, pp. 378–392, 1971.

———, and D. A. Sprott: Application of Likelihood Methods to Models Involving Large Numbers of Parameters, *J. Roy. Stat. Soc.*, ser. B, vol. 32, pp. 175–208, 1970.

——— and ———: Corrigendum, *J. Roy. Stat. Soc.*, ser. B, vol. 34, pp. 124–125, 1972.

——— and ———: Marginal and Conditional Likelihood, *Sankhyā*, ser. A, vol. 35, pp. 311–328, 1973.

Kao, J. H. K.: A Graphical Estimation of Mixed Weibull Parameters in Life-Testing Electron Tubes, *Technometrics*, vol. 1, pp. 389–407, 1959.

Lawless, J. F.: Conditional Confidence Interval Estimation for the Parameters of the Weibull Distribution, *Utilitas Mathematica*, vol. 2, pp. 71–87, 1972.

———: Conditional Versus Unconditional Confidence Intervals for the Parameters of the Weibull Distribution, *Jour. Amer. Stat. Assoc.*, vol. 68, pp. 665–669, 1973.

———: Approximations to Confidence Intervals for Parameters in the Extreme Value and Weibull Distributions, *Biometrika*, vol. 61, pp. 123–129, 1974.

Lieblein, J., and M. Zelen: Statistical Investigation of the Fatigue Life of Deep Groove Ball Bearings, *Journal of Research, National Bureau of Standards*, vol. 57, pp. 273–316, 1956.

Lindley, D. V.: Discussion on the Paper by Dr. Dawid, Prof. Stone and Dr. Zidek, *J. Roy. Stat. Soc.*, ser. B, vol. 35, pp. 218–219, 1973.

———: Comments on the Paper by Fraser and Mackay, on the Equivalence of Standard Inference Procedures, in Harper and Hooker (eds.), "Foundations of Probability Theory, Statistical Inference, and Statistical Theories of Science," vol. II, D. Reidel Publishing Company, Dordrecht-Holland, pp. 47–62, 1976.

Mann, N. R.: Point and Interval Estimation Procedures for the Two-Parameter Weibull and Extreme-Value Distributions, *Technometrics*, vol. 10, pp. 231–256, 1968.

———, R. E. Schafer, and N. D. Singpurwalla: "Methods for Statistical Analysis of Reliability and Life Data," John Wiley and Sons, New York, 1974.

Pitman, E. J. G.: Location and Scale Parameters, *Biometrika*, vol. 30, pp. 391–421, 1939.

———: Tests of Hypothesis Concerning Location and Scale Parameters, *Biometrika*, vol. 31, pp. 200–215, 1939.

Sprott, D. A.: Normal Likelihoods and Their Relation to Large Sample Theory of Estimation, *Biometrika*, vol. 60, pp. 457–465, 1973.

———, and J. D. Kalbfleisch: Examples of Likelihoods and Comparison with Point Estimates and Large Sample Approximations, *Jour. Amer. Stat. Assoc.*, vol. 64, pp. 468–484, 1969.

Thoman, D. R., L. J. Bain, and C. E. Antle: Inferences on the Parameters of the Weibull Distribution, *Technometrics*, vol. 11, pp. 445–460, 1969.

Weibull, W.: A Statistical Distribution Function of Wide Applicability, *Jour. of App. Mechanics*, vol. 18, pp. 293–297, 1951.

CHAPTER
THREE
NECESSARY METHODS

In Chap. 2 we examined statistical inference for data from a system with location-scale parameters. Our starting point was an inference base $\mathscr{I} = (\mathscr{M}, \mathscr{D})$ consisting of a location-scale model $\mathscr{M}$ for the compound system and the data $\mathscr{D}$ recording the observed response vector.

In Sec. 2-1 various core-reduction methods were mentioned for the location-scale inference base. Some of these methods were *necessary* methods, methods predicated by the definitions for the model and data in the inference base; the remainder were methods based on principles, some of a spectrum of principles proffered in inference theories and examined in detail in Chap. 4.

In this chapter we examine necessary methods for analyzing inference bases. Then in Chap. 4 we examine methods based on introduced principles.

For the discussion of necessary methods our starting point is an inference base $\mathscr{I} = (\mathscr{M}, \mathscr{D})$ consisting of a model $\mathscr{M}$ and data $\mathscr{D}$. The model $\mathscr{M} = (\Omega; \ldots)$ was defined in Sec. 1-1-2 and the inference base in Sec. 1-1-3. The model and data, as initially assembled to form an inference base, may contain arbitrary elements, elements beyond those specified by the definitions just cited. In this chapter we examine necessary reduction methods—methods for eliminating the arbitrary elements in an inference base as presented.

3-1 ON THE PARAMETER SPACE

In certain rather special applications an observed response may effectively reduce the parameter space. In some cases, then, this reduction gives a smaller statistical model—with the result that an inference base can often be substantially reduced. Let $\mathscr{I} = (\mathscr{M}, \mathscr{D})$ be an inference base with model $\mathscr{M}$ and data $\mathscr{D}$.

3-1-1 Preliminaries

A response **y** may be consistent with only some of the parameter values in the parameter space Ω of the model as given. For a response value **y** we define the *index set* $D(\mathbf{y})$ for that value **y** to be the set

$$D(\mathbf{y}) = \{\theta : f(\mathbf{y}|\theta) > 0\} \tag{3-1}$$

of parameter values for which the density is greater than zero; we use $f(\mathbf{y}|\theta)$ to designate the response density derived from the model $\mathcal{M}$. The set $D(\mathbf{y})$ is the set of possible parameter values for the data **y**.

Consider the simple example of data y from the model $\mathcal{M}$ for the uniform $(\theta \pm \frac{1}{2})$ distribution with θ in $\mathbb{R}$. For the index set we obtain

$$D(y) = \{\theta : f(y|\theta) > 0\} = (y - \tfrac{1}{2}, y + \tfrac{1}{2})$$

Thus from the inference base $(\mathcal{M}, y)$ we *know* that θ is in $(y \pm \frac{1}{2})$.

The preceding suggests that we use the simpler model with θ restricted to the interval $(y \pm \frac{1}{2})$. Such a "model," however, is clearly inappropriate; note that the simpler "model" has sample points whose index sets are not contained in the range of the "model." We explore this further and seek a proper correspondence between parameter sets and sample space sets.

3-1-2 Some Definitions

It can happen that the sets $D(\mathbf{y})$ form a partition of Ω. Or more generally it can happen that the sets $D(\mathbf{y})$ overlap in various ways but still remain subordinate to some nontrivial partition of Ω. Consider some details.

Let $\mathscr{P} = \{P\}$ be a partition of Ω, the finest partition that does not break up any set $D(\mathbf{y})$. Specifically: $\mathscr{P} = \{P\}$ is the finest partition of Ω such that each P satisfies

$$D(\mathbf{y}) \subset P \quad \text{or} \quad D(\mathbf{y}) \cap P = \phi \tag{3-2}$$

for all $D(\mathbf{y})$. Note: the finest partition can be obtained by considering partitions that satisfy (3-2) and then intersecting the sets of the partitions.

Now for each value **y** we define the *options set* $P(\mathbf{y})$ to be the element of the partition $\mathscr{P}$ that contains $D(\mathbf{y})$:

$$D(\mathbf{y}) \subset P(\mathbf{y}) \tag{3-3}$$

Of course, $\mathscr{P}$ is the finest partition of Ω with this property (3-3).

For the simple uniform example in Sec. 3-1-1 the sets $D(\mathbf{y})$ overlap without a break on the line $\Omega = \mathbb{R}$; accordingly we have $P(\mathbf{y}) = \Omega$ and the partition $\mathscr{P} = \{\mathbb{R}\}$ is trivial.

Now let $\mathscr{S}(\mathbf{y})$ be the sample space corresponding to the parameter values in the set $P(\mathbf{y})$:

$$\mathscr{S}(\mathbf{y}) = \bigcup_{\theta \in P(\mathbf{y})} \{\mathbf{y}' : f(\mathbf{y}'|\theta) > 0\}$$

$$= \{\mathbf{y}' : D(\mathbf{y}') \subset P(\mathbf{y})\}$$
$$= \{\mathbf{y}' : P(\mathbf{y}') = P(\mathbf{y})\}$$
$$= P^{-1} P(\mathbf{y})$$

Note that the various sets $\mathscr{S}(\mathbf{y})$ are the preimage sets of the function P; accordingly the sets $\mathscr{S}(\mathbf{y})$ form a partition of the initial sample space for the model. For some discussion of mappings and preimages in a statistical context see, for example, Fraser (1976, pp. 100–104).

Consider the simple uniform example but now with the restricted parameter space $\Omega = (-\infty, -1) \cup (1, \infty)$. The sets $D(\mathbf{y})$ are as before but restricted where needed to the set Ω. The finest partition satisfying (3-2) is $\mathscr{P} = \{(-\infty, -1), (1, \infty)\}$. Thus $P(y) = (-\infty, 1)$ or $(1, \infty)$ according as $y < 0$, or $y > 0$.

3-1-3 Model Reduction

Consider the model $\mathscr{M}$ as discussed in Sec. 3-1-1. Let P be a subset of Ω and let $\mathscr{M}_P$ be the model $\mathscr{M}$ but with the parameter space restricted to the set P. Of course, $\mathscr{M}_\Omega = \mathscr{M}$. We will be interested in $\mathscr{M}_P$ for P an element of the partition $\mathscr{P}$ described in Sec. 3-1-2.

We are now in a position to state the first necessary reduction method.

RM$_1$: Necessary reduction—parameter space The inference base $(\mathscr{M}, \mathbf{y})$ necessarily produces the inference base $(\mathscr{M}_{P(\mathbf{y})}, \mathbf{y})$.

As indicated in Sec. 3-1-2 this will be nontrivial only in cases where the partition $\mathscr{P}$ is nontrivial ($\mathscr{P} \neq \{\Omega\}$).

From an observed $\mathbf{y}$ we know that θ is in the index set $D(\mathbf{y})$ and thus in the options set $P(\mathbf{y})$; $P(\mathbf{y})$ is the smallest parameter range that provides a consistent model for $\mathbf{y}$. The reduction method RM$_1$ eliminates arbitrary elements from the inference base $(\mathscr{M}, \mathbf{y})$ and produces $(\mathscr{M}_{P(\mathbf{y})}, \mathbf{y})$; the parameter space Ω has been restricted to the options set $P(\mathbf{y})$.

Now consider the proof of the necessary reduction method RM$_1$. For this we recall the definitions for the model and the inference base in Secs. 1-1-2 and 1-1-3, and we show that the model eliminates arbitrary elements from the inference base as initially presented. The inference base restricts θ to the index set $D(\mathbf{y})$. The options set is the smallest parameter range P for which the sample space points have index sets contained in P; see formula (3-3) and the adjacent comments. The model $\mathscr{M}_{P(\mathbf{y})}$ is the smallest model consistent with the observed response $\mathbf{y}$; the distributions of $\mathscr{M}$ outside $\mathscr{M}_{P(\mathbf{y})}$ are arbitrary. It follows that $(\mathscr{M}, \mathbf{y})$ necessarily produces $(\mathscr{M}_{P(\mathbf{y})}, \mathbf{y})$.

3-1-4 A Regression Model Example

Consider the linear model $\mathscr{M}$ as given by

$$\mathbf{y} = X\beta + \mathbf{e} \tag{3-4}$$

where X is an $n \times r$ matrix of rank r, $\boldsymbol{\beta}$ is the vector of r regression coefficients, and $\mathbf{e}$ has a distribution with

$$E(\mathbf{e}) = 0 \qquad \text{VAR}(\mathbf{e}) = \sigma^2 Q$$

where Q is a known inner-product matrix of rank p. This model has been examined by Rao (1976) and more extensively by Zyskind et al. (1971).

The model (3-4) suggests a variation-based model $\mathscr{M}_V$ as in Sec. 1-2-4. For our present purposes, however, we treat (3-4) as just a way of describing a response model as presented in Sec. 1-2-1.

For our model (3-4) we have assumed that X has full column rank r, that is, the column vectors of X are linearly independent. For the case with linearly dependent columns we could find a substitute X matrix with linearly independent columns. Of course, the original linearly dependent columns might correspond to different input variables, and the dependence has occurred through lack of control of the input variables; this means a lack of identifiability for some of the original parameters. The substitute X matrix corresponds to a reexpression of those parameters that are identifiable—thus we restrict ourselves to the notationally tidier case in which X has full column rank r.

We consider the model $\mathscr{M}$ for the rather special case in which the inner-product matrix Q has rank $p < n$, and we assume that $\mathbf{e}$ has a continuous distribution with a nonzero density function on the p-dimensional subspace determined by the singular inner-product matrix; for our purposes here it is convenient to think in terms of normal distributions.

We first investigate the p-dimensional subspace for the distribution of $\mathbf{e}$. For this we determine the VARiance of the linear compound $\mathbf{l}'\mathbf{e}$:

$$\text{VAR}(\mathbf{l}'\mathbf{e}) = E(\mathbf{l}'\mathbf{e}\mathbf{e}'\mathbf{l}) = \sigma^2 \mathbf{l}'Q\mathbf{l}$$

This VARiance is zero for all vectors $\mathbf{l}$ in $\mathscr{L}^\perp(Q)$, the orthogonal complement of $\mathscr{L}(Q)$; it follows that the distribution of $\mathbf{e}$ lies exclusively in the p-dimensional subspace $\mathscr{L}(Q)$ of $\mathbb{R}^n$.

Now consider an inference base $(\mathscr{M}, \mathbf{y}^0)$ involving a model of the rather special type just described. The model as it stands is somewhat more general than our notation so far allows: the model has a density function but it is a density on a contour that varies with the parameter. The definitions, however, can be extended in a natural way and we obtain the index set

$$D(\mathbf{y}) = \{(\boldsymbol{\beta}, \sigma): \mathbf{y} - X\boldsymbol{\beta} \in \mathscr{L}(Q)\}$$

We need some additional notation. Let $\mathscr{L}(N)$ be the intersection of the regression subspace and the error subspace:

$$\mathscr{L}(N) = \mathscr{L}(X) \cap \mathscr{L}(Q) \tag{3-5}$$

where N has s linearly independent column vectors. The interesting cases will have $s < r$; we now examine these cases.

Let $\mathscr{L}(X^*)$ and $\mathscr{L}(Q^*)$ be the orthogonal complements of $\mathscr{L}(N)$ in the spaces $\mathscr{L}(X)$ and $\mathscr{L}(Q)$ respectively:

$$\mathcal{L}(X^*) = \mathcal{L}(X) \cap \mathcal{L}^\perp(N) \qquad \mathcal{L}(Q^*) = \mathcal{L}(Q) \cap \mathcal{L}^\perp(N) \tag{3-6}$$

We assume that X^* and Q^* have $r-s$ and $p-s$ linearly independent column vectors. Computer programs are readily constructed to determine matrices N, X^*, and Q^*.

Any possible $\mathbf{y}$ for the model $\mathcal{M}$ can then be written uniquely as

$$\mathbf{y} = N\mathbf{a}(\mathbf{y}) + X^*\mathbf{c}(\mathbf{y}) + Q^*\mathbf{d}(\mathbf{y})$$

relative to the basis (N, X^*, Q^*) where $\mathbf{a}(\mathbf{y})$, $\mathbf{c}(\mathbf{y})$, $\mathbf{d}(\mathbf{y})$ are the regression vectors of $\mathbf{y}$ on N, on X^*, and on Q^*, respectively. Correspondingly, for the location vector $X\boldsymbol{\beta}$ of the model we have

$$X\boldsymbol{\beta} = N\boldsymbol{\alpha} + X^*\boldsymbol{\gamma}$$

where $(\boldsymbol{\alpha}, \boldsymbol{\gamma})$ is equivalent to $\boldsymbol{\beta}$.

We now determine the index set $D(\mathbf{y})$, but do so in terms of the new coordinates $\boldsymbol{\alpha}, \boldsymbol{\gamma}, \sigma$:

$$\begin{aligned} D(\mathbf{y}) &= \{(\boldsymbol{\alpha}, \boldsymbol{\gamma}, \sigma): N[\mathbf{a}(\mathbf{y}) - \boldsymbol{\alpha}] + X^*[\mathbf{c}(\mathbf{y}) - \boldsymbol{\gamma}] + Q^*\mathbf{d}(\mathbf{y}) \in \mathcal{L}(N, Q^*)\} \\ &= \{(\boldsymbol{\alpha}, \boldsymbol{\gamma}, \sigma): \boldsymbol{\gamma} = \mathbf{c}(\mathbf{y})\} \\ &= \mathbb{R}^s \times \mathbf{c}(\mathbf{y}) \times \mathbb{R}^+ \end{aligned} \tag{3-7}$$

Note that such sets form a partition of the parameter space $\Omega = \mathbb{R}^r \times \mathbb{R}^+$; thus we have that the options set $P(\mathbf{y})$ is equal to the index set $D(\mathbf{y})$ in (3-7). The simple interpretation of the preceding is that an observed response $\mathbf{y}^0$ completely determines $r-s$ of the regression coefficients:

$$\begin{aligned} \gamma_1 &= c_1(\mathbf{y}^0) \\ &\vdots \\ \gamma_{r-s} &= c_{r-s}(\mathbf{y}^0) \end{aligned} \tag{3-8}$$

We now record the form of the model $\mathcal{M}_{P(\mathbf{y})}$ in which the original parameter space Ω is restricted to the options set $P(\mathbf{y}^0)$. Using the new parameters we have

$$\mathbf{y} = N\boldsymbol{\alpha} + X^*\mathbf{c}(\mathbf{y}^0) + \mathbf{e}$$

or, equivalently,

$$\mathbf{y} - X^*\mathbf{c}(\mathbf{y}^0) = N\boldsymbol{\alpha} + \mathbf{e} \tag{3-9}$$

where $\mathbf{e}$ is distributed as described after formula (3-4). Note that this model has only s regression parameters.

The necessary method RM_1 thus produces the inference base

$$(\mathcal{M}_{P(\mathbf{y}^0)}, \mathbf{y}^0) \tag{3-10}$$

involving only s regression parameters and the error standard deviation σ.

Rao (1976) and Zyskind et al. (1971) concentrated their attention on finding estimates with special properties; they did not make explicit use of this initial necessary reduction. This reinforces our view that estimation per se is a very special terminal decision procedure. It is difficult to see how any broadly based approach to inference could overlook the preliminary parameter space reduction just described.

If an applied statistician in a certain application was able to calculate certain regression coefficients without error, he or she would certainly be very quick to do so and would feel considerable satisfaction; one certainly could not pretend that one did not know the values of the particular regression coefficients. For such parameters the reduction method is basically nonstatistical. The formal necessary method RM_1 places this reduction method in the larger context of statistical inference.

3-2 ON THE SAMPLE SPACE

Sometimes a statistical model can contain a component probability space, a component that does not involve the parameter of the model; for this we recall our Sec. 1-1-2 requirement that the components of a statistical model be *descriptive*—and thus correspond to objective components of the particular investigation. In addition, sometimes the data can give the observed value on the probability space. The *probabilistic* requirement in Sec. 1-1-2 then isolates the conditional statistical model—eliminating the remainder of the initial model as arbitrary.

We first survey some basic results on probability conditioning. We then consider an inference base $\mathscr{I} = (\mathscr{M}, \mathscr{D})$ and investigate reduction based on a component probability space.

3-2-1 Probability Space Conditioning

For the moment consider a *probability space* as a model for some real world system; by contrast a statistical model has free parameters corresponding to unknowns of a system being investigated. In its simplest form a probability space model has a sample space $\mathscr{S}$, a σ-algebra $\mathscr{A}$ of events, and a probability measure P. For convenience we continue with our Chap. 1 assumption that probability is given by density functions and that these *are* density functions in a limiting sense. For the present let $\mathscr{M}$ designate the probability model. The model $\mathscr{M}$ describes characteristics of the system under repeated performances, and, of course, it also describes a realization from the system provided the realization is effectively concealed from the investigator.

Now consider a realization from the system in a situation where there is available partial information concerning the realization; more informally, we say that the realization is partly concealed from the investigator. If the information C concerning the realization has the form of an observed value of a well-defined function, then from probability theory the proper model is the conditional model, say $\mathscr{M}_C$, given the observed value of the function.

The interpretation of the conditional model parallels that for the original model. The model $\mathscr{M}_C$ describes repeated performances, not of the original system but of a selected system: the original system is repeatedly operated until a value occurs satisfying the condition C to some reasonable approximation; this response is then a value from the selected system.

For some discussion on the application and interpretation of conditional probability see Fisher (1961), Fraser (1971, 1976, pp. 161–162), and Brenner and Fraser (1977).

In summary: the model $\mathscr{M}$ provides the probability description for a realization that is effectively concealed from the investigator. The model $\mathscr{M}_C$ provides the probability description for a realization that is concealed except for the information C—on the assumption that the information has the special form mentioned above.

3-2-2 Component Probability Spaces

Now consider an inference base $\mathscr{I} = (\mathscr{M}, \mathscr{D})$ and suppose the model $\mathscr{M}$ as presented has a component probability space.

Our definition of the model $\mathscr{M}$ in Sec. 1-1-2 required that it be descriptive, exhaustive, and probabilistic. In particular, we then have that the component probability space properly describes objective variables of the investigation. Thus, the probability space is not some construct that records constant totals of other parameter-dependent probabilities, a topic we examine in Sec. 4-2.

Consider the data $\mathscr{D}$. In this section we examine the case where the data $\mathscr{D}$ fully identifies the value on the probability space. Then in Sec. 3-3 we examine a related case where there is partial identification of a realization. The probabilistic requirements discussed in Secs. 1-1-2 and 3-2-1 then determine the form of the model.

Now for some notation. We suppose that the data $\mathscr{D}$ identifies a value on the component probability space. Let $\mathscr{S}(\mathscr{D})$ be the restriction of the original space in accord with the identified value on the component probability space. And let $\mathscr{M}^{\mathscr{S}(\mathscr{D})}$ be the original model but with probabilities replaced by conditional probabilities given $\mathscr{S}(\mathscr{D})$.

We thus obtain the following second necessary reduction method.

RM$_2$: Necessary reduction—sample space The inference base $(\mathscr{M}, \mathscr{D})$ necessarily produces the inference base $(\mathscr{M}^{\mathscr{S}(\mathscr{D})}, \mathscr{D})$.

From the data $\mathscr{D}$ we know the value on the component probability space. The reduction method RM$_2$ eliminates arbitrary elements from the inference base $(\mathscr{M}, \mathscr{D})$ as presented and produces $(\mathscr{M}^{\mathscr{S}(\mathscr{D})}, \mathscr{D})$ in which the distributions are the conditional distributions given the value on the probability space.

We now consider a spectrum of examples. One interesting example, however, that we omit here, is the location-scale model with known distribution form, a special case of Sec. 2-2. We omit it because it, together with the more general case having a parametric family for distribution form, is amenable to the reduction method in Sec. 3-3 and will be examined there.

3-2-3 Random Choice of Measuring Instrument

Consider a familiar example illustrating the preceding method. Two instruments are available for measuring a physical quantity θ. The instruments are different, but have known distribution properties for measuring a physical quantity. The example involves an initial random choice of measuring instrument, say with equal probabilities, and then the subsequent measurement of θ with the chosen instrument.

We consider the example at face value, even though it has certain unrealistic characteristics. Certainly responsible investigators would know their instruments and would choose between them on the basis of the special properties known to them. It is hard to imagine a reason for random choice of instrument, but this is not our concern here.

Let $\mathcal{M}^{(1)}$ be the model for the appropriate multiple observations with the first instrument and $\mathcal{M}^{(2)}$ be the model for the second instrument. For the random choice of model we have a probability space with probability $\frac{1}{2}$ at instrument 1 and probability $\frac{1}{2}$ at instrument 2. Let us represent the composite model in the convenient symbolic form:

$$\mathcal{M} = \tfrac{1}{2}\mathcal{M}^{(1)} + \tfrac{1}{2}\mathcal{M}^{(2)}$$

A realized value for the composite model has the form (i, m), where i indicates the instrument chosen and m represents the resulting measurements with that instrument. The immediately available inference base then has the form $(\mathcal{M}, (i, m))$. We now apply the necessary method RM_2 and obtain the reduced inference base $(\mathcal{M}^{(i)}, m)$. This inference base concerning θ is specific to the measuring instrument actually used.

The preceding reduction conforms to a reasonable scientific viewpoint—that an investigator would use the performance properties of the instruments actually used together with data from these instruments. The example has been considered from a somewhat different viewpoint in the literature; see, for example, Cox (1958).

3-2-4 Sample Surveys and Experimental Design

Consider a sample survey involving simple random sampling. Let $\Pi = \{c_1, \ldots, c_N\}$ designate the population and let $\mathbf{y} = (y_1, \ldots, y_n)'$ designate a sample of n from Π. We obtain a random sample from the population by choosing with equal probability one of the $N^{(n)}$ different sequences of n elements from Π. As special notation we might use the index set $\{1, \ldots, N\}$ to designate the population and use a sequence of n different integers from that set to designate the sample. We then obtain a random sample by choosing with equal probability one of the $N^{(n)}$ sequences of n integers from $\{1, \ldots, N\}$.

One might think at first sight that the necessary reduction RM_2 was applicable to this sample survey situation; the method would produce the conditional model given the chosen sample.

From a very practical viewpoint we can see that RM_2 is not applicable, for we note that conditionally there is no randomness left—there is no conditional model. Indeed, the purpose of the randomization was precisely to build a statistical model. The use of the index set is thus somewhat misleading.

From a theoretical viewpoint we can see that RM_2 is not applicable. An investigator would be quite satisfied if the randomization could be handled entirely out of sight, handled by someone reliable but inaccessible—and in some sense this is the case. We do *not* know the particular permutation of the unknown population values because we do not know, of course, what these values are. Accordingly we view the index set as something quite arbitrary—used only to accomplish the permutation of the N unknown values in the population. From this viewpoint we do *not* have an observed value on the probability space of such permutations.

A somewhat similar situation arises with the randomization of experimental design: treatments are randomly assigned to experimental units. We randomize the treatments to units because we do not know the units—we do not know how they would respond. We randomize precisely to obtain a reliable statistical model.

In an application we know the permutation of treatments to index numbers for the experimental units but do not know the permutation of treatments to the unknown response behavior of the units.

Thus we see from practical and theoretical viewpoints that the necessary reduction RM_2 is not applicable for the randomization of sampling and experimental design.

3-2-5 Random Choice of Sample Size

Consider some particular system under investigation and suppose there is an initial random choice of the number of performances n, the sample size. Let p_n be the known probability function for this random choice.

We consider the example at face value, although its unusual features are quite similar to those for random choice of instrument in Sec. 3-2-3. We do suggest, however, that a responsible investigator would choose the sample size deliberately and by design.

Let $\mathcal{M}^n$ be the composite model for n performances. The probability space for sample size has probability p_n for sample size n. Accordingly, let us represent the combined model for sample size and multiple performances in the convenient symbolic form:

$$\mathcal{M} = \sum_{n=1}^{\infty} p_n \mathcal{M}^n$$

A realized value for the combined model has the form (n, r_n), where n designates sample size and r_n designates the response from n performances.

The immediately available inference base then has the form

$$\left(\sum_{n=1}^{\infty} p_n \mathcal{M}^n, (n, r_n) \right)$$

We now apply the necessary reduction method RM_2 and obtain the reduced inference base

$$(\mathcal{M}^n, r_n)$$

This inference base is specific to the particular sample size used. The example has been considered from a somewhat different viewpoint in the literature; see, for example, Cox and Hinkley (1974, p. 32).

3-2-6 Sample from a Mixed Population

Two distinct populations A_1 and A_2 with relative sizes q_1 and q_2 are actually intermingled and are not readily distinguished; we have $q_1 + q_2 = 1$. A physical quantity θ can be assessed by the reaction probability $p_1(\theta)$ with a member of population A_1 and by reaction probability $p_2(\theta)$ with a member of population A_2; the functions p_1 and p_2 are known.

A random sample of size n is chosen from the combined population. The members of the individual populations are identified giving totals n_1 and n_2 for populations A_1 and A_2, respectively. The individuals are tested yielding the following data array:

Population	Reactions	No reactions	Totals
A_1	n_{11}	n_{12}	n_1
A_2	n_{21}	n_{22}	n_2

The model for the sample of size n is multinomial with probabilities as follows:

Population	Reaction	No reaction
A_1	$q_1 p_1(\theta)$	$q_1(1 - p_1(\theta))$
A_2	$q_2 p_2(\theta)$	$q_2(1 - p_2(\theta))$

We have introduced enough notation at this point without introducing additional notation for a formal presentation of the inference base. The results, however, of the necessary reduction method RM_2 are clear. The reduced model involves a sample of size n_1 from population A_1 giving the binomial $(n_1, p_1(\theta))$ and a sample of size n_2 from population A_2 giving an independent binomial $(n_2, p_2(\theta))$. Subsequent analysis would then involve the product of binomials rather than the four-way multinomial.

For the examples in Secs. 3-2-3 and 3-2-5 we were able to present no realistic reasons for the presence of a probability space—no good reasons for randomly choosing the measuring instrument or the sample size. The present example, however, is quite different. The probability space randomization describes the sampling process from the mixed populations, and this sampling is an essential part of obtaining information about the populations.

The necessary reduction of this example was discussed in Fraser (1973) and used to resolve dilemmas that were present in earlier considerations of such examples. Some anticipation of the necessary reduction may be found in Basu (1964). Kalbfleisch (1975) gives some alternate discussion of the method in Fraser (1973).

3-2-7 Random Choice of Input Variable

Consider a random system with response

$$y = \alpha + \beta x + e$$

that depends linearly on an input variable x; we assume that the variation e has a distribution that is known or known in form, or is more general, as discussed in Sec. 1-2 for the location-scale model.

We now suppose that values for the input x are obtained from a distribution with known density $f(x)$ and that this distribution is separate from and antecedent to the random system just described. This distribution for x could be describing some deliberate randomization as in the examples in Secs. 3-2-3 and 3-2-5, or it could be describing some related system that produces x values in accord with $f(x)$; as such it is closer to the example in Sec. 3-2-6.

Let $\mathcal{M}$ designate the compound model covering both the random choice of n input values and the regression for the n response values. And let $\mathcal{M}^{\times}$ designate the regression model alone for the response vector $\mathbf{y} = (y_1, \ldots, y_n)'$ given the input vector $\mathbf{x} = (x_1, \ldots, x_n)'$.

The immediately available inference base then has the form

$$(\mathcal{M}, (\mathbf{x}, \mathbf{y}))$$

The necessary reduction method RM_2 gives the reduced base

$$(\mathcal{M}^{\times}, \mathbf{y})$$

This inference base is specific to the input values $x_1, \ldots, x_n$ actually used for the initial random system.

3-3 FACTORIZATION

With certain statistical models the data may provide only partial information concerning the realization on the basic sample space of the model. If the information concerning the realization has the form of an observed value of a well-defined function then the appropriate marginal and conditional models are required. This gives a separation or factorization of the inference base.

3-3-1 The Method

In the preceding section we considered a model with a component probability space. The data fully identified the realization on the probability space. Correspondingly the inference base used the conditional model given the realization on the probability space.

In the present section we consider a statistical model in which the basic sample space is not directly observable; and correspondingly the data only partially identify the realization on the basic sample space. In particular we consider the case where the data information is equivalent to the observed value of a well-defined function on the basic sample space.

This differs from the preceding section in two ways. The distribution on the relevant sample space *may* depend on the parameter. The realized value on the relevant space is known to lie on a particular contour *and* there is no information as to where it lies on the contour.

Consider the inference base $\mathscr{I} = (\mathscr{M}, \mathscr{D})$. We suppose that the basic sample space is not fully observable. We also suppose that the data information $\mathscr{D}$ is equivalent to the observed value D^0 of a well-defined function D on the basic sample space; in particular there is no differential information as to where the realization is on the contour specified by $D = D^0$.

For notation let $\mathscr{M}_D$ be the marginal model for the function D and let $\mathscr{M}^D$ be the conditional model given a value D for the function. Then from Secs. 1-1-2 and 3-2-1 we have the following third necessary reduction method:

RM$_3$: Necessary reduction—factorization The inference base $(\mathscr{M}, \mathscr{D})$ necessarily produces the inference bases $(\mathscr{M}_D, D^0)$ and $(\mathscr{M}^{D^0}, \mathscr{D})$.

Let λ in Λ be the parameter for the marginal model $\mathscr{M}_D$; the parameter λ may be the whole of the original parameter θ, or some reduction $\lambda = \lambda(\theta)$ on θ. The inference base $(\mathscr{M}_D, D^0)$ has an observed value D^0 and a model $\mathscr{M}_D$ for possible values for the function D; the parameter is λ.

The inference base $(\mathscr{M}^{D^0}, \mathscr{D})$ has the data $\mathscr{D}$ and the conditional model given the value D^0 for the function D. The data information $\mathscr{D}$, however, gives no differential information as to where the realization is, given the contour specified by $D = D^0$. Typically there are parameters unknown in value that prevent the identification of the realization on the contour $D = D^0$—they not only prevent the identification of the realization but provide no differential information concerning the realization. We consider this in more detail in Chap. 7. The factorization of method RM$_3$ follows from the probabilistic requirement in Sec. 1-1-2.

3-3-2 The Location-Scale Model

For a location-scale system consider the variation-based model

$$\mathscr{M}_V^n = (\Omega; \mathbb{R}^n, \mathscr{B}^n, \mathscr{V}^n, \mathscr{T}) \tag{3-11}$$

in Sec. 1-2-4. We examine the inference base $(\mathscr{M}_V^n, \mathbf{y}^0)$.

From Secs. 2-1-1 and 2-1-2 we see that the data give the value

$$\mathbf{d}(\mathbf{z}) = \mathbf{d}(\mathbf{y}^0) = \mathbf{d}^0 \tag{3-12}$$

for the function $\mathbf{d}(\mathbf{z})$ on the variation space; the variation space has the distributions $\mathscr{V}^n$. From the equation (2-9) it is plausible that the data give no differential information as to the value of $[\bar{z}, s(\mathbf{z})]$; we will give a precise verification of this in Chap. 7.

The marginal model describing the observed $\mathbf{d}(\mathbf{z})$ is available from formula (2-15); as noted earlier the space for $\mathbf{d}(\mathbf{z})$ is the unit sphere in $\mathscr{L}^1(1)$. We avoid extra notation and abbreviate the model as

$$\mathscr{M}_\mathbf{d} = (\Lambda; \{h_\lambda(\mathbf{d})\}) \tag{3-13}$$

The corresponding inference base is

$$(\mathscr{M}_\mathbf{d}, \mathbf{d}^0) \tag{3-14}$$

The conditional model describing the unobservable $[\bar{z}, s(\mathbf{z})]$ is available from formula (2-17); let $\mathscr{V}^*$ designate the class of densities. The unknowns standing between $[\bar{z}, s(\mathbf{z})]$ and $[\bar{y}, s(\mathbf{y})]$ are given by the equations in (2-9); we can reexpress these as a transformation T^* from $[\bar{z}, s(\mathbf{z})]$ to $[\bar{y}, s(\mathbf{y})]$:

$$T^*: \begin{cases} \bar{y} = \mu + \sigma\bar{z} \\ s(\mathbf{y}) = \sigma s(\mathbf{z}) \end{cases} \tag{3-15}$$

Thus we have the model

$$\mathscr{M}^{\mathbf{d}^0} = \{\Lambda \times \{(\mu, \sigma)\}; \mathbb{R} \times \mathbb{R}^+, \mathscr{B}^2, \mathscr{V}^*, \mathscr{T}^*\} \tag{3-16}$$

where $\mathscr{T}^* = \{T^*\}$ is the class of location-scale transformations (3-15). The corresponding inference base is

$$(\mathscr{M}^{\mathbf{d}^0}, [\bar{y}, s(\mathbf{y})]) \tag{3-17}$$

The preceding inference base has a distribution with parameter λ that describes the realization $[\bar{z}, s(\mathbf{z})]$ on the basic variation space. The data $[\bar{y}, s(\mathbf{y})]$ gives no information as to where the realization $[\bar{z}, s(\mathbf{z})]$ is located on the variation space; the unknown $[\mu, \sigma]$ stands between the realization and the available data $[\bar{y}, s(\mathbf{y})]$.

The necessary reduction method RM_3 applied to the initial inference base then gives the separation

$$(\mathscr{M}_\mathbf{d}, \mathbf{d}^0); (\mathscr{M}^{\mathbf{d}^0}, [\bar{y}, s(\mathbf{y})])$$

discussed in Sec. 2-1. The separation was subsequently used for the terminal methods in Sec. 2-2.

3-4 BY REEXPRESSION

In Chap. 1 we presented the inference base as the essential material from the specification and the performances of the system. As part of this we remarked that the notation used to present the inference base may unintentionally introduce arbitrary elements. In this section we consider a method for eliminating such arbitrary elements of notation.

Of course, the model and data as initially presented in an inference base may contain material seemingly of substance and yet beyond that determined by the definitions in Sec. 1-1. The methods in Secs. 3-1, 3-2, and 3-3 are concerned with eliminating such seemingly substantial but actually arbitrary elements. In this section the method is concerned more with eliminating the arbitrary elements of notation.

Our approach to eliminating these arbitrary elements is the conventional mathematical approach—the use of the invariant group of a mathematical structure. In Sec. 1-2-2 we discussed a class of transformations on a space that *was closed under the formation of products and the formation of inverses*. A class of transformations with this property is called a *group—a transformation group*. In this section we use transformation groups to eliminate arbitrary elements from the inference base.

3-4-1 Invariant Group of an Inference Base

Consider an inference base $\mathscr{I} = (\mathscr{M}, \mathscr{D})$ and let Ω be the parameter space of the model $\mathscr{M} = (\Omega; \ldots)$.

We examine transformations on the response space, the space of the response variables of the system under investigation. We view a transformation on the response space as providing a *reexpression* or alternate mode of expression for the response.

We might initially think of the class of all invertible transformations on the response space. In Chap. 1, however, we restricted our attention to models that have density functions, densities that are densities in a limiting sense. Accordingly, we restrict our attention here to transformations that respect densities. We thus consider the class of one-one transformations on the response space that are continuously differentiable each way; these transformations are called *diffeomorphisms*. The composition of two diffeomorphisms is a diffeomorphism and the inverse of a diffeomorphism is a diffeomorphism. Thus the class of diffeomorphisms is a group. In this section we examine the group G of diffeomorphisms on the response space of the problem under investigation.

Consider an element g in the group G of diffeomorphisms. A response variable $\mathbf{y}$ in the initial mode of expression becomes a response variable $g\mathbf{y}$ in the alternate mode of expression. A response value $\mathbf{y}^0$ in the given mode of expression becomes a value $g\mathbf{y}^0$ in the alternate mode of expression. Or, equivalently, data $\mathscr{D}$ in the initial mode become data $g\mathscr{D}$ in the alternate mode.

Now consider how a reexpression g affects the model $\mathscr{M} = (\Omega; \ldots)$. A parameter value θ indexes one of the possible distributions for the response in the initial mode of expression and of course indexes the corresponding distribution in the alternate mode of expression. In particular, the "true value" of the parameter on Ω designates the "true distribution" for the response in the initial mode of expression and also indexes the corresponding "true distribution" in the alternate mode of expression. Thus there is no transformation on the parameter space Ω—just a change of distribution for each θ value as dictated by the reexpression transformations.

The model $\mathscr{M}$ records Ω together with the various indexed distributions for the given mode of expression. We let $g\mathscr{M}$ designate the model for the reexpression g of the response variable; $g\mathscr{M}$ records the same Ω together with the various indexed distributions for the alternate mode of expression.

We emphasize particularly that $\mathscr{M}$ and $g\mathscr{M}$ have the same Ω and that a parameter value θ designates a distribution for the initial mode of expression and designates the corresponding distribution in the alternate mode of expression.

An inference base $(\mathscr{M}, \mathscr{D})$ in the initial mode of expression becomes an inference base $(g\mathscr{M}, g\mathscr{D})$ in the alternate mode of expression. We examine the following class consisting of the various modes of reexpression for the inference base:

$$\{(g\mathscr{M}, g\mathscr{D}): g \in G\} \tag{3-18}$$

These provide different ways of looking at the inference base as initially presented.

Sometimes we will find a nontrivial reexpression that does not change the model. Let $G_\mathscr{M}$ be the set of transformations $g \in G$ that *leave the model unchanged*:

$$G_\mathscr{M} = \{g : g\mathscr{M} = \mathscr{M}, g \in G\} \tag{3-19}$$

$G_\mathscr{M}$ is closed under products and inverses; $G_\mathscr{M}$ is a group, called the *invariant group* of $\mathscr{M}$; this usage of the term invariant group differs from another usage that does not take account of the indexing class Ω.

Now consider the class (3-18) of reexpression for the inference base. The initial model $\mathscr{M}$ is associated with various data values $g\mathscr{D}$; we write

$$G_\mathscr{M}\mathscr{D} = \{g\mathscr{D} : g \in G_\mathscr{M}\} \tag{3-20}$$

for the set of data values corresponding to the initial model $\mathscr{M}$.

We introduce here a simple property of a group of transformations on a space. Consider a group H on a space $\mathscr{S}$. The group H carries a point s into the *orbit*

$$Hs = \{hs : h \in H\} \tag{3-21}$$

of all images under the transformations. As a group is closed under products and inverses we see easily that the orbit Hs can be generated by applying H to any element of the orbit. We then have that the orbits form a partition $\mathscr{P}$ of the space $\mathscr{S}$:

$$\mathscr{P} = \{Hs : s \in \mathscr{S}\} \tag{3-22}$$

This partition $\mathscr{P}$ is often designated $\mathscr{S}/H$, called $\mathscr{S}$ modulo the group H. In conclusion we note that Hs gives a mapping from $\mathscr{S}$ onto $\mathscr{P} : s \to Hs$. Any function one-one equivalent to this mapping is called a *maximal invariant function* and it of course indexes the orbits Hs on the space $\mathscr{S}$.

Now consider the data set

$$G_\mathscr{M}\mathscr{D} = \{g\mathscr{D} : g \in G_\mathscr{M}\} \tag{3-23}$$

that is associated with the model $\mathcal{M}$ in the class (3-18). Or more generally consider the various data sets $G_\mathcal{M}\mathcal{D}$ as we consider various possible data $\mathcal{D}$. Let m be some maximal invariant function that indexes the orbits of the invariant group $G_\mathcal{M}$. Then the various possible data sets $G_\mathcal{M}\mathcal{D}$ are one-one equivalent to the various values $m(\mathcal{D})$ of the function m.

We have used the transformation G to determine the arbitrary elements in the inference base $(\mathcal{M}, \mathcal{D})$. Among the reexpressions recorded in (3-18) we find the model $\mathcal{M}$ associated not with $\mathcal{D}$ but with the data set $G_\mathcal{M}\mathcal{D}$ or equivalently with the value $m(\mathcal{D})$ of the function m. Let $\mathcal{M}_*$ designate the marginal model for the maximal invariant function m. Then the definition of the inference base in Sec. 1-1 gives

$$(\mathcal{M}_*, m(\mathcal{D}))$$

We thus obtain the following fourth necessary reduction method.

RM$_4$: Necessary reduction—reexpression The inference base $(\mathcal{M}, \mathcal{D})$ necessarily produces the inference base $(\mathcal{M}_*, m(\mathcal{D}))$.

The transformations in G provide various ways of looking at an inference base. As part of this we find that the model $\mathcal{M}$ has associated data $G_\mathcal{M}\mathcal{D}$ or equivalently associated data $m(\mathcal{D})$. Accordingly the inference base without the arbitrary elements is just $(\mathcal{M}_*, m(\mathcal{D}))$. For this it is appropriate to recall our definition of the model in Sec. 1-1-2—that all the relevant characteristics of the system are included—and to note then that $G_\mathcal{M}$ leaves the model unchanged.

3-4-2 Simple Normal Example

Consider a response y that is known to be normally distributed with unknown location μ and known scaling σ_0. Let $\mathbf{y}^0$ be the data for a sample of n.

We examine an inference base using the minimum response-based model from Sec. 1-2-1:

$$\mathcal{M}_R = (\Omega = \{\mu\}; \mathbb{R}^n, \mathcal{B}^n, \mathcal{F}) \qquad (3\text{-}24)$$

where

$$\mathcal{F} = \left\{ (2\pi\sigma_0^2)^{-n/2} \exp\left[-\frac{1}{2\sigma_0^2} \Sigma(y_i - \mu)^2 \right] : \mu \in \mathbb{R} \right\}$$

is the class of response density functions.

We take the model $\mathcal{M}_R$ at face value: a parameter space, a response space, a class of events, and a class of densities indexed by the parameter space. As a realistic model for applications we might well have additional essential ingredients, e.g., explicit identification of the variation, or distinctiveness of the coordinates, or individual length measure for each coordinate. However, we take (3-24) at face value and examine the inference base

$$(\mathcal{M}_R, \mathbf{y}^0) \qquad (3\text{-}25)$$

The likelihood function from a response value **y** is

$$L(\mathbf{y};\mu) = c \exp\left[-\frac{1}{2\sigma_0^2}\Sigma(y_i - \mu)^2\right]$$

$$= c \exp\left[-\frac{n}{2\sigma_0^2}(\bar{y} - \mu)^2\right] \tag{3-26}$$

As a function on the space $\mathbb{R}^n$ this likelihood map (see, for example, Fraser, 1976, p. 334) is equivalent to the simple functions Σy_i, or $\bar{y}$, called the likelihood statistic.

Now consider the group $G_\mathcal{M}$ of transformations that do not change the model $\mathcal{M}$. Under a diffeomorphism in $G_\mathcal{M}$ the density function does not change and in particular the dependence of this on the parameter μ does not change. Thus a transformation that leaves $\mathcal{M}$ unchanged in particular leaves the likelihood unchanged and thus must be a transformation within contours of the simple functions Σy_i or $\bar{y}$.

The conditional distribution given $\bar{y}$ is a rotationally symmetric $(n-1)$ dimensional normal centered on the intersection with the **1** vector. By contraction we can view this as a uniform distribution on a $(n-1)$ dimensional unit ball. With $n \geq 3$ the ball has dimension greater than or equal to 2 (a disc or more), and the group that leaves the uniform distribution invariant is transitive on that ball (carries any point into any other point on that ball). Accordingly, for $n \geq 3$ the group $G_\mathcal{M}$ is transitive on the contours of the function $\bar{y}$ or, equivalently, the function $\bar{y}$ is a maximal invariant; this discussion briefly surveys the key components for a more formal and detailed proof.

The distribution of the maximal invariant $\bar{y}$ is normal $(\mu, \sigma_0/\sqrt{n})$; the model is

$$\mathcal{M}_* = (\Omega = \{\mu\}; \mathbb{R}, \mathcal{B}, \mathcal{F}_*) \tag{3-27}$$

where

$$\mathcal{F}_* = \left\{\left(\frac{2\pi\sigma_0^2}{n}\right)^{-1/2} \exp\left[-\frac{n}{2\sigma_0^2}(\bar{y} - \mu)^2\right] : \mu \in \mathbb{R}\right\}$$

The reduction method RM_4 applied to the inference base (3-25) with $n \geq 3$ necessarily produces the following inference base:

$$(\mathcal{M}_*, \bar{y}^0) \tag{3-28}$$

This inference base involves only the sample average. Recall, however, the minimum ingredients for the initial model (3-24).

3-5 BY REEXPRESSION; A PARAMETER COMPONENT

In Sec. 3-4 we used the transformation group on the response space to eliminate arbitrary elements in the inference base. In this section we suppose that attention is entirely restricted to a parameter component of the inference base and we again use the transformation group to eliminate the arbitrary elements.

3-5-1 Invariant Group for a Parameter Component

Consider an inference base $\mathscr{I} = (\mathscr{M}, \mathscr{D})$. Let Ω be the parameter space for the model $\mathscr{M}$ and suppose that attention is entirely restricted to the parameter component $\phi = h(\theta)$ where h is a given function on Ω.

We continue with our restriction to models involving density functions and, accordingly, consider the group G of diffeomorphisms on the response space. We then examine the following class of reexpressions for the inference base:

$$\{(g\mathscr{M}, g\mathscr{D}) : g \in G\} \tag{3-29}$$

Now consider the preceding with attention entirely restricted to the parameter component $\phi = h(\theta)$. Typically, some transformations in G will leave the model $\mathscr{M}$ unchanged; and often some further transformations will leave the model unchanged *with respect to the parameter component* $\phi = h(\theta)$. Consider this in more detail. A transformation g leaves $\mathscr{M}$ unchanged with respect to ϕ if the models $\mathscr{M}$ and $g\mathscr{M}$ have for each ϕ value the same set of distributions but the indexing of these distributions can be different; thus g can induce a permutation of distributions with parameter values as long as the permutations are entirely within the contours of $h(\theta) = \phi$.

We might be tempted to say that such a transformation g changes the θ value but not the ϕ value. But this would be wrong and very misleading. The parameter values and the parameter space do not change under a transformation g in G. Rather, a transformation alters the distributions associated with θ values; and for the transformation g just discussed the distributions are altered, permuted along the contours of the parameter function $h(\theta) = \phi$.

Let $\bar{G}_\mathscr{M}$ designate the set of transformations g that leave the model unchanged in the ϕ-specific sense just described. With our attention restricted to the parameter component ϕ, the models $\mathscr{M}$ and $g\mathscr{M}$ are equivalent for g in $\bar{G}_\mathscr{M}$. Accordingly, we identify them for our present purposes and write $\mathscr{M}$ as the representative.

Now consider the class (3-29) of reexpressions for the inference base. The model $\mathscr{M}$ is associated with data $g\mathscr{D}$ for each g in $\bar{G}_\mathscr{M}$. We write

$$\bar{G}_\mathscr{M} \mathscr{D} = \{g\mathscr{D} : g \in \bar{G}_\mathscr{M}\} \tag{3-30}$$

for the set of data values that are associated with $\mathscr{M}$. Let $\bar{m}$ be the maximal invariant function that indexes the orbits of the invariant group $\bar{G}_\mathscr{M}$. Then the various possible data sets $\bar{G}_\mathscr{M} \mathscr{D}$ are one-one equivalent to the various values $\bar{m}(\mathscr{D})$ of the function $\bar{m}$.

We have used the transformation group to determine the arbitrary elements in the inference base. We have, of course, the model $\mathscr{M}$, but for data we have just the value $\bar{m}(\mathscr{D})$ for the function $\bar{m}$. Let $\mathscr{M}_*$ designate the marginal model for the function $\bar{m}$, the maximal invariant with respect to the invariant group $\bar{G}_\mathscr{M}$. Then the definition of inference base in Sec. 1-1 gives

$$(\mathcal{M}_*, \bar{m}(\mathcal{D})) \tag{3-31}$$

We thus obtain the following fifth necessary reduction method.

RM$_5$: Necessary reduction—parameter component With attention restricted to the parameter component $\phi = h(\theta)$, the inference base $(\mathcal{M}, \mathcal{D})$ necessarily produces the inference base $(\mathcal{M}_*, \bar{m}(\mathcal{D}))$.

3-5-2 Normal Example

Consider a response y that is known to be normally distributed with unknown location μ and unknown scaling σ. Let $\mathbf{y}^0$ be the data for a sample of n. We suppose that attention is entirely restricted to the parameter component σ, the second coordinate projection for $\theta = (\mu, \sigma)$.

As in Sec. 3-4-2 we again examine a minimum response-based model:

$$\mathcal{M}_R = (\Omega = \{(\mu, \sigma)\}; \mathbb{R}^n, \mathcal{B}^n, \mathcal{F}) \tag{3-32}$$

where

$$\mathcal{F} = \left\{ (2\pi\sigma^2)^{-n/2} \exp\left[-\frac{1}{2\sigma^2} \Sigma (y_i - \mu)^2 \right] : \begin{array}{l} \mu \in \mathbb{R} \\ \sigma \in \mathbb{R}^+ \end{array} \right\}$$

is the class of response density functions. Again we take the model $\mathcal{M}_R$ at face value; recall the remarks in Sec. 3-4-2. We now examine the inference base

$$(\mathcal{M}_R, \mathbf{y}^0) \tag{3-33}$$

with attention entirely restricted to the parameter component σ.

As a preliminary step we consider the invariant group $G_{\mathcal{M}}$ for the model itself. The methods in Sec. 3-4-2 can be followed; some needed likelihood material is available from Sec. 4-1-3. We then have that for $n \geq 4$ (not 3) the invariant group $G_{\mathcal{M}}$ has the maximal invariant $[\bar{y}, s(\mathbf{y})]$.

Now consider the restriction to the parameter component σ; the function h gives the projection of $\theta = (\mu, \sigma)$ onto the second coordinate. We enlarge the group $G_{\mathcal{M}}$ to the group $\bar{G}_{\mathcal{M}}$ by now allowing new pairings of distributions to parameter values, but within the contours of the projection function h. Some obvious additional transformations are those of the location group

$$H = \{[a, 1] : a \in \mathbb{R}\} \tag{3-34}$$

where from Sec. 1-2-1 we have the location transformations

$$[a, 1]\mathbf{y} = a\mathbf{1} + \mathbf{y}$$

and we have closure under composition and inverses. Fortunately with our choice of H we have that $[a, 1]$ provides a transformation of the orbits of $G_{\mathcal{M}}$, carrying one orbit intact to become another orbit. This manifests itself as

$$[a, 1][\bar{y}, s(\mathbf{y})] = [\bar{y} + a, s(\mathbf{y})] \tag{3-35}$$

The group $\bar{G}_\mathcal{M}$ will contain transformations generated from $G_\mathcal{M}$ and H. It has the maximal invariant

$$\bar{m}(\mathbf{y}) = s(\mathbf{y}) \tag{3-36}$$

The distribution of the maximal invariant $s(\mathbf{y})$ is σ-chi$(n-1)$; the model is

$$\mathcal{M}_* = (\Omega^* = \{\sigma\}; \mathbb{R}^+, \mathcal{B}, \mathcal{F}_*) \tag{3-37}$$

where

$$\mathcal{F}_* = \left\{ \frac{A_{n-1}}{(2\pi)^{(n-1)/2}} \exp\left(-\frac{s^2}{2\sigma^2}\right) \left(\frac{s}{\sigma}\right)^{n-2} \frac{1}{\sigma} : \sigma \in \mathbb{R}^+ \right\}$$

where $A_f = 2\pi^{f/2}/\Gamma(f/2)$ is the surface volume of the unit sphere in $\mathbb{R}^f$. $\mathcal{F}_*$ can be obtained by a simple rewrite of the rescaled gamma distribution of $s^2(\mathbf{y})$.

With attention restricted to the component σ the reduction method RM$_5$ applied to the inference base (3-33) with $n \geq 4$ necessarily gives the inference base

$$(\mathcal{M}_*, s(\mathbf{y}^0)) \tag{3-38}$$

This inference base involves only the sample residual length $s(\mathbf{y}^0)$. Recall, however, the minimum ingredients for the initial model (3-32).

It is of interest that the present necessary method does not work for the case when attention is restricted to the parameter μ.

REFERENCES AND BIBLIOGRAPHY

Basu, D.: Recovery of Ancillary Information, *Sankhyā*, ser. A, vol. 26, pp. 3–16, 1964.
Brenner, D., and D. A. S. Fraser: "When is a Class of Functions a Function?" Department of Statistics, University of Toronto, 1977.
Cox, D. R.: Some Problems Connected with Statistical Inference, *Ann. Math. Stat.*, vol. 29, pp. 357–372, 1958.
———, and D. V. Hinkley: "Theoretical Statistics," Chapman and Hall, London, 1974.
Fisher, R. A.: Sampling the Reference Set, *Sankhyā*, ser. A, vol. 23, pp. 3–8, 1961.
Fraser, D. A. S.: Events, Information Processing, and the Structural Model, in V. P. Godambe and D. A. Sprott (eds.), "Proc. Symposium on the Foundations of Statistical Inference," Holt, Rinehart and Winston of Canada, Toronto and Montreal, pp. 32–55, 1971.
———: The Elusive Ancillary, in D. G. Kabe and R. P. Gupta (eds.), "Multivariate Statistical Inference," North-Holland Publishing Company, Amsterdam and London, 1973.
———: Necessary Analysis and Adaptive Inference, *Jour. Am. Stat. Assoc.*, vol. 71, pp. 99–113, 1976.
———: "Probability and Statistics: Theory and Applications," Duxbury Press, North Scituate, Mass., 1976.
Kalbfleisch, J. D.: Sufficiency and Conditionality, *Biometrika*, vol. 62, pp. 251–259, 1975.
Rao, C. R.: Estimation of Parameters in a Linear Model, *Ann. Stat.*, vol. 4, pp. 1023–1037, 1976.
Zyskind, G., O. Kempthorne, A. Mexas, P. Papaidannov, and J. Seely: "Linear Models, Statistical Information and Statistical Inference," Study-Aerospace Research Laboratories, Project no. 7071, available from Clearing House, U.S. Department of Commerce, Springfield, Va., 22151, 1971.

CHAPTER
FOUR
DENSITY ALLOCATION METHODS

In Chap. 3 we examined necessary reduction methods, methods that eliminated arbitrary elements from the inference base. In this chapter we examine reduction methods that require an inference principle. The inference principles range from the very widely held sufficiency principle to the relatively uncommon weak ancillarity principle.

We also examine the support for the inference principles. We find that the support for some of the principles is rather weak, resting on a few appealing examples. Indeed, for one of the principles we find that all the appealing examples properly belong with one of the necessary methods in the preceding chapter.

We will see that the substance of the methods involves grouping together points on the response space to satisfy some criterion in terms of the density function—in other words, *allocating* points with associated density to satisfy some criterion. Accordingly, we call the methods in this chapter *density allocation* methods.

In Sec. 4-1 we consider the weak likelihood principle, otherwise called the sufficiency principle. In Sec. 4-2 we consider the ancillarity principle. And in the remaining sections we consider several principles that involve a mix or weakening of the preceding principles.

The reduction methods in this chapter are targeted primarily on inference bases that involve the response-based model $\mathcal{M}_R$ in Sec. 1-2-1. They can of course be applied to inference bases that use a variation-based model but the support for the needed inference principle may or may not be available in the altered context.

Accordingly, in this chapter we discuss the reduction methods in terms of a response-based model

$$\mathcal{M} = (\Omega; \mathcal{S}, \mathcal{A}, \mathcal{F})$$

where $\mathcal{S}$ is the response space and

$$\mathcal{F} = \{f(\cdot|\theta): \theta \in \Omega\}$$

is the class of response density functions. A wider application of a method may be in terms of a model with additional ingredients—but the support for the needed principle may no longer be available and the method itself may be quite inappropriate in the context.

The variables and parameters will typically be vectors but for the general discussions here we will write simply y, θ, for example, without the use of bold face.

4-1 SUFFICIENCY REDUCTION

In this section we examine a reduction method that uses the sufficiency or weak likelihood principle. This principle has wide acceptance in the literature.

The analysis of sufficiency seems to proceed most naturally by using properties of the likelihood function and likelihood map. Accordingly we give a preliminary discussion of these likelihood concepts.

4-1-1 Likelihood Function

For this we need the definition of the likelihood function. Some familiarity with likelihood has been assumed for the location-scale example in Sec. 2-2 and for the examples in Secs. 3-4 and 3-5. For some recent discussion and development see Fraser (1976, secs. 8-1, 8-4, 8-5).

For our discussions here, we use the definition of likelihood that is appropriate and fruitful for most areas of statistical inference—in fact, the original definition as found in Fisher (1922). The reader is cautioned that in most statistical texts, likelihood is used in a very limited way and the definition, as suitable for this limited use, does not conform to that needed for general inference or to that implicit in the original presentation of the concept. We do assume the availability of general discussions on likelihood as, for example, in Fraser (1976, secs. 8-1, 8-4, 8-5) and accordingly present here a relatively brief survey.

At a response value y the model $\mathcal{M}$ assigns a probability density $f(y|\theta)$, which depends on the parameter θ in Ω. If we allow, as is common, that no particular supporting measure has special significance, then we have that the probability density depends on θ, but is defined up to an arbitrary multiplicative constant:

$$L(y;\cdot) = cf(y|\cdot) \tag{4-1}$$

where c is an arbitrary positive multiplicative constant. This is called the *likelihood function* as obtained from the response value y. In this, we view y as taking a particular value and treat (4-1) as a real-valued function over Ω that plots $L(y;\theta)$ at the point θ in Ω.

More formally we define as follows:

Definition The likelihood function $L(y;\cdot)$ from the response value y is the equivalence class

$$L(y;\cdot) = \{cf(y|\cdot): c \in \mathbb{R}^+\} \qquad (4\text{-}2)$$

of similarly shaped functions of θ.

This definition avoids the direct use of a generic positive constant c.

The set of real-valued functions on Ω is designated by $\mathbb{R}^\Omega$; it is a vector space. The functions in the class (4-2) differ from each other by a positive multiplicative constant. Accordingly the class (4-2) is a *ray* from the origin of the vector space passing through the function $f(y|\cdot)$; alternatively, it can be described as the positive half of a one-dimensional subspace of $\mathbb{R}^\Omega$.

Consider $L(y;\cdot)$ for various response values y. As a *function* on the sample space $\mathscr{S}$ it maps a point y into a ray in $\mathbb{R}^\Omega$, the likelihood function for that point y; we call this function the *likelihood map*.

Now suppose we form sets or contours on the response space $\mathscr{S}$ in the following manner: we allocate to a contour all the response values y that have a particular likelihood function. And suppose we do this for each of the possible likelihood functions. Then we have formed a partition on the sample space $\mathscr{S}$, specifically the preimage partition of the likelihood map.

Now let $s(\cdot)$ be some simple convenient function on the sample space $\mathscr{S}$ that indexes the sets of the partition just formed; we call such a function the *likelihood statistic*. The likelihood statistic, then, is any response-space function that is one-one equivalent to the likelihood map.

4-1-2 Sufficiency

We now summarize some results concerning sufficiency. The concept of sufficiency is also due to Fisher (1920) and is approximately two years older than the likelihood concept. The present form of the concept of sufficiency originally due to Fisher is called B-sufficiency in a recent survey of generalizations of the concept by Barndorff-Nielsen (1971).

For the response model $\mathscr{M}$, we now define a sufficient statistic.

Definition The function $t(\cdot)$ is sufficient if the conditional distribution given t does not depend on the parameter θ.

For more detailed notation we let $r(\cdot)$ be a complementary function so that y is one-one equivalent to $[t(y), r(y)]$. We suppose that dt and dr designate suitable volume measures for t and r and we suppose that $h(t|\theta)$ and $g(r|t)$ are marginal and conditional density functions for t and for r given t. We then note the following:

Definition The function $t(\cdot)$ is sufficient if the joint density of t and r can be factored as

$$h(t|\theta)\, g(r|t)$$

where the conditional density $g(r|t)$ is independent of θ.

Now let us examine the likelihood function along contours of the statistic $t(\cdot)$:

$$\begin{aligned} L(y;\cdot) &= \{cf(y|\cdot) : c \in \mathbb{R}^+\} \\ &= \{cg(r|t)\, h(t|\cdot) : c \in \mathbb{R}^+\} \\ &= \{ch(t|\cdot) : c \in \mathbb{R}^+\} \end{aligned} \qquad (4\text{-}3)$$

If we tolerate the generic arbitrary positive constant c we can abbreviate the preceding as

$$L(y;\theta) = cf(y|\theta) = cg(r|t)\, h(t|\theta) = ch(t|\theta)$$

where the constant (with respect to θ) $g(r|t)$ has been absorbed into the arbitrary multiplicative constant c. Note from (4-3) that the likelihood function does not depend on r. Thus all the points on a given contour of $t(\cdot)$ have the same likelihood function—but the contour may not include all points with the particular likelihood function.

We now define a *minimal sufficient statistic*.

Definition A sufficient statistic $s(\cdot)$ is a minimal sufficient statistic if for any other sufficient statistic $t(\cdot)$ there exists a function q such that $s(y) = q[t(y)]$.

In effect the definition says that $s(\cdot)$ is a reduction on any other sufficient statistic, or equivalently that $s(\cdot)$ has a coarser preimage partition than any other sufficient statistic. Thus in accord with the preceding paragraph we are led to expect the minimal sufficient statistic to be the statistic for which each contour assembles all the points with a particular likelihood function. This is formalized in the following theorems.

Theorem 4-1 The likelihood statistic $s(\cdot)$ is a sufficient statistic for $\mathcal{M}$.

Theorem 4-2 The likelihood map is the minimal sufficient statistic; the likelihood statistic $s(\cdot)$ is the minimal sufficient statistic.

For a proof of Theorem 4-1 see, for example, Fraser (1976, p. 337). Theorem 4-2 follows trivially from our discussion on assembling points with a common likelihood function.

Our present organization of the material on likelihood and sufficiency makes the very intimate connection between the concepts seem almost trivial. The connection, however, was not noted publicly at statistics meetings until the early 1960s. Interestingly, Fisher's writings clearly contained the implicit connection—as early as the 1930s (Fisher, 1934).

4-1-3 Some Simple Examples

Consider the response model for a sample from the normal (μ, σ) distribution. The density function is

$$f(\mathbf{y}|\mu, \sigma) = (2\pi\sigma^2)^{-n/2} \exp\left[-\frac{1}{2\sigma^2}\sum(y_i - \mu)^2\right]$$

$$= (2\pi\sigma^2)^{-n/2} \exp\left[-\frac{s^2(\mathbf{y})}{2\sigma^2} - \frac{n(\bar{y} - \mu)^2}{2\sigma^2}\right] \quad (4\text{-}4)$$

and the likelihood function is

$$L(\mathbf{y}; \mu, \sigma) = c(\sigma^2)^{-n/2} \exp\left[-\frac{s^2(\mathbf{y})}{2\sigma^2} - \frac{n(\bar{y} - \mu)^2}{2\sigma^2}\right] \quad (4\text{-}5)$$

From $[\bar{y}, s(\mathbf{y})]$ we are able to determine the particular likelihood function by substituting in the preceding expression. Conversely, if we start with a particular likelihood we can find, say, the parameter value $(\hat{\mu}, \hat{\sigma}^2)$ maximizing likelihood,

$$\hat{\mu} = \bar{y} \qquad \hat{\sigma}^2 = \frac{s^2(\mathbf{y})}{n}$$

and thus determine the value of $[\bar{y}, s(\mathbf{y})]$. It follows that $[\bar{y}, s(\mathbf{y})]$ is one-one equivalent to the likelihood map and is thus the likelihood statistic. It then follows from Theorem 4-2 that $[\bar{y}, s(\mathbf{y})]$ is the minimal sufficient statistic. In general, however, the maximum likelihood estimator may not be one-one equivalent to the likelihood statistic.

The preceding is a special case of the exponential model

$$f(y|\theta) = k(\theta) \exp\left[\sum_1^r \psi_j(\theta) t_j(y)\right] h(y) \quad (4\text{-}6)$$

The model for a sample from the preceding has the same basic form:

$$f(\mathbf{y}|\theta) = k^n(\theta) \exp\left\{\sum_{j=1}^r \psi_j(\theta)\left[\sum_{i=1}^n t_j(y_i)\right]\right\} \Pi h(y_i) \quad (4\text{-}7)$$

accordingly, it suffices to consider a density function in the notation of expression (4-6).

The functions $1, \psi_1(\theta), \ldots, \psi_r(\theta)$ may be linearly dependent. If so, one function can be expressed linearly in terms of the others, substituted in (4-6), and the expression rearranged to obtain one less ψ function; iteration then gives an expression (4-6) in which the functions $1, \psi_1(\theta), \ldots, \psi_r(\theta)$ are linearly independent and have a smaller value for r.

74 INFERENCE AND LINEAR MODELS

Now consider the likelihood function, or more conveniently the logarithm of the likelihood function

$$l(y;\theta) = \ln L(y;\theta) = \ln cf(y|\theta)$$
$$= 1 \cdot a + \ln c(\theta) + \psi_1(\theta)a_1(y) + \cdots + \psi_r(\theta)a_r(y)$$

Note that we use $a = \ln c$ as an arbitrary real constant and have absorbed $\ln h(y)$ into that constant.

Clearly $(a_1(y), \ldots, a_r(y))$ determines the likelihood function. Conversely, the linear independence of the ψ's shows that the likelihood function determines $(a_1(y), \ldots, a_r(y))$. It follows that $(a_1(y), \ldots, a_r(y))$ is the likelihood statistic and is the minimal sufficient statistic. For some further discussion see, for example, Fraser (1976, Sec. 8-5).

4-1-4 Weak Likelihood Principle

Consider the response model $\mathcal{M}$ and let $s(\cdot)$ be the likelihood statistic, the minimal sufficient statistic.

In Sec. 4-1-1 we saw that by allocating to a particular contour all the points with a particular likelihood function we obtained the contours of the likelihood statistic. A contour of the likelihood statistic thus consists of points for which the density function has the same dependence on θ. If the density function—the core of the response-based model—does not distinguish points along any contour of the likelihood statistic, then why should any analysis based on the model distinguish among points along a contour of the likelihood statistic? Or, why use data beyond the value of the likelihood statistic? The lack of any obvious answer provides what may be called the likelihood argument for the principle described below.

In Sec. 4-1-2 we saw that the conditional distribution of the location of the response along any contour of the minimal sufficient statistic $s(\cdot)$ does not depend on θ. Thus, if a technician in the laboratory has given us the value of the minimal sufficient statistic $s(\cdot) = s$ and is prepared to give us the value of the complementary function $r(\cdot)$, we would be encountering what could be called post-randomization—the opportunity to obtain a value from a *known* probability distribution. Why should such a value be obtained? Why should the data include anything beyond the value of the minimal sufficient statistic? The lack of any obvious answer provides the sufficiency argument for the principle described below.

S: Weak likelihood (sufficiency) principle For an inference base $(\mathcal{M}, y)$, the data y should be replaced by $s(y)$, the value of the likelihood (minimal sufficient) statistic.

The sufficiency support deriving from Sec. 4-1-2 is the traditional and widely accepted endorsement for the principle. The weak likelihood support deriving from this section is more recent and more closely tied to the views in the preceding chapter; we feel that it provides more cogent and substantial support for the principle. Note, however, that the two supporting arguments are appropriate only with the response-based model. Thus the arguments as given would provide little if any support for use of the principle with a variation-based model.

We are now in a position to present the first density allocation method of reduction. Let $\mathscr{M}$ be a minimum response-based model, $s(y)$ be the likelihood (minimal sufficient) statistic, and $\mathscr{M}_*$ be the marginal model for $s(y)$. We then have the following method:

RM_{01}: **Density-allocation reduction—likelihood** The weak likelihood principle S and the inference base $(\mathscr{M}, y)$ necessarily produce the inference base $(\mathscr{M}_*, s(y))$.

The proof is straightforward. Principle S replaces the data y with the data $s(y)$. The definition of the model in Sec. 1-1-2 then replaces $\mathscr{M}$ by $\mathscr{M}_*$ giving the inference base $(\mathscr{M}_*, s(y))$.

4-1-5 An Example

Consider a response-based model for the location-scale system examined in Chap. 2. The usual procedures for deriving tests and confidence intervals for two parameters μ and σ require in general a preliminary reduction so that the data are essentially two dimensional. For the present problem we then ask—when is the likelihood statistic two dimensional?

For the cases in which the initial response density is positive on the real line the normal model in Sec. 4-1-3 is the only model yielding a two-dimensional likelihood statistic. And for the cases in which the endpoints of the distribution can vary, the uniform and the left-facing and right-facing exponentials are the only models yielding a two-dimensional likelihood statistic. See, for example, Fraser (1976, p. 344) together with Dynkin (1951) or Ferguson (1962).

If we then restrict our attention to models appropriate to common and reasonable applications we eliminate the variable carrier models and are left with just the normal model examined in Sec. 4-1-3. Thus the applicability of reduction method RM_{01} to applied statistical models is very limited—handling just the *normally* distributed case of the location-scale model.

In conclusion, it is perhaps worth mentioning that in sampling from a location-scale distribution on the real line the sufficiency reduction is typically from the initial vector $(y_1, \ldots, y_n)$ just to the order statistic $(y_{(1)}, \ldots, y_{(n)})$.

4-2 ANCILLARITY REDUCTION

In our *constructive* approach to sufficiency in the preceding section we saw that contours of the sufficient statistic were formed by allocating to a contour all the points that have the same dependence on θ. In this section we proceed in an opposite direction and allocate to a contour points with different dependences on

76 INFERENCE AND LINEAR MODELS

θ but allocate in such a way that the dependences average out along the contour: the resulting function is called an *ancillary statistic*. In this section we examine a reduction method involving an ancillary statistic and a related ancillarity principle. This principle, often called the *conditionality principle*, has special appeal for some theoreticians.

4-2-1 The Ancillary Statistic

Consider a response-based model $\mathcal{M}$ as discussed at the beginning of this chapter. For the method in this section we now define an ancillary statistic.

Definition A function $a(\cdot)$ is ancillary if its marginal distribution does not depend on the parameter θ.

Let $r(\cdot)$ be a complementary function so that y is one-one equivalent to $(a(y), r(y))$. Then using the notation and assumptions in Sec. 4-1-2 we obtain the following:

Definition The function $a(\cdot)$ is ancillary if the joint density of a and r can be factored as

$$h(a)g(r|a, \theta)$$

where the marginal density h is independent of the parameter θ.

The concept of an ancillary statistic was presented by Fisher (1925) and developed in subsequent papers. It was largely neglected by the statistical profession until Buehler (1959) and Wallace (1959). This original form of ancillarity due to Fisher is called B-ancillarity in a recent survey of generalizations by Barndorff-Nielsen (1971).

How can we construct an ancillary statistic? Consider a contour of a function $a(y)$ which is to be an ancillary statistic. Along the contour, points typically will have probabilitiy densities with various dependences on θ; the integration of the density along the contour, however, will average out the differing dependences giving a marginal probability for the contour that does not depend on θ. Thus, points are lumped together on a contour so that the total probability for the contour is θ-independent.

For a simple example we turn to a discrete distribution and examine a four-way multinomial with a single real parameter. The example comes from genetics, as is indicated by the labels for the following arrays:

Probabilities

	A	aa
B	$(2 + \theta)/4$	$(1 - \theta)/4$
bb	$(1 - \theta)/4$	$\theta/4$

Data

	A	aa
B	y_{11}	y_{12}
bb	y_{21}	y_{22}

The distribution of $(y_{11}, y_{12}, y_{21}, y_{22})$ is multinomial $[n, (2 + \theta)/4, (1 - \theta)/4, (1 - \theta)/4, \theta/4]$.

The sample space for $(y_{11}, \ldots, y_{22})$ typically has a large number of points. We can, however, see the salient characteristics of the example by looking at the special case $n = 1$; this amounts to a single observation and an indicator function for the "observed" cell. We then have just four sample points and they effectively correspond to the four cells in the left array of probabilities.

How can we construct an ancillary statistic for this model? Starting in the upper left-hand corner with its probability $(2 + \theta)/4$, we can look for some other point with a compensating probability. The lower left-hand or the upper right-hand corner has a compensating probability. Suppose we choose the lower left-hand corner. A contour of an ancillary is then the first column; a second contour is given by the second column. Thus the indicator function for columns is ancillary:

A	aa
$(2 + \theta)/4$	$(1 - \theta)/4$
$(1 - \theta)/4$	$\theta/4$
3/4	1/4

For a general value of n the corresponding ancillary is given by the column total in the earlier array; we have that the column totals are multinomial $(n, 3/4, 1/4)$, in effect binomial $(n, 3/4)$ or binomial $(n, 1/4)$ depending on the particular column used.

In the preceding paragraph we could equally well have chosen the upper right-hand corner to pair with the initial top left corner. We would then have obtained a constant total for the probabilities in each row. For a general value of n the corresponding ancillary would be given by row totals in the original array; we would have that the row totals are multinomial $(n, 3/4, 1/4)$.

For this four-way multinomial with a single real parameter we have found two very different ancillaries. The ancillary idea of balancing out θ differences has some appeal but is rather arbitrary. The duplicity we have just observed points up this arbitrary aspect and points to serious difficulties for the ancillary approach.

The preceding example is a simplified version of examples in Basu (1964) and Fraser (1973). The present version in a genetics context may be found in Fisher (1956, p. 47) and Barndorff-Nielsen (1971).

4-2-2 Ancillary Reduction

Consider the response model $\mathcal{M}$ and let $a(\cdot)$ be an ancillary statistic.

In Sec. 4-2-1 we saw that the dependence of the density on θ typically varied along a contour of the statistic $a(\cdot)$ and yet the total or marginal density for the ancillary $a(\cdot)$ was θ-independent.

One interpretation that can be projected on an ancillary statistic is that of pre-

78 INFERENCE AND LINEAR MODELS

randomization: A value a is obtained from the *known* probability distribution for the function $a(\cdot)$; the remainder of the data is then viewed as an observation from the conditional distribution given $a(y) = a$, a distribution that is θ-dependent. Viewing the distribution of $a(\cdot)$ as providing a form of pre-randomization, we can ask why the value a should matter other than in determining the conditional distribution that can be used for assessing the remainder of the data. The lack of any quick and easy answer provides support for the ancillarity principle described below.

Let $a(\cdot)$ be an ancillary with marginal model $\mathcal{M}_*$ and let $\mathcal{M}^a$ be the conditional model for y given $a(y) = a$. Note that the model $\mathcal{M}_*$ does not have parameters; it is a *mathematical construct*—a probability space model marginalized from the initial model by averaging θ out of θ-dependent probabilities.

A: Ancillarity principle For an inference base $(\mathcal{M}, y)$ the model $\mathcal{M}$ should be replaced by $\mathcal{M}^{a(y)}$, the conditional model given an ancillary statistic value $a(y)$.

The pre-randomization argument presented above is the typical kind of support for the principle. In fact, however, the common acceptance of the principle rests almost entirely on a number of examples with prima facie appeal.

We are now in a position to present the second density allocation method of reduction:

RM$_{02}$: Density allocation reduction—ancillarity The ancillarity principle A and the inference base $(\mathcal{M}, y)$ necessarily produce the inference base $(\mathcal{M}^{a(y)}, r(y))$.

The proof is straightforward. Principle A replaces the model $\mathcal{M}$ with the model $\mathcal{M}^{a(y)}$. The definition of the model in Sec. 1-1-2 then replaces the data y by the data given $a(y)$; with the notation earlier in this section this amounts to replacing y by $r(y)$. The resulting inference base is $(\mathcal{M}^{a(y)}, r(y))$.

4-2-3 Some Examples

First consider the four-way multinomial example considered in Sec. 4-2-1:

	A	aa				A	aa	
B	$(2+\theta)/4$	$(1-\theta)/4$	3/4	B	y_{11}	y_{12}	$y_{1\cdot}$
bb	$(1-\theta)/4$	$\theta/4$	1/4	bb	y_{21}	y_{22}	$y_{2\cdot}$
	3/4	1/4			$y_{\cdot 1}$	$y_{\cdot 2}$	n

We have noted that the column totals $(y_{\cdot 1}, y_{\cdot 2})$ form an ancillary $a(\mathbf{y})$ with the multinomial $(n, 3/4, 1/4)$ distribution. The conditional distribution given this ancillary may be described as follows: y_{11} is binomial $[y_{\cdot 1}, (2+\theta)/3]$ and independently y_{12} is binomial $(y_{\cdot 2}, 1 - \theta/3)$. Principle A and method RM_{02} then give an inference base in which y_{11} and y_{12} are the data, and the model involves the product of the preceding independent binomial distributions; this is an analysis within columns.

In Sec. 4-2-1 we also noted that the row totals $(y_1 \cdot y_2 \cdot)$ form an ancillary with the multinomial $(n, 3/4, 1/4)$ distribution. Principle A and method RM_{02} then give an inference base in which the rows are analyzed as independent binomials; this is an analysis within rows.

Thus principle A is self-contradictory. The principle says to do something. But the something can be different things that contradict each other.

The multinomial example is, of course, discrete. Consider a somewhat similar example involving continuous variables. Let (y_1, y_2) be bivariate normal with mean $(0, 0)$ and VARiance matrix

$$\begin{pmatrix} 1 & \rho \\ \rho & 1 \end{pmatrix}$$

a matrix with a parameter ρ in $(-1, +1)$.

The first coordinate projection function y_1 is ancillary: it has the normal $(0, 1)$ distribution. The conditional distribution given y_1 says that y_2 is normal $(\rho y_1, \sqrt{1-\rho^2})$. Principle A and method RM_{02} then give an inference base in which an observed y_2 is analyzed against the normal $(\rho y_1, \sqrt{1-\rho^2})$ model.

In a parallel way, however, we see that the second coordinate projection function y_2 is ancillary. This leads to an inference base in which an observed y_1 is analyzed against the normal $(\rho y_2, \sqrt{1-\rho^2})$ model.

Again, we see the self-contradictory nature of principle A.

4-2-4 Some Discussion

The examples just discussed do not paint a very attractive scene for the ancillarity principle A and the method RM_{02}. However, the presentation usually given for this principle and method (for example, Cox and Hinkley, 1974) has several very appealing examples as well as the two rather unattractive examples just examined.

The appealing examples are all concerned with systems that typically admit more detailed models than the minimum response models we have been discussing in this chapter. In fact, these examples, as in Cox and Hinkley (1974), are all examples that were mentioned in Sec. 3-2 and for which we used the more detailed model appropriate to the applications. These other examples are: the random choice of measuring instrument; the random sample size; and the location or location-scale example. These more attractive examples were thus drawing their strength from the more detailed contexts in Chap. 3 and are thus inappropriate for illustrating the principle and method in this section.

One example mentioned in Sec. 3-2 but not discussed until it was examined more generally in Sec. 3-3 is the location-scale example. We discuss it briefly now. Suppose we consider the response-based model in Sec. 1-2-1 and examine the analysis in Sec. 2-1 for the special case that the distribution form is known. For the model we have the class

$$\mathscr{F} = \{\sigma^{-n}\Pi f(\sigma^{-1}(y_i - \mu)): \mu \in \mathbb{R}, \sigma \in \mathbb{R}^+\}$$

of response distributions where f is the known density describing the distribution form.

The function $\mathbf{d(y)}$ given by

$$\mathbf{d(y)} = s^{-1}(\mathbf{y})(\mathbf{y} - \bar{y}\mathbf{1})$$

has a distribution (2-15) that is independent of $\theta = (\mu, \sigma)$; thus the function $\mathbf{d(y)}$ is ancillary. The ancillarity principle A and the method RM_{02} then produce an inference base involving the value of $[\bar{y}, s(\mathbf{y})]$ with the conditional model (2-16). The analysis in Chap. 2 proceeded in this manner and it corresponds closely to the analysis suggested in an early Fisher paper (1934) concerning ancillaries; the more general case with a parametric family for distribution form was not examined in Fisher's papers.

Recall, however, that with the proper model, the variation-based $\mathscr{M}_V$, the necessary reduction method RM_3 in Sec. 3-3 directly gives the procedures discussed in Chap. 2.

4-3 SUFFICIENCY-ANCILLARITY REDUCTION

Sufficiency leads to a marginal model; ancillarity leads to a conditional model. In a few special cases a statistic can be both sufficient and ancillary—sufficiency for one parameter component and ancillarity for an independent parameter component. In this section we examine this combination of sufficiency and ancillarity for a response model $\mathscr{M}$.

4-3-1 Sufficiency-Ancillarity

We now examine a blend of sufficiency and ancillarity, a blend that uses a statistic with sufficiency properties for one parameter component and ancillarity properties for the complementary parameter component. For this, suppose that the parameter θ in Ω can be reexpressed as (ψ, ϕ) on a Cartesian product $\Omega_2 \times \Omega_1$. Thus $(\psi(\theta), \phi(\theta))$ provides a one-one reparameterization onto a product space. The parameters ψ and ϕ are thus independent in a reasonable way, which we call *Cartesian independence*.

On the sample space, now consider a function $t(\cdot)$ and let $r(\cdot)$ be a complementary function so that y is one-one equivalent to $[t(y), r(y)]$. We suppose that dt and dr designate suitable volume measures for t and r, and we suppose that $h(t|\theta)$ and $g(r|t, \theta)$ are marginal and conditional density functions for t and for r given t. We now define a sufficient-ancillary statistic.

Definition A function $t(\cdot)$ is sufficient-ancillary for (ψ, ϕ) if the marginal distribution of t depends only on ψ and the conditional distribution given t depends only on ϕ where the parameter components ψ and ϕ are Cartesian independent.

Thus a statistic t is sufficient-ancillary if the joint density of t and r can be factored as

$$h(t|\psi)g(r|t, \phi) \qquad (4\text{-}8)$$

where the parameter (ψ, ϕ) belongs to the Cartesian product $\Omega_2 \times \Omega_1$.

This separation of variables and parameters was proposed in Fraser (1956) as sufficiency (ψ). Some separation of variables but without the required Cartesian independence of parameters may be found in Olshevsky (1940) and implicitly in Neyman (1935). An emphasis on the ancillary component may be found in Sandved (1967). In a recent survey Barndorff-Nielsen (1971) has used the term S-sufficiency and S-ancillarity, depending on the parameter component under discussion. Basu (1977) more recently has reattributed the sufficiency-ancillarity definition to Olshevsky and Neyman but has overlooked the absence of the second half of the definition concerned with Cartesian independence.

Consider a simple example. Let y_1 and y_2 be independent Poisson (θ_1) and Poisson (θ_2). We introduce new parameters replacing (θ_1, θ_2) by (ψ, ϕ) where

$$\psi = \theta_1 + \theta_2 \qquad \phi = \frac{\theta_1}{\theta_1 + \theta_2} \qquad (4\text{-}9)$$

The function $t = y_1 + y_2$ has the Poisson (ψ) distribution and y_1 given t has the binomial (t, ϕ) distribution. It follows then that t is sufficient-ancillary for (ψ, ϕ).

How can we construct a sufficient-ancillary statistic? Consider a separation of the parameter θ into the Cartesian product of the parameters ψ and ϕ, as discussed earlier in this section. Suppose we allocate to a contour or assemble on a contour all the points for which the variation in the likelihood function is entirely in terms of the parameter ϕ; thus y_2 is on the same contour as y_1 if and only if the ratio

$$\frac{L(y_2; \psi, \phi)}{L(y_1; \psi, \phi)} \qquad (4\text{-}10)$$

is independent of ψ and thus involves just ϕ and, of course, y_1 and y_2. This is a direct positive construction procedure paralleling that in Sec. 4-1-1.

Now let $t(\cdot)$ be a function that indexes the contours just constructed and let $r(\cdot)$ be a complementary function that indexes points along the contours. We also suppose that density functions are available for these new coordinates t and r giving

$$h(t|\phi, \psi)g(r|t, \phi)$$

The construction condition for the contours of $t(\cdot)$ ensures that ψ is absent from the second factor; indeed, it ensures that the contours of $t(\cdot)$ are the largest contours with this ψ-independence. From this we see that there is sufficiency-ancillarity if and only if the parameter ϕ is absent from the marginal distribution for the constructed function $t(\cdot)$.

If there are functions that are sufficient-ancillary for (ψ, ϕ) then the construction procedure gives one such function, a particular one. There may, however, be more than one, depending on how the pre- and post-randomization discussed in Secs. 4-1 and 4-2 are separated between marginal and conditional variables.

4-3-2 Sufficiency-Ancillarity Reduction

Consider the response model $\mathcal{M}$ and let $t(\cdot)$ be a sufficient-ancillary statistic for (ψ, ϕ).

If attention is restricted to the parameter ψ then $t(\cdot)$ has some of the properties of a sufficient statistic, for if a technician in a laboratory has given us the value of the statistic $t(\cdot) = t$ and is prepared to give us the value of the complementary function $r(\cdot)$, then we would be encountering what we have called post-randomization—the opportunity here to obtain a value from an *unknown* probability distribution, a distribution, however, that is completely independent of the parameter ψ of concern. Thus for inference concerning ψ why should such a value be obtained? Why should the data record anything beyond the value of the statistic $t(\cdot)$? The lack of any obvious answer provides support for the sufficiency aspect of the following principle. A discussion similar to that of Sec. 4-2-2 provides support for the ancillarity aspect of the definition.

Note that the preceding argument would not be applicable using the Olshevsky version. Consider the Poisson example in Sec. 4-3-1 with θ_1 restricted to an interval (θ_1', θ_1''). Then the possible values of ϕ would satisfy

$$\theta_1'/\psi < \phi < \min\{\theta_1''/\psi, 1\}$$

and the distribution of the complementary function $r(\cdot)$ would not be independent of the parameter ψ.

The preceding discussion provides support for a sufficiency-ancillarity principle. For this, let $t(\cdot)$ be sufficient-ancillary for (ψ, ϕ); let $\mathcal{M}_*$ be the marginal model for $t(\cdot)$ with parameter ψ; and let $\mathcal{M}^t$ be the conditional model for y given $t(y) = t$ with parameter ϕ. We then have the following sufficiency-ancillarity principle.

SA: Sufficiency-ancillarity principle For an inference base $(\mathcal{M}, y)$, the data y should be replaced by $t(y)$ for inferences concerning ψ, and the model $\mathcal{M}$ should be replaced by $\mathcal{M}^{t(y)}$ for inferences concerning ϕ.

We are now in a position to present the third density allocation method of reduction.

RM$_{03}$: Density allocation reduction—sufficiency-ancillarity The sufficiency-ancillarity principle SA and the inference base $(\mathcal{M}, y)$ necessarily produce the inference base $(\mathcal{M}_*, t(y))$ for inference concerning ψ and the inference base $(\mathcal{M}^{t(y)}, r(y))$ for inference concerning ϕ.

The proof routinely follows the pattern in Secs. 4-1 and 4-2.

4-3-3 Some Examples

First consider the example from Sec. 4-3-1. Let y_1 and y_2 be independent Poisson (θ_1) and Poisson (θ_2). We have noted that $t = y_1 + y_2$ is sufficient-ancillary for (ψ, ϕ) where $\psi = \theta_1 + \theta_2$ and $\phi = \theta_1/(\theta_1 + \theta_2)$.

Now suppose that inferences are to be made separately for ψ and ϕ. For the parameter ψ principle SA and method RM$_{03}$ give the inference base with Poisson (ψ) distribution and data t. And for the parameter ϕ principle SA and method RM$_{03}$ give the inference base with binomial (t, ϕ) distribution and data y_1.

A somewhat similar example arises with the three-way multinomial. Let $(y_1, y_2, n - y_1 - y_2)$ be multinomial (n, pp_1, pq_1, q) where $p + q = 1$ and $p_1 + q_1 = 1$. The function $t = y_1 + y_2$ has the binomial (n, p) distribution and y_1 given t has the binomial (t, p_1) distribution. It follows then that t is sufficient-ancillary for (p, p_1).

Now suppose that inferences are to be made separately for p and p_1. For the parameter p the principle SA and the method RM$_{03}$ give the inference base with binomial (n, p) distribution and data t. And for the parameter p_1, the principle SA and the method RM$_{03}$ give the inference base with binomial (t, p_1) distribution and data y_1.

The preceding examples are both discrete. Continuous examples are much more difficult to find. A continuous albeit nonparametric example may be found in Fraser (1956).

4-3-4 Some Discussion

The example involving independent Poisson distributions has some appealing features. Consider some physical context involving a possibly nonhomogeneous Poisson process. Let t be the total count for a specified time interval and let y_1 and y_2 be the counts for first and second component intervals. Interest could be centered on the average rate for the full specified time interval; this is given by the parameter $\psi = \theta_1 + \theta_2$. Or, separately from the preceding, interest could be centered on the homogeneity of the process—how θ_1 and θ_2 relate to the lengths of the corresponding time intervals of observation.

Our Poisson example has now been enlarged to include obvious objective aspects of the physical situation. And not surprisingly we find that methods from Chap. 3 are available; specifically we can use the method in Sec. 3-2 for part of the analysis discussed in Sec. 4-3-3. *If ψ has some specified value*, then the total count is a characteristic of the full time interval of observation and it has a known distribution. The reduction method RM_2 then specifies that inference concerning ϕ be made in terms of y_1 and the binomial (t, ϕ) distribution. This inference base for ϕ, however, is the same for each specified value of ψ. Thus since the italicized "if" clause holds for some ψ, we obtain necessarily the reduced inference base in Sec. 4-3-3 for the parameter component ϕ.

In conclusion, we should note that the sufficient-ancillary statistic need not be unique. Principle SA then has some potential for the self-contradictory properties discussed in Sec. 4-2-3.

4-4 WEAK SUFFICIENCY AND ANCILLARITY

The sufficiency-ancillarity concept in the preceding section requires a very sharp separation of Cartesian-independent parameters, one for a marginal distribution and the other for a conditional distribution. In this section we relax the requirements and obtain two different extensions: *weak sufficiency* and *weak ancillarity*.

These principles are not widely acknowledged in the literature. The search for such extended principles is no doubt rooted in the failure of the earlier principles to come to grips with the simple separation of μ-inference and σ-inference for that simplest of models, the normal (μ, σ). Even at a considerable price in terms of loss of support, these extended principles only partially handle the very simple normal problem.

4-4-1 Weak Sufficiency and Ancillarity

Consider a response-based model $\mathscr{M}$ and suppose the parameter θ in Ω can be reexpressed as (ψ, ϕ) in Ω^*, which need not be a Cartesian product as required in Sec. 4-3-1.

Now consider a sample space function $t(\cdot)$ and a complementary function $r(\cdot)$ and suppose that density functions are available as in Sec. 4-3-1. The concept of sufficiency-ancillarity requires the following separation of parameters in the density function:

$$h(t\,|\,\psi)g(r\,|\,t, \phi) \tag{4-11}$$

Consider a first relaxation called weak sufficiency.

Definition The function $t(\cdot)$ is weakly sufficient for ψ if the factorization has the form

$$h(t\,|\,\psi)g(r\,|\,t, \phi, \psi) \tag{4-12}$$

and the conditional distribution $g(r\,|\,t, \phi, \psi)$ provides no direct information concerning ψ.

Two interpretations for "no direct information" will be discussed below. Consider also a second relaxation called weak ancillarity.

Definition The function $t(\cdot)$ is weakly ancillary for ϕ if the factorization has the form

$$h(t\,|\,\phi,\psi)g(r\,|\,t,\phi) \tag{4-13}$$

and the marginal distribution $h(t\,|\,\phi,\psi)$ provides no direct information concerning ϕ.

We now examine two interpretations for "no direct information."

A type of weak ancillarity was proposed by Cox (1958). For this, the marginal density $h(t\,|\,\phi,\psi)$ is said to provide no direct information (NDI-1) concerning ϕ if at each point t the likelihood ratio for any two ϕ values can be duplicated by two ψ values; specifically, if for each t, ϕ_1, ϕ_2, ψ there exists ϕ, ψ_1, ψ_2 such that

$$\frac{L(y;\phi_1,\psi)}{L(y;\phi_2,\psi)} = \frac{L(y;\phi,\psi_1)}{L(y;\phi,\psi_2)}$$

A parallel definition can be used for the conditional density $g(r\,|\,t,\phi,\psi)$ in (4-12) to give weak sufficiency.

Another type of weak ancillarity was proposed by Barndorff-Nielsen (1971) and is called *M-ancillarity*. For this, the marginal density $h(t\,|\,\phi,\psi)$ is said to provide no direct information (NDI-2) concerning ϕ if for each value of ϕ the density family $h(t\,|\,\phi,\psi)$ is *universal*; specifically, if for each value of ϕ and value for t there is a ψ value that places maximum density at the particular t value,

$$h(\cdot\,|\,\phi,\psi) \leq h(t\,|\,\phi,\psi)$$

In a sense a t value can be fully "explained" by variation in ψ alone.

A type of weak sufficiency was proposed by Barndorff-Nielsen (1971) and called *M-sufficiency*; it used NDI-2 for the conditional distribution.

In Secs. 4-2 and 4-3 we have mentioned some of the difficulties and self-contradictory properties of the principles of ancillarity and of sufficiency-ancillarity. The possibilities for unsatisfactory properties are far greater for the two extensions: weak sufficiency and weak ancillarity.

The extended definitions have been proposed, it seems, in order to handle some very simple problems that are not amenable to the methods presented earlier in this chapter; an example is given in Sec. 4-4-2. The wider use of the method, beyond such very simple transparent problems, seems to face, however, many very unsatisfactory possibilities. We feel that the case for the extended definition is so weak that it is inappropriate to bother with more than the one simple example to follow.

4-4-2 An Example

Consider a response model for a sample $(y_1, \ldots, y_n)$ from the normal (μ, σ) distribution. The weak likelihood (sufficiency) principle S can be used to support a reduction to $[\bar{y}, s(\mathbf{y})]$; we then have that $\bar{y}$ is normal $(\mu, \sigma/\sqrt{n})$ and, independently, that $s(\mathbf{y})$ has the chi $(n-1)$ distribution scaled by the factor σ.

The sufficiency-ancillarity principle in Sec. 4-3 does not lead to separate inferences for μ and for σ. We examine the usefulness of the present principles, weak sufficiency and weak ancillarity.

First consider the parameter μ. The function $\bar{y}$ is sometimes referred to as being sufficient for μ, but this does not conform to the definition in Sec. 4-3 and in a quite general sense is perhaps very misleading, for with σ unknown there is no reasonable way that $\bar{y}$ alone can be viewed as "sufficient" for inferences concerning μ: its reliability is available from the complementary $s(\mathbf{y})$. Our present principles do not lead us to the use of $\bar{y}$ alone for inferences concerning μ.

Now consider the parameter σ. The function $s(\mathbf{y})$ does seem to be a rather natural function for inference concerning σ. First we note that $s(\mathbf{y})$ is weakly sufficient for σ using either NDI-1 or NDI-2. Then we note that $\bar{y}$ is weakly ancillary for σ using either NDI-1 or NDI-2. These properties are sometimes used as support for $s(\mathbf{y})$ for inference concerning σ. Note that the two properties, weak sufficiency and weak ancillarity, sit side by side largely because of the statistical independence of $\bar{y}$ and $s(\mathbf{y})$.

The presence of a nice example somehow tends to lend strength to tentatively proffered principles—here weak sufficiency and weak ancillarity. Serious cautions, however, are indicated. The normal has so many natural simplicities and symmetries that just about anything not totally misdirected will work. The niceness of the example does underlie something more substantial, for we can recall that for inference concerning the parameter σ the normal model is amenable to the necessary reduction method RM_5 as in Sec. 3-5-2.

Certainly reduction concepts are needed. The case for weak sufficiency and weak ancillarity, however, is at present very very *weak*. Indeed the concepts, away from the very simple example, may be quite misleading and damaging.

REFERENCES AND BIBLIOGRAPHY

Barnard, G. A., and D. A. Sprott: A Note on Basu's Examples of Anomalous Ancillary Statistics, in V. P. Godambe and D. A. Sprott (eds.), "Proc. Symposium on the Foundations of Statistical Inference," Holt, Rinehart, and Winston of Canada, Toronto and Montreal, pp. 163–176, 1971.

Barndorff-Nielsen, O.: "On Conditional Inference," Mimeographed Report, Aarhus University, Denmark, 1971.

———: On $\mathscr{M}$-ancillarity, *Biometrika*, vol. 60, pp. 447–455, 1973.

———: Factorization of Likelihood Functions for Full Exponential Families, *Jour. Roy. Stat. Soc.*, ser. B, vol. 38, pp. 37–44, 1976.

———— and H. K. Kvist: Note on Exponential Families and $\mathscr{M}$-ancillarity, *Scand. Jour. Statist.*, vol. 1, pp. 36–38, 1974.
Basu, D.: The Family of Ancillary Statistics, *Sankyā*, ser. A, vol. 21, pp. 247–256, 1959.
————: Recovery of Ancillary Information, *Sankhyā*, ser. A, vol. 26, pp. 3–16, 1964.
————: On the Elimination of Nuisance Parameters, *Jour. Amer. Stat. Assoc.*, vol. 72, 355–366, 1977.
Birnbaum, A.: On the Foundations of Statistical Inference, *Jour. Amer. Stat. Assoc.*, vol. 57, pp. 269–306, 1962.
————: On Durbin's Modified Principle of Conditionality, *Jour. Amer. Stat. Assoc.*, vol. 65, pp. 402–403, 1970.
Buehler, R. J.: Some Validity Criteria for Statistical Inferences, *Ann. Math. Stat.*, vol. 30, pp. 845–863, 1959.
Cox, D. R.: Some Problems Connected with Statistical Inference, *Ann. Math. Stat.*, vol. 29, pp. 357–372, 1958.
————: The Choice between Alternative Ancillary Statistics, *Jour. Roy. Stat. Soc.*, ser. B, vol. 33, pp. 251–255, 1971.
———— and D. V. Hinkley, "Theoretical Statistics," Chapman and Hall, London, 1974.
Dynkin, E. B.: Necessary and Sufficient Statistics for a Family of Probability Distributions, *Uspehi, Math. Nauk (N.S.)* vol. 6, no. 1 (41), pp. 68–90, 1951. (Also, selected *Transl. Math. Stat. Prob.*, pp. 17–40).
Edwards, A. W. F.: *Likelihood*, Cambridge University Press, Cambridge, 1972.
Ferguson, T. S.: Location and Scale Parameters in Exponential Families of Distributions, *Ann. Math. Stat.*, vol. 33., 986–1001, 1962.
Fisher, R. A.: A Mathematical Examination of the Method of Determining the Accuracy of an Observation by the Mean Error, and by the Mean Square Error, *Monthly Notices of the Royal Astronomical Society*, vol. LXXX, no. 8, pp. 758–770, 1920; also as paper 2 in Fisher (1950).
————: On the Mathematical Foundations of Theoretical Statistics, *Phil. Trans. Roy. Soc. London*, ser. A, vol. 222, pp. 309–368, 1922; also as paper 10 in Fisher (1950).
————: Theory of Statistical Estimation, *Proc. Camb. Phil. Soc.*, vol. 22, pt. 5, pp. 700–725, 1925; also as paper 11 in Fisher (1950).
————: Two New Properties of Mathematical Likelihood, *Proc. Roy. Soc.*, ser. A, vol. 144, pp. 285–307, 1934; also as paper 24 in Fisher (1950).
————: "Contributions to Mathematical Statistics," John Wiley and Sons, New York, 1950.
————: "Statistical Methods and Scientific Inference," Oliver and Boyd, London, 1956.
Fraser, D. A. S.: Sufficient Statistics with Nuisance Parameters, *Ann. Math. Stat.*, vol. 27, pp. 838–842, 1956.
————: On the Sufficiency and Likelihood Principles, *Jour. Amer. Stat. Assoc.*, vol. 58, pp. 641–647, 1963.
————: On Sufficiency and the Exponential Family, *Jour. Roy. Stat. Soc.*, ser. B, vol. 25, pp. 115–123, 1963.
————: Sufficiency for Regular Models, *Sankhyā*, ser. A, vol. 27, pp. 137–144, 1966.
————: Sufficiency and Conditional Sufficiency, *Sankhyā*, ser. A, vol. 28, pp. 145–150, 1966.
————: Sufficiency for Selection Models, *Sankhyā*, ser. A, vol. 28, pp. 329–334, 1966.
————: Sufficiency or Conditional Sufficiency, *Sankhyā*, ser. A, vol. 29, pp. 239–244, 1967.
————: The Elusive Ancillary, in D. G. Kabe and R. P. Gupta (eds.), "Multivariate Statistical Inference," North-Holland Publishing Company, Amsterdam and London, 1973.
————: "Probability and Statistics: Theory and Applications," Duxbury Press, North Scituate, Mass., 1976.
Hájek, J.: On Basic Concepts of Statistics, *Proc. Fifth Berkeley Symposium*, vol. 1, pp. 139–162, 1967.
Halmos, P. R., and L. J. Savage: Application of the Radon-Nikodym Theorem to the Theory of Sufficient Statistics, *Ann. Math. Stat.*, vol. 20, pp. 225–241, 1949.
Lehman, E. L., "Testing Statistical Hypothesis," John Wiley and Sons, New York, 1966.
Neyman, J.: Sur un Teorma Concernente le Cosidette Statistiche Sufficienti, *Giorn. Ist. Ital. Att.*, vol. 6, pp. 320–334, 1935.

Olshevsky, L.: Two Properties of Sufficient Statistics, *Ann. Math. Stat.*, vol. II, pp. 104–106, 1940.

Sverdrup, E.: The Present State of the Decision Theory and the Neyman-Pearson Theory, *Rev. Inst. Int. Statist.*, vol. 34, pp. 309–333, 1966.

Sandved, E.: A Principle for Conditioning on an Ancillary Statistic, *Skand. Aktuar*, vol. 50, pp. 39–47, 1967.

———: Ancillary Statistics in Models without and with Nuisance Parameters, *Skand. Aktuar.*, vol. 55, pp. 81–91, 1972.

Wallace, D. L.: Conditional Confidence Level Properties, *Ann. Math. Stat.*, vol. 30, pp. 864–876, 1959.

CHAPTER
FIVE

TERMINAL METHODS OF INFERENCE

In Chaps. 3 and 4 we examined various methods for the reduction and simplification of an inference base. In this chapter we examine the subsequent components of inference, the common *terminal methods* of inference; these are the methods for summarizing and presenting the output inferences from a reduced inference base.

The terminal inferences are viewed in the larger sense of a full spectrum of results, not one or two results from a single method. Statistical inference involves this full spectrum.

5-1 TESTS OF SIGNIFICANCE

Perhaps the most elementary and basic inference method is that of testing or assessing a hypothesis that specifies the value of the parameter or the value of some component parameter.

Consider an inference base

$$(\mathcal{M}, \mathcal{D}) \tag{5-1}$$

and let H_0 be a hypothesis that specifies the value of the parameter in Ω or specifies the value of some component parameter. The hypothesis H_0 produces an alleged model $\mathcal{M}_0$, which in the first case would be a probability model and in the second case would be a statistical model using the restricted parameter. Thus H_0 gives the alleged inference base

$$(\mathcal{M}_0, \mathcal{D}) \tag{5-2}$$

A test of significance, then, is an examination or assessment of the alleged inference base (5-2) to see if it is realistic and plausible, or if it is inherently self-contradictory and the original inference base (5-1) is needed.

An assessment of an inference base is really an assessment of the data in relation to the model. Could the data reasonably have come from a system described by the alleged model? If the alleged model $\mathcal{M}_0$ is a probability model —a statistical model without a parameter—then the assessment amounts to seeing whether the data could reasonably have come from a particular distribution. More generally, the assessment amounts to seeing whether the data could reasonably have come from a statistical model, that is, some collection of probability models.

Such an assessment is in part concerned with unusual, rare, or unlikely values. Such values are typically those that are distant from where the distribution or model is concentrating probability. In the case of a fully specified probability distribution, an experienced worker in probability should be able to make a direct assessment of an observed value in relation to a distribution—whether it is unusual or extreme and to what degree. In the more general case, an experienced worker may still be able to make some direct assessment—whether a value is unusual or extreme for all the possibilities in the alleged model.

In Secs. 2-2-2, 2-2-3, and 2-3-2 we examined tests of significance for a location parameter μ and a scale parameter σ. The preliminary reduction, there, gave an inference base that permitted relatively straightforward assessments, and for each test we had an observed value to assess against a single distribution that was determined by the hypothesis.

As part of these assessments, we calculated an observed level of significance—*the probability of as great or greater departure from the centre of the distribution.* For the numerical example in Sec. 2-3-2, however, we recorded the probability of such departure just for the particular tail on which the observed value occurred. This is highly informative in itself, and with the near-symmetrical distribution involved, can be doubled to give the preceding departure probability to reasonable accuracy.

Of course, if the distribution under the hypothesis is far from symmetrical, then it is rather unclear how to compare distances on the two tails and thus how to calculate the probability of as great or greater departure from the centre of the distribution. What is needed is some measure of departure from the centre, a *measure of discrepancy.* For the earlier example, we used absolute distance from the "obvious" centre. More generally, a choice of a measure of discrepancy is an arbitrary input to the inference analysis, and can be based on an ordering from the center of the distribution, or on various probability or likelihood criteria that we do not investigate here. Often the one-tailed probability seems quite appropriate for an assessment—provided the context clearly indicates that it *is* a one-tailed probability.

An extreme possibility arises if the distribution specified by the hypothesis has a hole or shallow spot in the centre. How should this be incorporated into an observed level of significance? With an asymmetric distribution we measured probability of departure on the particular tail. With a hole in the centre and an observed value in the hole we would certainly record what happened; a suitable measure of discrepancy might be rather difficult to come up with and would certainly be quite specific to the particular distribution involved.

For tests of significance based on normal models, almost all reasonable approaches lead to the same assessment calculations. This pleasant state for the normal with its many natural symmetries tends to raise expectations that there can be some absolute measures of discrepancy, suitable for all models. In fact, such absolute measures do not seem to be available and, indeed, there seems to be little need for them; this is partly indicated by the approaches we mentioned for asymmetric distributions and for distributions with a "hole in the centre."

An extreme value for an alleged or hypothesized distribution might in fact be an extreme value for all the other distributions indicated by the full model. In that case the data may be casting doubts on the full model itself. If likelihood is available from a spectrum of models, then various comparisons are possible using likelihood ratios and likelihood functions (see, for example, Sec. 2-3-1).

Our discussion so far has been concerned largely with an observed value on the real line. The preliminary reduction, however, may produce an observed value that is still multivariate. The example just mentioned from Sec. 2-3-1 was of this type, and only likelihood assessment seemed available there. In other problems, generalized likelihood ratios and some techniques from hypothesis-testing theory may be useful and give distances and measures of discrepancy that are one dimensional rather than many dimensional.

For multiparameter problems we may often be able to separate the parameters and examine them in some natural sequence. We explore some possibilities in this direction in Sec. 7-4 and Chap. 11.

The approach in this section has been deliberately informal. It has emphasized the direct assessment of data and has tried to avoid tendencies toward premature formulation and overformulation.

5-2 CONFIDENCE REGIONS

Confidence intervals and confidence regions are the most important and widely used method of inference. Trends do occur and other methods may arise, but we must acknowledge the ubiquitous nature of confidence procedures.

5-2-1 Confidence Method

Consider a reduced or simplified inference base

$$(\mathcal{M}, \mathcal{D}) \tag{5-3}$$

and let $\phi = \phi(\theta)$ be a parameter of special interest for a particular purpose in statistical inference.

The confidence method can be presented most easily in terms of the tests of significance discussed in Sec. 5-1. Suppose that for each value of the parameter ϕ, we have a $1 - \alpha$ test of significance for testing that parameter value, that is, we can form a set of response values that leads to acceptance of ϕ with probability $1 - \alpha$ if ϕ is the true value of the parameter. This set of response values is called the *acceptance region* for ϕ. We can then see whether the observed response is in the acceptance region for ϕ. The values of the parameter that lead to acceptance in the preceding way form a $1 - \alpha$ *confidence region for the parameter* $\phi = \phi(\theta)$.

A $1 - \alpha$ confidence region as just presented has the property that in repeated performances, the confidence region will cover the true parameter value with probability $1 - \alpha$, whatever that parameter value is. The preceding is the standard textbook interpretation and we do not elaborate on it here. For some recent discussion on confidence regions, see Fraser (1976, Sec. 10-5, pp. 579–584).

The close connection with tests of significance suggests that to the degree that reasonable tests of significance are available so also are reasonable confidence regions. The reduction of the model and the inference base by the methods in Chaps. 3 and 4 removes much of the arbitrariness from the formation of confidence regions. There still remains a range of possibilities corresponding mostly to the range of possibilities for the measure of discrepancy; recall the example in Sec. 2-3.

Most statistical texts tend to focus on 95 and on 99 percent confidence intervals. Certainly for a particular end use an investigator could be interested in a particular confidence level. For inference more generally, however, it seems appropriate to record a spectrum of confidence intervals at a variety of confidence levels. For example, the computer printout for the analysis in Sec. 2-3 gives confidence intervals at the levels 90, 95, 99, and 99.9 percent; in fact a broader spectrum would seem preferable but one needs to weigh the benefits against the bulk of computer printout that can be received.

The intervals at different levels for the example in Sec. 2-3 form a nested family of intervals. This nesting occurs with nice examples, and frequently with real parameters. For the case where the nesting occurs for a full range of confidence levels we have what can be called a *confidence distribution* for the parameter. Such nested intervals and confidence distributions are available in a natural way for the variation-based models introduced in Sec. 1-2-4; we return to this in detail in Sec. 7-3. At this stage we view a confidence distribution solely as a tool for obtaining confidence intervals.

In this book we emphasize the importance of having a wide spectrum of confidence intervals, hopefully intervals that nest one within another. It is a small step then to a full spectrum of intervals or a confidence distribution. We choose to avoid this small step as it seems to attract an unnatural amount of attention from some areas of statistics and, in turn, deflects attention from the serious and substantial topics under discussion.

What meaning should we attach to a confidence distribution? Certainly such a distribution means what all of the component confidence intervals mean. And in some contexts it is possible to attach further meaning to the distribution; specifically, with variation-based models we can view the distribution as having a probability meaning based on direct probability statements concerning the variation. We do note, here, that such extensions are not of central concern for this book, and we will merely record a few results of mathematical interest in appropriate supplements.

5-2-2 Some Examples

The standard interpretation for confidence regions was mentioned briefly in the preceding Sec. 5-2-1. It is a fairly strong, easily understood interpretation, and provides backbone for standard statistical theory in applications.

In this section we use several simple examples to investigate certain unusual features that can arise in all but the special situations—those that use the variation-based model introduced in Chap. 2.

The first two examples show that we should treat the term confidence with caution and mild scepticism: the quoted percentage may differ from the realities of an application.

> **Example 5-1** Recall the example in Sec. 3-2-3 involving the random choice of measuring instrument; we now consider this example entirely in the context of a response-based model and initially ignore the obvious availability—indeed necessity—for the stronger variation-based model. To simplify the calculations let the first instrument have an error distribution that is uniform $(-1, +1)$ and the second instrument have an error distribution that is uniform $(-5, +5)$; we assume as before that the measuring instrument is randomly chosen with equal probabilities from the two available.
>
> Now consider the preceding for a single observation: we have the inference base $(\mathcal{M}, y)$ where y is the observed measurement. A 90 percent confidence interval for the location θ is $(y - 4, y + 4)$.
>
> Note an anomalous feature of the preceding interval. If the investigator records the particular instrument used (how could he neglect to do so?) and if he finds it to be the first instrument, then he can assert categorically that $(y - 4, y + 4)$ brackets θ; he has 100 percent confidence. Indeed, he can assert this without any use of probability or statistics.
>
> Alternatively, if he finds it to be the second instrument, then he can assert that the interval $(y - 4, y + 4)$ brackets the true θ with confidence only 80 percent.
>
> Clearly, disregarding the information concerning the instrument used gives anomalous results. Indeed, it violates basic scientific principles.

Now consider a more substantial example illustrating the same anomalous results, but in a more detailed and less transparent context.

> **Example 5-2** Consider a sample (y_1, y_2) from the uniform $(\theta - \tfrac{1}{2}, \theta + \tfrac{1}{2})$ distribution with θ in $\mathbb{R}$. Again, we deliberately depart from the variation-based

methods as indicated in Chap. 2 and start with

$$t = \frac{y_1 + y_2}{2} = \frac{y_{(1)} + y_{(2)}}{2} \tag{5-4}$$

as an estimator for the parameter θ. The density for t is easily derived and we have

$$f(t - \theta) = \begin{cases} 2 - 4|t - \theta| & |t - \theta| < \tfrac{1}{2} \\ 0 & \text{otherwise} \end{cases} \tag{5-5}$$

We can, of course, test any value for θ by seeing if an observed t is reasonable for the distribution (5-5) using the particular θ value. Correspondingly, we can form a confidence interval by choosing a range of "acceptable" values for $t - \theta$ and then finding the θ values that are acceptable given the observed t value.

A 75 percent probability interval for $t - \theta$ is $(-\tfrac{1}{4}, \tfrac{1}{4})$; this is the central 75 percent interval for the deviation $t - \theta$. The resulting 75 percent confidence interval for θ from a value t is

$$C(t) = \{\theta : t - \theta \in (-\tfrac{1}{4}, \tfrac{1}{4})\}$$
$$= (t - \tfrac{1}{4}, t + \tfrac{1}{4}) \tag{5-6}$$

The preceding 75 percent confidence interval has some rather unusual features that should make it very unattractive for applications. Indeed, consider the confidence interval for cases in which the sample range is $|y_2 - y_1| > \tfrac{1}{2}$. A short integration for the uniform distribution on $(\theta - \tfrac{1}{2}, \theta + \tfrac{1}{2}) \times (\theta - \tfrac{1}{2}, \theta + \tfrac{1}{2})$ shows that this range condition occurs with probability 25 percent. Now if y_2 differs from y_1 by more than $\tfrac{1}{2}$, then using the fact that $|y_i - \theta| < \tfrac{1}{2}$ for each i we find that $(y_1 + y_2)/2 = t$ must be within $\tfrac{1}{4}$ of the true θ and thus $(t - \tfrac{1}{4}, t + \tfrac{1}{4})$ must cover the true θ:

$$P\left(\left|\frac{y_1 + y_2}{2} - \theta\right| < \tfrac{1}{4} \,\middle|\, |y_2 - y_1| > \tfrac{1}{2}\right) = 100\% \tag{5-7}$$

Thus the confidence for $(t - \tfrac{1}{4}, t + \tfrac{1}{4})$ is 100 percent. In an application with $|y_2 - y_1| > \tfrac{1}{2}$ a practicing statistician would be able to say categorically that $(t \pm \tfrac{1}{4})$ brackets the true parameter value. And yet the interval as developed is an ordinary "75 percent confidence interval."

In a compensating way, however, if $|y_2 - y_1|$ is very small, say approximately zero, then the conditional probability of covering the true θ is approximately 50 percent:

$$P\left(\left|\frac{y_1 + y_2}{2} - \theta\right| < \tfrac{1}{4} \,\middle|\, |y_2 - y_1| \doteq 0\right) \doteq 50\% \tag{5-8}$$

In an application with such a value for $|y_2 - y_1|$, a practicing statistician would clearly be overstating the case to attach 75 percent confidence to $(t \pm \tfrac{1}{4})$.

The intermediate cases are easily examined using the conditional distribution for t given the difference $d = |y_2 - y_1|$,

$$g(t \mid d, \theta) = \begin{cases} \dfrac{1}{1-d} & |t - \theta| < (1-d)/2 \\ 0 & \text{otherwise} \end{cases} \qquad (5\text{-}9)$$

and the marginal distribution for d,

$$h(d) = \begin{cases} 2 - 2d & 0 < d < 1 \\ 0 & \text{otherwise} \end{cases} \qquad (5\text{-}10)$$

We can view the coverage probabilities as being as high as 100 percent and as low as 50 percent—but averaging out with respect to the distribution for d to the nominal value 75 percent. However we choose to view this phenomenon, we do need to acknowledge the absoluteness of the 100 percent statement given by (5-7). Clearly we do not need probability theory to obtain the statement (5-7): if $|y_i - \theta| < \tfrac{1}{2}$ for $i = 1, 2$ and if $|y_2 - y_1| > \tfrac{1}{2}$, then by ordinary deductive logic we *know* that $|t - \theta| < \tfrac{1}{4}$. Thus it follows by elementary logic.

In conclusion, we can note that the reduction method RM_2 from Chap. 3 used with the obvious and appropriate variation-based model gives the conditional confidence levels, e.g., as in (5-7) and (5-8). The example was discussed from a different viewpoint in Welch (1939).

We now consider a third example that avoids the obvious faults that lead to unusual coverage probabilities for a given nominal confidence level. We examine it in terms of the response-based model, but avoid the faults by conforming to the pattern required with the appropriate variation-based model. The example shows how a casual choice of confidence intervals—that somehow inadvertently ties in with response values—can produce the same kind of anomalous results as those found in the preceding examples.

Example 5-3 For this example we turn to a discrete distribution. Let y and θ take values on the integers 0, 1, 2, 3, treated as integers modulo 4, and let the distribution of y be uniform on $\theta - 1, \theta, \theta + 1$ (modulo 4).

There are three possible values for $z = y - \theta$, namely 3, 0, 1; each of these values has the same probability $\tfrac{1}{3}$. A 66.7 percent acceptance region for tests or confidence intervals can be formed by selecting two of the three values. If we select the values $z = 0, 1$, we obtain the following 66.7 percent confidence procedure:

Observation y	Confidence region for θ
0	$\{3, 0\}$
1	$\{0, 1\}$
2	$\{1, 2\}$
3	$\{2, 3\}$

(5-11)

If we use a different pair of z values, then we obtain a different 66.7 percent confidence procedure, just differing in an obvious way from the preceding. There are, of course, $\binom{3}{2} = 3$ different ways of selecting two z values from the three possible values and, accordingly, there are three different procedures of the general kind just described.

Now suppose we have an investigator who inadvertently or carelessly chooses his regions from the preceding three procedures, and does so in a way that ties in with response values in the following pattern:

Observation y	Region chosen for θ
0	$\{1, 3\}$
1	$\{0, 1\}$
2	$\{1, 3\}$
3	$\{0, 3\}$

(5-12)

For an investigator who chooses a region consistently in this way we can easily determine the long-run performance probabilities:

Value of θ	Probability of covering θ, in percentage
0	66.7
1	100
2	0
3	100

(5-13)

Thus we have the same anomalous coverage probabilities—some above and some below the nominal 66.7 percent confidence level.

Note that each of the three mentioned procedures has a regular coverage probability of 66.7 percent and does not have the anomalies in the earlier example. However, an inadvertent or unfortunate selection of intervals tied in with response values produces again the anomalous, sometimes high, sometimes low, coverage probabilities. Thus, if there is selection among various reasonable procedures there can be anomalous and disturbing possibilities.

A fourth example shows that unusual coverage values can all be *above* the nominal confidence level. This property for the coverage values does, however, overlook certain possibilities and these possibilities correspond to a "low coverage" value—or to large losses within a betting framework for assessing probabilities.

Example 5-4 Now suppose that θ can take integer values and that $y = \theta - 1$, $\theta + 1$ with equal probability $\frac{1}{2}$. A 50 percent confidence interval for θ can be obtained as the point $y + 1$, or as the point $y - 1$.

We have seen that selection among procedures can lead to high and low coverage probabilities. Suppose for the present example that an investigator chooses the interval $y - 1$ if $y > 0$, the interval $y + 1$ if $y \leq 0$; in a sense the investigator is betting inward towards zero. The coverage probabilities are then as follows:

Value of θ	Probability of covering θ, in percentage
0	100
1	100
Otherwise	50

Thus it seems that the investigator's procedure is at least 50 percent effective, and in special cases is 100 percent effective.

A randomized version of this procedure spreads the $\theta = 0, 1$ advantages to all the possible integer values for θ: use a discrete distribution with positive probability at each integer; randomly obtain a value from this discrete distribution; and then follow the preceding procedure adjusted so that "betting" is inward toward the value just obtained.

Asserted probabilities and confidence intervals are sometimes assessed in terms of betting between two supposedly equal participants. For the preceding confidence procedure with an asserted 50 percent confidence, the person betting in favor of coverage would have a net possibility of gain: he or she would win for sure if $\theta = 0, 1$ and would break even otherwise.

In a way it is always possible to be 100 percent correct for a particular parameter value; use intervals that always contain that parameter value when it is a possibility. This is the case with the initial nonrandomized procedure. The nonfinite nature of the integers on the real line allows the investigator to choose intervals that include 0 (also 1) whenever 0 (and 1) are possibilities, and yet still have a protected 50 percent interval otherwise.

The initial procedure can be viewed as betting that θ is inward toward the centre 0 on the line. Betting inward would seem to have the possibility of missing values far out. In a sense this is the case.

The assessment procedure in terms of betting allows the person placing bets to decide where and how to bet depending on assessment of the betting situation. Correspondingly, for two equal participants, we need to allow the person accepting bets to decide whether or not to accept individual bets. It turns out that with a sequence of θ values that tends to ∞, a person betting on the preceding confidence procedure can lose heavily. Thus the procedure with its asserted 50 percent confidence level does not have protected money-making properties for the bettor; certain possibilities can lead to very large losses.

For some discussion of betting assessments for such confidence intervals see Fraser (1977), in which the preceding is discussed in detail.

We thus return in effect to the Example 5-3 situation. If selection is made among a spectrum of good confidence procedures, then the high-low coverage probabilities are possible. Whether such a selection of intervals from a spectrum has real significance for applications is not clear. Central intervals, meaningful in most contexts, are a natural and sensible approach to inference concerning real-valued parameters.

5-3 LIKELIHOOD

The likelihood function from an observed response provides by itself an assessment of the various possible values for the parameter. This provides a widely available although somewhat primitive method of inference.

5-3-1 Observed Likelihood Function

Consider an inference base $(\mathcal{M}, \mathcal{D})$. The model $\mathcal{M}$ can be any of those examined in Sec. 1-2. For most of the discussion in this section, however, it suffices to have the minimum response-based model discussed in Sec. 1-2-1. Let $f(y|\theta)$ be the density for the observable response y given the parameter θ in Ω.

The likelihood function from a response value was defined in Sec. 4-1-1. It presents the probability for the response value as a function of the possible θ for the model

$$L(y;\theta) = cf(y|\theta) \tag{5-14}$$

The arbitrary constant c allows for an arbitrarily sized neighbourhood at the response value. Informally, an observed likelihood is the *probability for what is observed as a function of the possible values for θ*; the formal definition is given by (4-2). The likelihood function was used in Sec. 4-1 largely in a mechanical way to obtain the weak likelihood (sufficiency) reduction.

The equivalence class of similarly shaped functions presented informally by (5-14) or formally by (4-2) describes the relative probability for y as a function of possible values for θ. A common way to handle an equivalence class is to choose a representative from the class based, say, on some general characteristic; for some recent and detailed discussion see Fraser (1976, p. 313f). A simple representative is obtained by standardizing with respect to the maximum,

$$L_1(y;\theta) = \frac{f(y|\theta)}{f(y|\hat{\theta})}$$

provided there is a finite maximum value for $f(y|\theta)$ at a point, say $\hat{\theta} = \hat{\theta}(y)$. This representative is sometimes called the relative likelihood function, but this is a misnomer. The likelihood function itself *is* relative, one θ value to another, and the preceding representative is just one of many ways of recording or presenting the relative properties of the likelihood function.

In this section we view the likelihood function as a direct and primitive means for assessing an observed response.

Definition The observed likelihood function records the probability for the observed response as a function of the parameter θ of the model.

We are, of course, assuming in this section that the model $\mathcal{M}$ for the system is given. Typically, then, there is no problem calculating the observed likelihood function; simplistically, just substitute the observed y in the function $f(\cdot|\cdot)$ and let the resulting y section be indeterminate with respect to a multiplicative constant c. With one real parameter a likelihood function can be plotted directly or by computer. With two real parameters the contours of the function can be plotted directly or by computer; also with computer graphics the full function can be displayed in perspective and prints produced in hard copy. With more than two real parameters, direct examination of the likelihood function becomes difficult; one straightforward possibility is to examine various sections through the likelihood function.

When we examine an observed likelihood function it is natural to ponder the likelihood functions we might have obtained—from other response values of the system. It is then natural, one step further, to ponder the patterns of likelihood functions we would obtain under the various distributions given by the parameter of the system—in other words, the *statistical model* for the likelihood function as the "response" of the system. Some indication of this from a pragmatic viewpoint may be found in Sec. 2-2-1 and also, diffusely, in Sec. 2-5.

The weak likelihood (sufficiency) principle S in Sec. 4-1-4 prescribed in effect that the initial data and model be replaced by the observed likelihood function and the model for possible likelihood functions. The principle is not presented explicitly in this way in Sec. 4-1-4, but in the earlier Sec. 4-1-2, we did record the equivalence of the likelihood map and the likelihood statistic when the model is available.

There is, however, a *strong likelihood principle* that effectively says forget about the model and use only the observed likelihood function.

L: Strong likelihood principle An inference base $(\mathcal{M}, y)$ should be replaced by the observed likelihood function $L(y; \cdot)$.

This principle goes much farther than any suggested in Chap. 4. By using this principle we no longer have a nontrivial inference base; we have the observed likelihood function as the data and we have a model that consists just of the indexing space Ω.

The development in this book does not in any way endorse the strong likelihood principle L. The methods in the preceding section require more than just the likelihood function, and our approach generally is that inference should be based on all the nonarbitrary ingredients in the inference base.

Sometimes, as in Sec. 2-2-1, an observed likelihood function may be available and yet the model for the likelihood may be essentially impossible to extract and use. In such cases we are then de facto using just the observed likelihood; but this is not by principle, but by pure force of circumstances. Recall, however, that in Sec. 2-2-1 we recommended the empirical or Monte Carlo examination of likelihood functions that can arise with a model. Typically this examination will be by computer simulation, and will give background information as to what likelihood functions arise under various parameter values. This background information would then allow a safer and more incisive use of an observed likelihood function.

In the profession there is a strong concern and unease with the strong likelihood principle; this largely derives from results in Birnbaum (1962) and related papers on inference principles. Birnbaum presents an argument that the weak likelihood principle S and the ancillarity principle A together imply the strong likelihood principle L. Some dissent concerning the validity of the argument has been registered, for example, by Fraser (1963). In addition, there has been some discomfort that seemingly acceptable principles S and A should imply a principle L, so seemingly unacceptable to most statisticians, except the committed Bayesians. The validity of the argument is one aspect, as mentioned above. Another is the questionable nature of principle A (see Sec. 4-2). We have noted in Sec. 4-2-4 that all the attractive examples supporting principle A are in fact examples of the necessary method RM_2 in Sec. 3-2, and that the unattractive residual examples are those left for principle A. Also, we have noted the self-contradictory nature of principle A. These difficulties leave principle A rather loose and unsatisfactory as an ingredient for an argument in support of the strong likelihood principle.

5-3-2 Techniques with Likelihood Functions

For multiparameter problems the direct assessment of a likelihood function may be rather difficult; we now examine some techniques for simplifying this assessment.

Consider a likelihood function with parameter θ, and suppose the parameter θ, perhaps a reexpression of the initial parameter, has components θ_1, θ_2.

The *profile likelihood* for the parameter component θ_1 is

$$\sup_{\theta_2} L(y; \theta_1, \theta_2) \tag{5-15}$$

This gives the profile or silhouette of the likelihood surface viewed along θ_1 sections.

The *section likelihood* for θ_2 given θ_1 is

$$\frac{L(y; \theta_1, \theta_2)}{\sup_{\theta'_2} L(y; \theta_1, \theta'_2)} \tag{5-16}$$

This is the θ_1 section of the likelihood surface.

Now suppose that the sample space function $t(\cdot)$ has some special relevance for the parameter component θ_1; in particular that it has a distribution independent of θ_2; this function could arise perhaps on the basis of a reduction method in Chap. 4. Let $h(t|\theta_1)$ be the marginal density function for t given θ_1. Then the *marginal likelihood function* for θ_1 is

$$ch(t|\theta_1) \tag{5-17}$$

This is, of course, a regular likelihood function, but calculated from the value t together with the corresponding model h.

Alternatively, suppose that the sample space function $r(\cdot)$ conditional on the preceding $t(y) = t$ has some special relevance for the parameter component θ_2; this function could arise on the basis of a reduction method in Chap. 4. Let $g(r|t, \theta_1, \theta_2)$ be the conditional density function for r given t, θ_2, and perhaps θ_1. Then the *conditional likelihood* for θ_2 given θ_1 is

$$cg(r|t, \theta_1, \theta_2) \tag{5-18}$$

as a function of θ_2. This is, of course, a regular likelihood function for θ_2, but calculated from r together with the conditional model for r given t; note that if θ_1 is involved in the preceding expression, then the expression as it stands is just a θ_1-section for the full likelihood from the conditional model.

The preceding presents a few indications of the wealth of possibilities for forming likelihood-type or likelihood-based functions. For example, paralleling formula (5-15) we could integrate out the parameter θ_2 using some weight or measure function, as indicated by some other considerations. Or with formulas (5-15) or (5-16) we could modulate with some weight function for the variable in question, again as indicated by some other considerations. Or we could explore any plausible mathematical operation on the initial likelihood or on likelihoods from component variables.

It is perhaps all too easy to call such functions "likelihood functions." But there may be little left of what can reasonably be called "likelihood." This brings us, of course, to many questions: What do such functions mean? How do they interrelate? Are they in some sense useful? Are they in fact misleading? Do they have significance or purpose?

5-3-3 Some Discussion

Consider a simple example bearing on the interrelationship of two of the techniques.

Example 5-5 Consider a sample $\mathbf{y} = (y_1, \ldots, y_n)'$ from the normal (μ, σ) distribution. The likelihood function from $\mathbf{y}$ is

$$L(\mathbf{y}; \mu, \sigma) = c(\sigma^2)^{-n/2} \exp\left[-\frac{1}{2\sigma^2}\sum(y_i - \mu)^2\right]$$

Suppose we focus attention on σ^2 and profile out the parameter μ following formula (5-15); we obtain

$$c(\sigma^2)^{-n/2} \exp\left[-\frac{1}{2\sigma^2}\sum(y_i - \bar{y})^2\right] \qquad (5\text{-}19)$$

This likelihood-type function depends on the response value **y** only through the sample space function $t(\mathbf{y}) = \Sigma(y_i - \bar{y})^2$.

Consider the marginal likelihood function for σ^2 based on the preceding function $t(\mathbf{y}) = \Sigma(y_i - \bar{y})^2$. The marginal density is

$$h(t\,|\,\sigma^2) = \Gamma^{-1}\left(\frac{n-1}{2}\right)\left(\frac{t}{2\sigma^2}\right)^{(n-1)/2 - 1} e^{-t/2\sigma^2} \frac{1}{2\sigma^2}$$

The corresponding likelihood function is

$$c(\sigma^2)^{-(n-1)/2} \exp\left[-\frac{1}{2\sigma^2}\sum(y_i - \bar{y})^2\right] \qquad (5\text{-}20)$$

Note the discrepancy between the expressions (5-19) and (5-20); they are different functions of σ^2. The first is obtained by a mathematical operation on a likelihood function and the second is the likelihood function from the variable needed for the first expression. Each expression might be treated as a likelihood function and yet only the second is a true likelihood function.

A much more extreme example arises if we have a large number n of normal samples of size 2 from different normal distributions with unknown means but common variance. The analog of formula (5-19) has the power of σ^2 equal to $-2n/2$ and the analog of formula (5-20) has the power equal to $-n/2$; this is a *big* difference or discrepancy.

These examples focus on a major problem or difficulty with likelihood-type functions: they may not be likelihood functions and yet be treated as likelihood functions because "likelihood" is used in the name. As a precaution we suggest the following principle:

Consistency principle—likelihood A function should be treated as a likelihood function only if it is the likelihood function from the variable on which it is based.

This principle would prescribe the marginal likelihood (5-20) in place of the profile "likelihood" (5-19).

At this stage the principle is offered more as a guide or general recommendation concerning the many possibilities for the techniques in Sec. 5-3-2. The techniques, without guidance from methods in preceding chapters, are, however, rather loosely based and exploratory.

The principle does focus on the marginal likelihood (5-17) as opposed to the other functions. Marginal likelihood was introduced in Fraser (1965, 1966) for particular variables that have some special significance (for example, see Sec. 2-2-1; also see Sec. 7-3-1).

Marginal likelihood can, of course, be calculated from any component variable. There can be little merit or safety in the examination of such a likelihood unless there is some justification for the choice of the particular component variable. In some cases the choice can be based on, say, the principles in Chap. 4. In other cases the choice may be necessary (see, for example, Secs. 2-2-1, 2-3-1, and the later Sec. 7-3-1).

For the conditional likelihood there is a special case of particular interest. Suppose that $\theta_2 = \theta$ and that θ_1 is trivial, and then consider the reduction method RM_{02} based on an ancillary $a(\cdot)$ in Sec. 4-2. The conditional likelihood based on the conditional model given $a(y) = a$ is of course just the original likelihood from the full response.

In general a conditional likelihood can be very misleading. For example, there may be a nontrivial likelihood coming from the conditioning variable, and it could be an important and significant part of the likelihood from the full response.

In this section we have considered the direct assessment of likelihood functions. We do not propose that only the likelihood function be examined, but there may be cases where only the likelihood function is available.

5-4 INFERENCE AND DECISIONS

We have developed the concept of an inference base and examined various methods for reduction and terminal inference. The inference base formalizes all the relevant model and data information concerning the unknowns of the system under investigation; and the various methods of inference are concerned with organizing, extracting, and presenting what the inference base implies concerning the unknowns of the system.

As part of concentrating our attention on inference we have avoided any consideration of terminal decisions, decisions specific to some particular use for a particular purpose at a particular time. Such decisions are, of course, based on the available information concerning the unknowns of the system, but they are not directly concerned with these unknowns. We do not examine this specialized area of terminal decisions—which could be called expedient statistics. Rather we focus our attention on organizing the available information; we focus on inference—what can be inferred from an investigation concerning the unknowns of a system.

We have not discussed any point-estimation techniques in the preceding sections and yet they are often found under the general heading of inference. We argue that point-estimation theory belongs with the study of terminal decision. Consider an example.

Example 5-6 Let $\mathbf{y} = (y_1, \ldots, y_n)'$ be a sample from the normal (μ, σ) distribution and suppose that a point estimate is wanted for the parameter μ. The common estimate for μ is the sample average $\bar{y}$. Presenting this value can be viewed as an expedient, not as an inference in the sense we have been developing.

An estimate, by itself, is a very minimal output from an inference base. It provides no indication of reliability or precision. The contention here is that such an output, a single point estimate, can only be meaningful in some particular context for some particular purpose. In a different context with a different purpose, a different estimate would typically be the *decision*—based, say, on a different measure of reliability or a different loss function. The tailoring of results to a minimum form for a specific use is a part of decision theory, and, as such, is a separate consideration from the inference methods examined in this book. For some recent comments see Fraser (1976, p. 284).

5-4-1 Combining Inference Bases

Consider two different investigations concerning an unknown of some system. By this we do not mean, for example, the results from two different blocks of a randomized block design; they are both part of the full design. Rather we are thinking of two distinct investigations, each of which yields an inference base:

$$(\mathcal{M}_1, \mathcal{D}_1) \qquad (\mathcal{M}_2, \mathcal{D}_2)$$

These inference bases are concerned with the same unknown. Accordingly $\mathcal{M}_1$ and $\mathcal{M}_2$ have the same parameter space Ω and the same true value of the parameter θ.

Certainly the two inference bases can be mathematically combined. The model would be the product model but with the common parameter space. The data would be the vector combination $(\mathcal{D}_1, \mathcal{D}_2)$.

Some plausible arguments can certainly be given for combining the inference bases, if only to examine the combined base in comparison with the individual bases.

The viewpoint in this book is that the component inference bases should be available. Indeed, the viewpoint is stronger—that the component bases are primary and that a decision, a type of terminal decision, is involved in forming the combined base. This becomes a rather serious matter in the following Bayesian context.

5-4-2 Bayesian Analysis

Consider the typical context for Bayesian analysis. For this we picture some observational or experimental investigation of a system. The available specification has produced a model that formalizes the relevant aspects of the system; the model is the set of possible descriptions for the system as needed for the particular performances. This definition clearly excludes preferences or shadings among the possibilities: the model is the *set* of possibilities.

A nonnegative numerical value placed on each of the possibilities is sometimes referred to as a sort of "prior likelihood" and a nonnegative measure placed on the set of possibilities can be referred to as a sort of "prior distribution." In subjective Bayesian analyses, these are organized as the considered impressions, judgments, and feelings of a particular investigator or statistician. Thus with different investigators we would have different prior likelihoods or different prior distributions.

One motivating force for the introduction of these prior assessments is the need to obtain an analysis of an inference base; we have referred to this earlier in Sec. 1-1-4. To the degree that available theory is unable to produce inference methods, an investigator can feel free to explore possible methods. But a method proffered is no better than the support available for it.

Now let us return to a particular investigator with prior likelihood or prior distribution. Also suppose we have an inference base recording what is available from the particular investigation. We then have two different sources of information concerning the parameter: the "prior" density or likelihood and the inference base. The former is a loose and personal thing far from the firm kind of information we had in mind for the discussions in Sec. 5-4-1. The viewpoint there is that the two sources of information should be separate and should both be available and that to combine them represents a type of terminal decision. For some recent discussions see Fraser (1976, pp. 576–578) and also Fraser (1972, 1974).

The case for keeping different sources of information separate can be highlighted by considering an example involving standard probability and statistical analysis. It is, however, a type of example often incorrectly classified under the heading of Bayesian analysis.

5-4-3 An Example

A certain population consists of two racial groups in the proportion 1 to 5. A certain characteristic θ, perhaps social, cultural, or economic, is known to be distributed as the normal (85, 10) in racial group A_1 and as the normal (115, 10) in racial group A_2. An assessment procedure is available for measuring θ in any individual and the measurements are normal $(\theta, 10)$.

For an individual sampled from the population, information is wanted concerning his characteristic θ. The individual is measured by the assessment procedure and the value $y = 110$ is obtained; what are the inferences concerning the θ value? The racial origin of the individual may or may not be available.

The individual The assessment procedure has given the observed value $y = 110$ from the normal $(\theta, 10)$ distribution. The routine methods from Sec. 5-2 and earlier sections give the following:

Confidence level, in percentage	Confidence interval
$68\frac{1}{4}$	$(110 \pm 10) = (100, 120)$
95	$(110 \pm 19.6) = (90.4, 129.6)$
99	$(110 \pm 25.8) = (84.2, 135.8)$

These and other confidence intervals can be summarized in the form of the confidence distribution: normal (110, 10) with density given by

$$\frac{1}{\sqrt{2\pi}\,10}\exp\left[-\frac{1}{200}(\theta-110)^2\right] \tag{5-21}$$

The population origin The individual was sampled from a population in which the characteristic θ is a mix of two normals:

$$\frac{1}{6}\frac{1}{\sqrt{2\pi}\,10}\exp\left[-\frac{1}{200}(\theta-85)^2\right]+\frac{5}{6}\frac{1}{\sqrt{2\pi}\,10}\exp\left[-\frac{1}{200}(\theta-115)^2\right] \tag{5-22}$$

Simple integration gives the following confidence intervals:

Confidence level, in percentage	Confidence interval
$68\frac{1}{4}$	(94.95, 123.76)
95	(74.63, 133.81)
99	(66.19, 140.12)

These and other confidence intervals can be summarized in the form of the confidence distribution: $(\frac{1}{6})$ normal (85, 10) + $(\frac{5}{6})$ normal (115, 10) with density given by (5-22). The mean of this distribution is 110 and the standard deviation is 15.

The racial origin The racial origin of the individual is available by special request: the individual belongs to racial group A_1. The individual thus came from a population in which the characteristic θ is normal (85, 10). The confidence intervals for θ are immediately available:

Confidence level, in percentage	Confidence interval
$68\frac{1}{4}$	(85 ± 10) = (75, 95)
95	(85 ± 19.6) = (65.4, 104.6)
99	(85 ± 25.8) = (59.2, 110.8)

These and other confidence intervals can, of course, be summarized in the form of the confidence distribution: normal (85, 10) with density given by

$$\frac{1}{\sqrt{2\pi}\,10}\exp\left[-\frac{1}{200}(\theta-85)^2\right] \tag{5-23}$$

The individual and the population origin We now examine the kind of inferences that would come from the combined inference base involving the assessment and the population origin. The prior density given by (5-22) multiplied by the likelihood (5-21) gives the following conditional density (normalized) given the data $y = 110$:

$$0.04273 \frac{1}{\sqrt{2\pi \times 50}} \exp\left[-\frac{(\theta - 97.5)^2}{100}\right] + 0.95727 \frac{1}{\sqrt{2\pi \times 50}} \exp\left[-\frac{(\theta - 112.5)^2}{100}\right]$$

(5-24)

The mean of this distribution is 111.86 and the standard deviation is 7.86. Some confidence intervals are:

Confidence level, in percentage	Confidence interval
$68\frac{1}{4}$	(104.48, 119.37)
95	(95.62, 126.23)
99	(88.78, 130.61)

The confidence distribution is unimodal with the peak at 112.43, approximately; the confidence density is given by (5-24).

The individual and the racial origin We now examine the kind of inferences that would come from the combined inference base involving the assessment and the racial origin. The prior density given by (5-23) multiplied by the likelihood (5-21) gives the following conditional density (normalized) given the data $y = 110$:

$$\frac{1}{\sqrt{2\pi \times 50}} \exp\left[-\frac{(\theta - 97.5)^2}{100}\right]$$

(5-25)

which is normal (97.5, 7.07). Some confidence intervals are:

Confidence level, in percentage	Confidence interval
$68\frac{1}{4}$	$(97.5 \pm 7.07) = (90.43, 104.57)$
95	$(97.5 \pm 13.86) = (83.64, 111.36)$
99	$(97.5 \pm 18.24) = (79.26, 115.74)$

These and other confidence intervals can, of course, be summarized in the form of the confidence distribution: normal (97.5, 7.07). Note that the central value here is 97.5, which is 12.5 units below the value in the direct assessment of the individual's characteristic θ.

In summary, note the following mean and standard deviations for the confidence distribution:

Inference base	Mean	SD
The individual	110	10
Population origin	110	15
Racial origin	85	10

Combined inference base	Mean	SD
The individual and the population origin	111.9	7.9
The individual and the racial origin	97.5	7.1

We have expressed the view that inference bases should be combined only for some appropriate terminal decision. There may not be any appropriate terminal decisions for the present problem that would require a combined inference base. We would, of course, remain then with the component inference bases.

REFERENCES AND BIBLIOGRAPHY

Birnbaum, A.: On the Foundations of Statistical Inference (with Discussion), *Jour. Amer. Stat. Assoc.*, vol. 57, pp. 269–326, 1962.

Fraser, D. A. S.: On the Sufficiency and Likelihood Principles, *Jour. Amer. Stat. Assoc.*, vol. 58, pp. 641–647, 1963.

————: On Local Inference and Information, *Jour. Roy. Stat. Soc.*, ser. B, vol. 26, pp. 253–260, 1964.

————: "Lecture Notes on Statistical Inference," University of Toronto, Toronto, 1965.

————: Structural Probability and a Generalization, *Biometrika*, vol. 53, pp. 1–9, 1966.

————: Data Transformations and the Linear Model, *Ann. Math. Stat.*, vol. 38, pp. 1456–1465, 1967.

————: Bayes, Likelihood, or Structural, *Ann. Math. Stat.*, vol. 46, pp. 777–790, 1972.

————: Comparison of Inference Philosophies, in G. Menges (ed.), "Information Inference and Decision," D. Reidel Publishing Company, Dordrecht, Holland, 1974.

————: "Probability and Statistics: Theory and Applications," Duxbury Press, North Scituate, Mass., 1976.

————: "Confidence, Posterior Probability, and the Buehler Example," *Ann. of Stat.*, vol. 5, pp. 892–898, 1977.

Welch, B. L.: On Confidence Limits and Sufficiency with Particular Reference to Parameters of Location, *Ann. Math. Stat.*, vol. 10, pp. 58–69, 1939.

CHAPTER
SIX
THE REGRESSION MODEL

In Chap. 2 we considered a real-valued response, with location and scale unknown but with distribution form known or known up to a shape parameter λ. For this we had already noted (Sec. 1-2-3) that the distribution form itself was directly observable, even with the unknown location and scaling. We were thus required to use the more descriptive variation-based model in Sec. 1-2-4.

Now suppose the system has input variables that can be changed and the investigator wants information on how such changes affect the response. Also suppose that the background information concerning the system and related systems specifies that the only effect from the input variables is on the general level or *location* of the response and is linear over the appropriate range for the input variables.

In this chapter we investigate the appropriate model and its analysis with data—the linear regression model and regression analysis.

6-1 CORE METHODS OF ANALYSIS

For notation let y designate the response and $x_1, \ldots, x_r$ designate the input variables. Some of the "input" variables x may actually be combinations of other input variables, thus allowing the usual polynomial and interactive regression models.

Also let f_λ be the density function describing the variation, with a possible shape parameter λ taking values in a space Λ. We suppose that f_λ has been suitably standardized as discussed in Sec. 1-2-2; e.g., the central 68.27 percent probability is in the interval $(-1, +1)$. Now let z designate the standardized variable for variation corresponding to the response variable y.

110 INFERENCE AND LINEAR MODELS

Let σ designate the unknown response scaling for the variation and let $\beta_1 x_1 + \cdots + \beta_r x_r$ designate the general level of the response, linear in the input variables. We thus obtain

$$y = \beta_1 x_1 + \cdots + \beta_r x_r + \sigma z \tag{6-1}$$

where z has the standardized distribution $f_\lambda(z)$ with λ in Λ. This is the common way for presenting the regression model, although often the combination σz is written as e and called the error.

For multiple performances of the system let

$$X = (\mathbf{x}_1 \ldots \mathbf{x}_r) = \begin{pmatrix} x_{11} \ldots x_{1r} \\ \cdots \cdots \cdots \\ x_{n1} \ldots x_{nr} \end{pmatrix} \tag{6-2}$$

designate the design matrix where the ith row records the values for the input variables $x_1, \ldots, x_r$ on the ith performance; we assume that X has rank $r < n$. Also let $\boldsymbol{\beta} = (\beta_1, \ldots, \beta_r)'$ designate the linear coefficients for the general response level, the *regression coefficients*, and let $\mathbf{y} = (y_1, \ldots, y_n)'$ and $\mathbf{z} = (z_1, \ldots, z_n)'$ record the n values for the response and the corresponding values for the standardized variation. We then obtain

$$\mathbf{y} - X\boldsymbol{\beta} + \sigma \mathbf{z} \tag{6-3}$$

where $\mathbf{z}$ has the sample distribution $f_\lambda(\mathbf{z}) = \Pi f_\lambda(z_i)$ with λ in Λ.

6-1-1 The Models

We now examine the various kinds of models as introduced in Sec. 1-2. For this we let $\Omega = \{(\boldsymbol{\beta}, \sigma, \lambda)\} = \mathbb{R}^r \times \mathbb{R}^+ \times \Lambda$ designate the full parameter space.

For the minimum *response-based model* we have

$$\mathcal{M}_R = (\Omega; \mathbb{R}^n, \mathcal{B}^n, \mathcal{F}) \tag{6-4}$$

where $\mathbb{R}^n$ is the response space, $\mathcal{B}^n$ is the Borel class in $\mathbb{R}^n$, and $\mathcal{F}$ is the class

$$\mathcal{F} = \left\{ \sigma^{-n} f_\lambda(\sigma^{-1}(\mathbf{y} - X\boldsymbol{\beta})) : \begin{array}{c} \boldsymbol{\beta} \in \mathbb{R}^r \\ \sigma \in \mathbb{R}^+ \end{array} \lambda \in \Lambda \right\} \tag{6-5}$$

Let $\mathbf{y}^0$ be the observed response resulting from the design matrix X; we then have the inference base

$$(\mathcal{M}_R, \mathbf{y}^0) \tag{6-6}$$

For the *variation-based model* we refer to the general discussion in Sec. 1-2-4. With multiple observations at given settings for the input variables the distribution for the variation is directly observable and identifiable, and must then be included in the model as an objective component. The model for the variation is

$$(\Lambda; \mathbb{R}^n, \mathcal{B}^n, \mathcal{V}) \tag{6-7}$$

where

$$\mathcal{V} = \{f_\lambda(\mathbf{z}) = \Pi f_\lambda(z_i) : \lambda \in \Lambda\} \tag{6-8}$$

The response **y** is a particular presentation of the variation **z**; we write
$$\mathbf{y} = X\boldsymbol{\beta} + \sigma\mathbf{z} = [\boldsymbol{\beta}, \sigma]\mathbf{z} \tag{6-9}$$
and use a minor extension of the notation in Sec. 1-2-2. Specifically we have
$$[\mathbf{b}, c]\mathbf{y} = X\mathbf{b} + c\mathbf{y}$$
$$[\mathbf{b}, c][\mathbf{b}^*, c^*] = [\mathbf{b} + c\mathbf{b}^*, cc^*]$$
$$[\mathbf{b}, c]^{-1} = [-c^{-1}\mathbf{b}, c^{-1}]$$
and thus see that the class of transformations
$$\mathcal{T} = \{[\boldsymbol{\beta}, \sigma]; \boldsymbol{\beta} \in \mathbb{R}^r, \sigma \in \mathbb{R}^+\} \tag{6-10}$$
is *closed under the formation of products and inverses*, and accordingly is a *group*— as remarked in the discussion at the beginning of Sec. 3-4.

We can now present the variation-based model $\mathcal{M}_V$:
$$\mathcal{M}_V = (\Omega; \mathbb{R}^n, \mathcal{B}^n, \mathcal{V}, \mathcal{T}) \tag{6-11}$$
This has essentially two components: the model (6-7) for the objective variation and the model (6-10) for all the possible presentations for the response from the variation. Let $\mathbf{y}^0$ be the observed response from the design matrix X; we then have the inference base
$$(\mathcal{M}_V, \mathbf{y}^0) \tag{6-12}$$

We do not specifically introduce notation for the pivotal version of the model. Rather, we refer to the discussion in Sec. 1-2-5 which relates the pivotal model $\mathcal{M}_P$ to the response-based model $\mathcal{M}_R$ and to the variation-based model $\mathcal{M}_V$. Apart from minor notational differences, the pivotal model is a response model $\mathcal{M}_R$ plus a pivotal function; the pivotal function gives pivotal distributions but without the objective requirement (Sec. 1-1-2) for ingredients of a statistical model.

6-1-2 Preliminary Reduction

Consider the inference base $\mathcal{I} = (\mathcal{M}, \mathcal{D})$ where $\mathcal{M}$ is a regression model $\mathcal{M}_R$, $\mathcal{M}_V$, or some blend thereof, and $\mathcal{D}$ is an observed response vector $\mathbf{y}^0$.

The inference base records the observed $\mathbf{y}^0$; let $\mathbf{z}^0$ be the corresponding realized value for the standardized variation **z**. We examine how much of $\mathbf{z}^0$ is identifiable from $\mathbf{y}^0$. The equation (6-9) gives
$$\mathbf{z}^0 = [\boldsymbol{\beta}, \sigma]^{-1}\mathbf{y}^0 = -X\sigma^{-1}\boldsymbol{\beta} + \sigma^{-1}\mathbf{y}^0 \tag{6-13}$$
for some unknown $\boldsymbol{\beta}, \sigma$; this identifies $\mathbf{z}^0$ as a point on the half $(r+1)$ space given by
$$\begin{aligned}\mathcal{L}^+(X; \mathbf{y}^0) &= \{X\mathbf{b} + c\mathbf{y}^0 : \mathbf{b} \in \mathbb{R}^r, c \in \mathbb{R}^+\} \\ &= \{[\mathbf{b}, c]\mathbf{y}^0 : \mathbf{b} \in \mathbb{R}^r, c \in \mathbb{R}^+\}\end{aligned} \tag{6-14}$$

which is half of the linear space $\mathscr{L}(X;\mathbf{y}^0)$; it is the half space subtended by $\mathscr{L}(X)$ and passing through the observed $\mathbf{y}^0$. Thus the space of *possible* values for $\mathbf{z}^0$ is not n dimensional but is essentially $(r+1)$ dimensional. We can present this more formally by saying that we have observed the value of the *function* $\mathscr{L}^+(X;\mathbf{z})$ carrying $\mathbf{z}$ to $\mathscr{L}^+(X;\mathbf{z})$:

$$\mathscr{L}^+(X;\mathbf{z}) = \mathscr{L}^+(X;\mathbf{z}^0) = \mathscr{L}^+(X;\mathbf{y}^0)$$

but do not have further information concerning the value of $\mathbf{z}^0$ on the identified half space.

Under the variation-based model $\mathscr{M}_V$ we have observed all but $(r+1)$ dimensions of the essential variation $\mathbf{z}$ for a performance of the compound system. Accordingly, by necessary method RM_3 in Sec. 3-3 we obtain the reduction to the *marginal model for what has been observed* and the *conditional model for the unobserved*; these provide respectively for inference concerning λ and for inference concerning (β, σ) given λ.

Under the response-based model $\mathscr{M}_R$ we can find the following grounds for using the marginal and conditional models. The introduction of the weak sufficiency principle in Sec. 4-4 can be used to support the marginal distribution for inference concerning λ; and *with attention restricted to* λ the necessary method RM_5 in Sec. 3-5 requires the use of the marginal distribution for inferences concerning λ. For inference given λ the introduction of the ancillarity principle in Sec. 4-2 can be used to support the conditional model for inference concerning (β, σ); however, recall from Secs. 4-2-3 and 4-2-4 that the ancillarity principle A is self-contradictory.

6-1-3 Suitable Coordinates

In the preceding section we have noted that an n-dimensional vector $\mathbf{z}$ should be examined in terms of where $\mathbf{z}$ lies in an $(r+1)$-dimensional region $\mathscr{L}^+(X;\mathbf{z})$ and also in terms of which $(r+1)$-dimensional region contains $\mathbf{z}$. For this it is, of course, convenient to have familiar coordinates, but we emphasize again that there is nothing absolute in a choice of coordinates. It is a device for talking about points that are there, in $\mathbb{R}^n$, already.

For the half $(r+1)$ space $\mathscr{L}^+(X;\mathbf{z})$ the vectors in X form a natural choice of r of the needed $(r+1)$ basis vectors. Let $\mathbf{d}(\mathbf{z})$ be a vector of unit length lying in $\mathscr{L}^+(X;\mathbf{z})$ and orthogonal to the vectors of X. We have

$$\mathbf{d}(\mathbf{z}) = s^{-1}(\mathbf{z})[\mathbf{z} - X\mathbf{b}(\mathbf{z})]$$
$$= [\mathbf{b}(\mathbf{z}), s(\mathbf{z})]^{-1}\mathbf{z} \qquad (6\text{-}15)$$

where

$$\mathbf{b}(\mathbf{z}) = (X'X)^{-1}X'\mathbf{z}$$

and $\qquad (6\text{-}16)$

$$s^2(\mathbf{z}) = |\mathbf{z} - X\mathbf{b}(\mathbf{z})|^2 = \mathbf{z}'[I - X(X'X)^{-1}X']\mathbf{z}$$

Note that $X\mathbf{b}(\mathbf{z})$ is the projection of $\mathbf{z}$ into $\mathscr{L}(X)$ and that $s^2(\mathbf{z})$ is the squared length of the residual. Thus

$$\mathbf{z} = X\mathbf{b}(\mathbf{z}) + s(\mathbf{z})\mathbf{d}(\mathbf{z}) = [\mathbf{b}(\mathbf{z}), s(\mathbf{z})]\mathbf{d}(\mathbf{z}) \qquad (6\text{-}17)$$

From this, we see that $\mathbf{z}$ has coordinates $[\mathbf{b}(\mathbf{z}), s(\mathbf{z})]$ with respect to the basis $(X, \mathbf{d}(\mathbf{z}))$ for the half space. Note that the possible values for $\mathbf{d}(\mathbf{z})$ generate the unit sphere in the orthogonal complement $\mathscr{L}^\perp(X)$ of $\mathscr{L}(X)$.

Again, we find it appropriate to emphasize that there is nothing absolute in any particular choice of coordinates—a choice just provides a means of saying where $\mathbf{z}$ is in $\mathbb{R}^n$. We could have used any of a wealth of possibilities. Such arbitrary choices have no effect on the inferences.

Now consider the equations

$$\mathbf{y} = X\beta + \sigma\mathbf{z}$$
$$\sigma^{-1}(\mathbf{y} - X\beta) = \mathbf{z} \qquad (6\text{-}18)$$

in terms of these coordinates. We see that $\mathbf{y}$ and $\mathbf{z}$ differ by a location-scale transformation (6-9) and thus lie on the same half plane

$$\mathscr{L}^+(X; \mathbf{z}) = \mathscr{L}^+(X; \mathbf{y}) \qquad (6\text{-}19)$$

or *orbit* under the group (6-10); accordingly, they have the same identifying basis vector

$$\mathbf{d}(\mathbf{z}) = \mathbf{d}(\mathbf{y}) \qquad (6\text{-}20)$$

We use the decomposition (6-17) for both $\mathbf{y}$ and $\mathbf{z}$ and substitute in the equation (6-18):

$$\begin{aligned}\mathbf{y} &= X\mathbf{b}(\mathbf{y}) + s(\mathbf{y})\mathbf{d}(\mathbf{y}) \\ &= X\beta + \sigma\mathbf{z} \\ &= X\beta + \sigma[X\mathbf{b}(\mathbf{z}) + s(\mathbf{z})\mathbf{d}(\mathbf{z})] \\ &= X[\beta + \sigma\mathbf{b}(\mathbf{z})] + \sigma s(\mathbf{z})\mathbf{d}(\mathbf{z})\end{aligned} \qquad (6\text{-}21)$$

This gives (with the orthogonality of X and $\mathbf{d}$)

$$\begin{aligned}\mathbf{b}(\mathbf{y}) &= \beta + \sigma\mathbf{b}(\mathbf{z}) \\ s(\mathbf{y}) &= \sigma s(\mathbf{z})\end{aligned} \qquad (6\text{-}22)$$

for the coordinates of X, and gives $\mathbf{d}(\mathbf{y}) = \mathbf{d}(\mathbf{z})$ which we had deduced earlier.

We now have convenient coordinates for the half $(r + 1)$ space as given by (6-20) and for points on the half space as given by (6-22).

6-1-4 Marginal and Conditional Distributions

The preliminary reduction in Sec. 6-1-2 leads us to consider the *marginal distribution* for the observed half space as given by $\mathbf{d}(\mathbf{y}) = \mathbf{d}(\mathbf{z})$ and the *conditional distribu-*

tion for points on the half space as given by $[\mathbf{b}(\mathbf{y}), s(\mathbf{y})]$ for $\mathbf{y}$ or by $[\mathbf{b}(\mathbf{z}), s(\mathbf{z})]$ with (6-22) for $\mathbf{z}$.

The initial distribution describing $\mathbf{y}$ and $\mathbf{z}$ is given by

$$\sigma^{-n} f_\lambda(\sigma^{-1}(\mathbf{y} - X\boldsymbol{\beta})) \, d\mathbf{y} = f_\lambda(\mathbf{z}) \, d\mathbf{z} \tag{6-23}$$

To obtain the marginal and conditional distributions we first need to make the change of variable

$$\mathbf{y} \leftrightarrow (\mathbf{b}(\mathbf{y}), s(\mathbf{y}), \mathbf{d}(\mathbf{y}))$$
$$\mathbf{z} \leftrightarrow (\mathbf{b}(\mathbf{z}), s(\mathbf{z}), \mathbf{d}(\mathbf{z})) \tag{6-24}$$

The substitution for the density function is straightforward. The substitution for the differential can be obtained easily by noting that $\mathbf{b}(\mathbf{z}), s(\mathbf{z}), \mathbf{d}(\mathbf{z})$ provide locally orthogonal coordinates for $\mathbf{z}$ in $\mathbb{R}^n$: for $\mathbf{b}(\mathbf{z})$ we have† Euclidean volume $|X'X|^{1/2} \, d\mathbf{b}$; for $s = s(\mathbf{z})$ we have Euclidean length ds; and for $\mathbf{d}(\mathbf{z})$ we have $s^{n-r-1} da$ where da is used for surface volume on the unit sphere in $\mathscr{L}^\perp(X)$ and thus $s^{n-r-1} da$ for surface volume on the sphere for $s\mathbf{d}(\mathbf{z})$ with radius s. This gives

$$d\mathbf{z} = |X'X|^{1/2} \, d\mathbf{b} \, ds \, s^{n-r-1} \, da \tag{6-25}$$

for the change of variable in the differential. Then by substitution in (6-23) we obtain

$$\sigma^{-n} f_\lambda(\sigma^{-1}[X(\mathbf{b} - \boldsymbol{\beta}) + s\mathbf{d}]) s^{n-r-1} |X'X|^{1/2} \, d\mathbf{b} \, ds \, da \tag{6-26}$$

for $\mathbf{b} = \mathbf{b}(\mathbf{y})$, $s = s(\mathbf{y})$, and $\mathbf{d} = \mathbf{d}(\mathbf{y})$; and correspondingly obtain

$$f_\lambda(X\mathbf{b} + s\mathbf{d}) s^{n-r-1} |X'X|^{1/2} \, d\mathbf{b} \, ds \, da \tag{6-27}$$

for $\mathbf{b} = \mathbf{b}(\mathbf{z})$, $s = s(\mathbf{z})$, and $\mathbf{d} = \mathbf{d}(\mathbf{z})$.

The *marginal distribution for* $\mathbf{d} = \mathbf{d}(\mathbf{y}) = \mathbf{d}(\mathbf{z})$ is obtained by integration from either (6-26) or (6-27):

$$h_\lambda(\mathbf{d}) \, da = \int_{\mathbb{R}^r} \int_{\mathbb{R}^+} f_\lambda(X\mathbf{b} + s\mathbf{d}) s^{n-r-1} |X'X|^{1/2} \, d\mathbf{b} \, ds \, da \tag{6-28}$$

This integration is usually not available in closed form but is accessible by a range of computer techniques.

The *conditional distribution for* $\mathbf{b}(\mathbf{y}) = \mathbf{b}$ *and* $s(\mathbf{y}) = s$ *given* $\mathbf{d}$ is obtained by division:

$$h_\lambda^{-1}(\mathbf{d}) \sigma^{-n} f_\lambda(\sigma^{-1}[X(\mathbf{b} - \boldsymbol{\beta}) + s\mathbf{d}]) s^{n-r-1} |X'X|^{1/2} \, d\mathbf{b} \, ds \tag{6-29}$$

The corresponding [by (6-22)] *conditional distribution for* $\mathbf{b}(\mathbf{z}) = \mathbf{b}$, $s(\mathbf{z}) = s$ *given* $\mathbf{d}$ is

$$h_\lambda^{-1}(\mathbf{d}) f_\lambda(X\mathbf{b} + s\mathbf{d}) s^{n-r-1} |X'X|^{1/2} \, d\mathbf{b} \, ds \tag{6-30}$$

† Suppose we have an orthonormal set V of vectors that generate $\mathscr{L}(X)$ and let $X = VT$ (where T is $r \times r$) express the X vectors in terms of this new basis; let $\mathbf{b}$ be coordinates with respect to X and $\mathbf{a}$ be coordinates with respect to V. Thus $X\mathbf{b} = VT\mathbf{b} = V\mathbf{a}$ gives $T\mathbf{b} = \mathbf{a}$; then Euclidean volume can be written $da = |T| db = |V'V|^{1/2} |T| db = |T'V'VT|^{1/2} db = |X'X|^{1/2} db$.

These conditional distributions are distributions on the $(r+1)$-dimensional half space $\mathscr{L}^+(X;\mathbf{y}) = \mathscr{L}^+(X;\mathbf{z})$. They are recorded here in terms of the choice of coordinates $[\mathbf{b}(\mathbf{y}), s(\mathbf{y})]$ and $[\mathbf{b}(\mathbf{z}), s(\mathbf{z})]$, but could equally have been recorded in terms of any other choice of coordinates.

The preceding marginal and conditional distributions are the particular distribution discussed at the end of Sec. 6-1-2.

6-1-5 Parameter Components

The response $\mathbf{y}$ and the objective variation $\mathbf{z}$ are related by the equation

$$[\beta, \sigma]^{-1}\mathbf{y} = \sigma^{-1}(\mathbf{y} - X\beta) = \mathbf{z} \tag{6-31}$$

For the variation-based model $\mathscr{M}_V$ this expresses the objective variation $\mathbf{z}$ in terms of the response $\mathbf{y}$. For the response-based model $\mathscr{M}_R$ it records a pivotal function for the response. For both cases we have noted in Secs. 6-1-2 and 6-1-3 that the equation separates as

$$\mathbf{d}(\mathbf{y}) = \mathbf{d}(\mathbf{z}) \tag{6-32}$$

for the observable part of $\mathbf{z}$ and

$$\sigma^{-1}[\mathbf{b}(\mathbf{y}) - \beta] = \mathbf{b}(\mathbf{z})$$
$$\sigma^{-1}s(\mathbf{y}) = s(\mathbf{z}) \tag{6-33}$$

for the unobservable part of $\mathbf{z}$. The relevant distributions are recorded in Sec. 6-1-4.

Now consider separately the two parameter components β and σ. The equation (6-33) can be rearranged so that β and σ are separated:

$$s^{-1}(\mathbf{y})[\mathbf{b}(\mathbf{y}) - \beta] = s^{-1}(\mathbf{z})\mathbf{b}(\mathbf{z}) = \mathbf{T}(\mathbf{z})$$
$$\sigma^{-1}s(\mathbf{y}) = s(\mathbf{z}) \tag{6-34}$$

This separation from (6-33) to (6-34) is *unique*, unique up to reexpression of the individual components in (6-34); we will examine this for the general case in Chap. 7.

We could rewrite (6-34) in a more familiar form involving the t statistic and the residual standard deviation. This has advantages for the numerical integration. We would, however, be picking up a variety of simple constants that unnecessarily complicate some of the distribution expressions to follow; accordingly, we derive the distributions for the components in (6-34) as they stand.

Consider the parameter β. We have the unique separation of the equations (6-33) giving the following components involving β:

$$s^{-1}(\mathbf{y})[\mathbf{b}(\mathbf{y}) - \beta] = \mathbf{T}(\mathbf{z}) = s^{-1}(\mathbf{z})\mathbf{b}(\mathbf{z}) \tag{6-35}$$

The distribution for this comes from the conditional distributions (6-29) and (6-30). These distributions, of course, produce the same distribution for the left- and right-hand sides of (6-34); the resulting distribution for $\mathbf{T} = \mathbf{T}(\mathbf{z})$ is easily derived:

$$g_\lambda^L(\mathbf{T}:\mathbf{d})\, d\mathbf{T} = \int_0^\infty h_\lambda^{-1}(\mathbf{d}) f_\lambda(s(X\mathbf{T} + \mathbf{d})) s^{n-1}\, ds\, |X'X|^{1/2}\, d\mathbf{T} \qquad (6\text{-}36)$$

In a similar way, if we consider a component of β we obtain a further separation of (6-35) and then obtain a corresponding distribution by integration of (6-36).

Now consider the parameter component σ. We have, of course, the unique separation of the equations (6-33) giving the following component involving σ:

$$\sigma^{-1} s(\mathbf{y}) = s(\mathbf{z}) \qquad (6\text{-}37)$$

The distribution for this comes from the conditional distributions (6-29) and (6-30). These distributions, of course, produce the same distribution for the left- and right-hand sides of (6-37); the resulting distribution for $s(\mathbf{z})$ is easily derived:

$$g_\lambda^S(s:\mathbf{d})\, ds = \int_{R^r} h_\lambda^{-1}(\mathbf{d}) f_\lambda(X\mathbf{b} + s\mathbf{d}) \, |X'X|^{1/2}\, d\mathbf{b}\, s^{n-r-1}\, ds \qquad (6\text{-}38)$$

6-2 TERMINAL METHODS OF ANALYSIS

In the preceding section we examined the core-reduction methods for the regression model inference base $\mathscr{I} = (\mathscr{M}, \mathbf{y}^0)$. We now examine some terminal inference methods that are based on these core-reduction methods.

From Sec. 6-1-2 we have a separation of the basic distribution into the marginal distribution for an identified component and a conditional distribution for an unidentified component. This separation was necessary for the variation-based model $\mathscr{M}_V$ and needed the introduction of special principles for the response-based model $\mathscr{M}_R$.

The marginal distribution for $\mathbf{d}(\mathbf{y}) = \mathbf{d}(\mathbf{z})$ is recorded in formula (6-28). The observed value of this function is $\mathbf{d}(\mathbf{y}) = \mathbf{d}(\mathbf{z}) = \mathbf{d}(\mathbf{y}^0)$ as obtained from the data $\mathbf{y}^0$ in the inference base. Let $\mathbf{d}^0$ designate $\mathbf{d}(\mathbf{y}^0)$.

The conditional distribution of $\mathbf{b}(\mathbf{y})$ and $s(\mathbf{y})$ or of $\mathbf{b}(\mathbf{z})$ and $s(\mathbf{z})$, in either case given $\mathbf{d}$, is recorded in formulas (6-29) and (6-30). For $\mathbf{b}(\mathbf{y})$ and $s(\mathbf{y})$ the observed values are $\mathbf{b}(\mathbf{y}^0)$ and $s(\mathbf{y}^0)$. For $\mathbf{b}(\mathbf{z})$ and $s(\mathbf{z})$ the realized values are unobservable; indeed, this necessitates the use of the conditional distribution for $\mathbf{b}(\mathbf{z})$ and $s(\mathbf{z})$.

With this separation of the distribution as a starting point we now consider various terminal methods of inference.

6-2-1 Inference: Shape λ

The marginal distribution (6-28) for $\mathbf{d}(\mathbf{y}) = \mathbf{d}(\mathbf{z})$ depends on the shape parameter λ. The observed value is $\mathbf{d}(\mathbf{y}) = \mathbf{d}(\mathbf{z}) = \mathbf{d}^0$. We consider inference concerning λ.

Under the variation-based model $\mathscr{M}_V$, we have a distribution $f_\lambda(\mathbf{z})$ for the objective variation, and $\mathbf{d}(\mathbf{z}) = \mathbf{d}^0$ is the only observable value available from the inference base. This gives a necessary reduction RM3 (see Sec. 3-3). Under the model $\mathscr{M}_R$

the function $\mathbf{d}(\mathbf{y}) = \mathbf{d}(\mathbf{z})$ is weakly sufficient (see Sec. 4-4). Also by focusing attention on λ we obtain the reduction to $\mathbf{d}(\mathbf{y})$ as a necessary reduction using RM_5 in Sec. 3-5.

The distribution for $\mathbf{d}(\mathbf{y}) = \mathbf{d}(\mathbf{z})$ is a distribution on the unit sphere generated by $\mathbf{d}$ in $\mathscr{L}^\perp(X)$; this sphere is an $(n - r - 1)$-dimensional manifold.

For any chosen $\mathbf{d}$ and λ the value of the density function $h_\lambda(\mathbf{d})$ requires an $(r + 1)$-dimensional integration; combinations of simulation methods and direct integration provide access to such values. For the observed $\mathbf{d} = \mathbf{d}^\circ$ we can then obtain the likelihood function

$$L(\mathbf{d}^\circ; \lambda) = ch_\lambda(\mathbf{d}^\circ) \qquad (6\text{-}39)$$

For one-, two-, and three-dimensional parameters λ various computer graphing techniques allow assessment of (6-39)—for cases where the preceding integrations can be completed.

The likelihood function (6-39) is a marginal likelihood function as introduced in Fraser (1965, 1967, 1968). Other derivations are possible provided the principles needed are chosen propitiously (see Sec. 2-2-1).

The likelihood function $L(\mathbf{d}^\circ; \lambda)$ for λ from the observed $\mathbf{d}(\mathbf{y}) = \mathbf{d}(\mathbf{z}) = \mathbf{d}^\circ$ allows some direct assessment of λ values in relation to the inference base. We would, of course, like to use the observed likelihood in relation to the model for that likelihood, but in general the distribution $h_\lambda(\mathbf{d})$ seems less than tractable; see Sec. 2-2-1.

6-2-2 Inference: Location β

We now consider inference for the location parameter β given a value for λ.

From Sec. 6-1-4 we have the separation of the basic distribution, which gives the conditional distribution (6-29) for $\mathbf{b}(\mathbf{y})$ and $s(\mathbf{y})$ and (6-30) for $\mathbf{b}(\mathbf{z})$ and $s(\mathbf{z})$; the value $\mathbf{d} = \mathbf{d}^\circ$ is used in these formulas. Then, from Sec. 6-1-5 we have the unique separation of β in the equation

$$s^{-1}(\mathbf{y})[\mathbf{b}(\mathbf{y}) - \beta] = s^{-1}(\mathbf{z})\mathbf{b}(\mathbf{z}) = \mathbf{T}(\mathbf{z}) \qquad (6\text{-}40)$$

The conditional distribution for $\mathbf{T}(\mathbf{z})$ with $\mathbf{d} = \mathbf{d}^\circ$ is given in (6-36).

Under the variation-based model $\mathcal{M}_V$ the necessary description for the unobservable $\mathbf{b}(\mathbf{z})$ and $s(\mathbf{z})$ is given by the conditional probability distribution (6-30). A value for β then identifies a contour of $\mathbf{T}(\mathbf{z})$, and the marginal distribution for $\mathbf{T}(\mathbf{z})$ in (6-36) is then the necessary basis for tests and confidence intervals for β. Under the model $\mathcal{M}_R$ the introduction of a weak ancillarity principle related to that in Secs. 4-2 and 4-4 gives some support to the conditional distribution (6-29); the *further* reduction to the distribution (6-36) for $\mathbf{T}(\mathbf{z})$ can be based on the necessary method RM_5 in Sec. 3-5.

Consider the hypothesis: $\beta = \beta_0$. On the assumption that $\beta = \beta_0$, the value of $\mathbf{T}(\mathbf{z}) = s^{-1}(\mathbf{z})\mathbf{b}(\mathbf{z})$ is observable:

$$\mathbf{T} = s^{-1}(\mathbf{z})\mathbf{b}(\mathbf{z}) = s^{-1}(\mathbf{y}^\circ)[\mathbf{b}(\mathbf{y}^\circ) - \beta_0] \qquad (6\text{-}41)$$

This observed value can be compared with the distribution (6-36), with $\mathbf{d} = \mathbf{d}^\circ$, and the given value of λ, to see whether it is a reasonable, high-density value, or a questionable value, or an impossible value out on the edge of the distribution where the density is essentially zero. The hypothesis can be assessed accordingly.

Now consider a confidence region for $\boldsymbol{\beta}$. The observed value $\mathbf{y}^0$ gives various values for

$$\mathbf{T} = s^{-1}(\mathbf{z})\mathbf{b}(\mathbf{z}) = s^{-1}(\mathbf{y}^0)[\mathbf{b}(\mathbf{y}^0) - \boldsymbol{\beta}] \tag{6-42}$$

depending on the values being considered for $\boldsymbol{\beta}$. Let A be a $1 - \alpha$ acceptance region for the distribution (6-36) for $\mathbf{T}$:

$$\int_A g_\lambda^L(\mathbf{T}:\mathbf{d}^0) \, d\mathbf{T} = 1 - \alpha \tag{6-43}$$

We then obtain the observed $1 - \alpha$ confidence region for $\boldsymbol{\beta}$:

$$C(\mathbf{y}^0) = \{\boldsymbol{\beta} : s^{-1}(\mathbf{y}^0)[\mathbf{b}(\mathbf{y}^0) - \boldsymbol{\beta}] \in A\}$$
$$= \mathbf{b}(\mathbf{y}^0) - s(\mathbf{y}^0)A$$

The preceding is, of course, a $1 - \alpha$ confidence region based on (6-40) and the conditional distribution (6-36). The random region has the form

$$C(\mathbf{y}) = \mathbf{b}(\mathbf{y}) - s(\mathbf{y})A(\mathbf{d})$$

where $A(\mathbf{d})$ is obtained from (6-43) with $\mathbf{d}$ replacing $\mathbf{d}^0$. This has conditional confidence $1 - \alpha$ given $\mathbf{d}(\mathbf{y}) = \mathbf{d}$ and thus has marginal confidence $1 - \alpha$.

6-2-3 Inference: Components of $\boldsymbol{\beta}$

Consider some component coordinates of the location parameter $\boldsymbol{\beta}$. Let $\boldsymbol{\beta}^*$ designate a particular set of p coordinates of the original $\boldsymbol{\beta}$. Then from (6-42) we have

$$\mathbf{T}^* = s^{-1}(\mathbf{z})\mathbf{b}^*(\mathbf{z}) = s^{-1}(\mathbf{y}^0)[\mathbf{b}^*(\mathbf{y}^0) - \boldsymbol{\beta}^*]$$

where $\mathbf{T}^*$, $\mathbf{b}^*$ record the coordinates of $\mathbf{T}$, $\mathbf{b}$ that correspond to those of $\boldsymbol{\beta}^*$. The distribution for $\mathbf{T}^*$ is obtained from (6-36) by integrating out the coordinates not represented in $\mathbf{T}^*$. Tests and confidence regions can then be formed following the pattern in Sec. 6-2-2.

In ordinary regression analysis the parameter components $\beta_1, \beta_2, \ldots, \beta_r$ are usually ordered from the most obviously present parameter β_1 to the least obviously present parameter β_r. In addition, it is typically the case that a non null value for a latter parameter, say β_r, implies a possible effect corresponding to preceding parameters. In accordance with this the parameters are usually tested sequentially: (1) test $\beta_r = \beta_{r,0}$; (2) if $\beta_r = \beta_{r,0}$, then test $\beta_{r-1} = \beta_{r-1,0}$; and so on. The sequential testing stops with a significant effect.

Consider the second test: if $\beta_r = \beta_{r,0}$, then test $\beta_{r-1} = \beta_{r-1,0}$. If the given $\beta_r = \beta_{r,0}$ is fully used then there would be pooling of the error variance, pooling that could inflate the error variance if, in fact, β_r was different from $\beta_{r,0}$. The common procedure is to be safe and not pool the error variance; in effect this amounts to testing $\beta_{r-1} = \beta_{r-1,0}$ without formally assuming $\beta_r = \beta_{r,0}$. In short, the lack of significance for β_r gives ground for testing β_{r-1}, but the test is performed in a *safe* manner that does not assume $\beta_r = \beta_{r,0}$ for the analysis—in other words, without pooling of error variance. For the general analysis without normal theory we are able to conform to this pattern of analysis; as we shall see in Chap. 7 this occurs because of some rather special properties of the presentations or transformations involved.

With a nonnormal distribution for variation it is important to note that an analysis of a variance table needs to record actual projections or projection coefficients and not the usual squared lengths.

6-2-4 Inference : Scale σ

We now consider inference for the scale parameter σ. Again as in Sec. 6-2-2 we have a separation of the distribution for $\mathbf{b}(\mathbf{y})$, $s(\mathbf{y})$ in (6-29) and for $\mathbf{b}(\mathbf{z})$, $s(\mathbf{z})$ in (6-30). We then have the unique separation of σ in the equation

$$\sigma^{-1}s(\mathbf{y}) = s(\mathbf{z})$$

The conditional distribution for $s(\mathbf{z})$ with $\mathbf{d} = \mathbf{d}^0$ is given by (6-38).

Tests and confidence intervals for σ are then available exactly in the pattern described in Sec. 2-2-3.

6-2-5 The Normal Case

For the location-scale analysis in Secs. 2-1 and 2-2 we did not derive the distributions for the special case of normal error but quoted an appropriate reference. For the regression model, however, it seems reasonable to record the substitutions. Consider the formulas in Sec. 6-1-4 with

$$f_\lambda(z) = (2\pi)^{-1/2} \exp\left(\frac{-z^2}{2}\right)$$

Note that now there is no free parameter λ. The conditional distribution (6-30) for $[\mathbf{b}(\mathbf{z}), s(\mathbf{z})]$ has the form

$$h_\lambda^{-1}(\mathbf{d})(2\pi)^{-n/2} \exp\left[-\tfrac{1}{2}(X\mathbf{b} + s\mathbf{d})'(X\mathbf{b} + s\mathbf{d})\right] s^{n-r-1} \, ds \, |X'X|^{1/2} \, d\mathbf{b}$$

$$= \frac{|X'X|^{1/2}}{(2\pi)^{n/2}} \exp(-\tfrac{1}{2}\mathbf{b}'X'X\mathbf{b}) d\mathbf{b} \, \frac{A_{n-r}}{(2\pi)^{(n-r)/2}} \exp\left(-\frac{s^2}{2}\right) s^{n-r-1} \, ds$$

where we have used the normalizing constant for the multivariate normal $[\mathbf{0}; (X'X)^{-1}]$ and for the chi $(n - r)$ distribution; recall the definition of A_f in Sec. 2-3-1 and note the correspondence with the gamma and chi-square normalizing constants. It follows that $\mathbf{b}$ is multivariate normal $[\mathbf{0}; (X'X)^{-1}]$ and independently s is chi $(n - r)$; the normalizing constant $h_\lambda(\mathbf{d}) = 1/A_{n-r}$ shows that $\mathbf{d}$ has a uniform distribution on the unit sphere in $\mathscr{L}^\perp(X)$.

The transformations or presentations (6-22) then show that $\mathbf{b}(\mathbf{y})$ is multivariate normal $[\boldsymbol{\beta}, \sigma^2(X'X)^{-1}]$ and independently $s(\mathbf{y})$ is σ-chi $(n - r)$. The ordinary normal analysis then follows necessarily, but on the basis of the necessary method RM_3

in Sec. 3-3. By contrast the usual presentation of the normal analysis requires the introduction of a sufficiency principle as examined in Sec. 4-1. For some discussion in the present framework see Fraser (1976, p. 476f).

6-3 REGRESSION WITH SERIAL CORRELATION

The sequence of response values in a regression analysis is often a sequence in time or in space. In accord with this, the design matrix may contain component vectors that allow for temporal or spatial trends. However, there can be other ways in which temporal and spatial effects can influence the response sequence. More specifically, the variation that affects the response may contain correlations— higher correlations between near responses and lower correlations between distant responses; this is called *serial correlation*. The reason could be that there is underlying variation that affects nearby responses but not more distant responses. In this section, we examine the regression model with serially correlated variation.

6-3-1 The Model

We consider the variation-based regression model $\mathcal{M}_V$ given by (6-11). In its abbreviated form, this model can be written

$$\mathbf{y} = X\boldsymbol{\beta} + \sigma \mathbf{z} \qquad (\boldsymbol{\beta}, \sigma) \text{ in } \mathbb{R}^r \times \mathbb{R}^+$$
$$\mathbf{z} \text{ has a distribution } f_\lambda(\mathbf{z}) \qquad \lambda \text{ in } \Lambda \tag{6-44}$$

We examine the inference base $(\mathcal{M}_V, \mathbf{y}^0)$.

For serial correlation we consider a distribution form $f_\lambda(\mathbf{z})$ on $\mathbb{R}^n$ that has a parameter component that adjusts the serial correlation. In this general form the model and the data analysis are exactly as discussed in preceding sections.

For our discussion here of serial correlation we examine the normal case. Specifically we let $f_\lambda(\mathbf{z})$ be a normal distribution that is standardized for each coordinate but does have correlations between variables; the integration in formula (6-28) can then be completed in closed form. For the normal distribution for variation we use

$$f_R(\mathbf{z}) = \frac{|R|^{-1/2}}{(2\pi)^{n/2}} \exp\left(-\tfrac{1}{2}\mathbf{z}'R^{-1}\mathbf{z}\right) \tag{6-45}$$

where R is a correlation matrix. Thus $\mathbf{z}$ is $N(\mathbf{0}; R)$. A common choice for the matrix R has a single real parameter ρ with $-1 < \rho < 1$ or $0 < \rho < 1$:

$$R = \begin{bmatrix} 1 & \rho & \rho^2 & \cdots & \rho^{n-1} \\ \rho & 1 & & & \vdots \\ \rho^2 & & \ddots & & \\ \vdots & & & & \rho \\ \rho^{n-1} & \cdots & & \rho & 1 \end{bmatrix} \tag{6-46}$$

With this correlation matrix the correlation drops off multiplicatively with the distance between the responses.

6-3-2 The Analysis

Sec. 6-1-4 records the basic distribution components for analyzing the regression model. As noted above we can, for the correlated normal, complete the necessary integration in closed form.

We make a few general comments before examining the details of the integration. The conditional distribution for **b** and s is concerned with the primary parameters β and σ; this distribution involves a chosen or given value for the secondary parameter λ, here R. The analysis for β and σ is modified regression analysis with the particular conditional distribution just mentioned and the modifications as indicated briefly at the end of Sec. 6-2-3.

The marginal distribution $h_R(\mathbf{d})$ for the orbit is available in closed form. Thus, we are able to do more than just likelihood analysis; specifically, we can make tests and form confidence regions.

We now derive the marginal density $h_R(\mathbf{d})$ for the orbit. We, of course, continue to use the reference point **d** on the unit sphere in $\mathscr{L}^\perp(X)$. From (6-28) with (6-45) we have

$$h_R(\mathbf{d}) = \int f_R(X\mathbf{b} + s\mathbf{d})s^{n-r-1}|X'X|^{1/2}\,d\mathbf{b}\,ds$$

$$= \int \frac{|R|^{-1/2}}{(2\pi)^{n/2}} \exp\left[-\tfrac{1}{2}(X\mathbf{b} + s\mathbf{d})' R^{-1}(X\mathbf{b} + s\mathbf{d})\right]s^{n-r-1}|X'X|^{1/2}\,d\mathbf{b}\,ds \quad (6\text{-}47)$$

The integration of a multivariate normal over r coordinates gives in effect the marginal distribution for the "remaining" coordinates. For this, it can be convenient to have orthonormal coordinates. Let P_1 be a matrix of r orthonormal column vectors that span $\mathscr{L}(X)$ and let P_2 be a matrix of $n - r$ orthonormal column vectors that span $\mathscr{L}^\perp(X)$; then in part,

$$\mathscr{L}(P_2) = \mathscr{L}^\perp(X) \qquad P_2'P_2 = I \quad (6\text{-}48)$$

Thus $\mathbf{b} = (X'X)^{-1}X'\mathbf{z}$ and $\mathbf{l} = P_2'\mathbf{z}$ together form a new set of n coordinates relative to the n basis vectors (X, P_2); here we do not use the orthonormal coordinates for $\mathscr{L}(X)$. Note that

$$\mathbf{l} = P_2's\mathbf{d} = sP_2'\mathbf{d} = s\mathbf{d}_*$$
$$\mathbf{d}_* = P_2'\mathbf{d} \quad (6\text{-}49)$$

and thus that $\mathbf{l} = s\mathbf{d}_*$ with just $n - r$ coordinates, gives a very convenient set of coordinates for **d**, a set of just $n - r$ coordinates that complements the coordinates **b**.

The marginal distribution for $\mathbf{l} = s\mathbf{d}_*$ is available from standard statistics textbooks; e.g., see Fraser (1976, pp. 177, 200). We have that **l** is $N(\mathbf{0}; P_2'RP_2)$; let $R_* = P_2'RP_2$ be the $(n - r) \times (n - r)$ variance matrix for **l**. This gives us the result of the location integration of **b** in (6-47); we have

$$h_R(\mathbf{d}) = \int_0^\infty \frac{|R_*|^{-1/2}}{(2\pi)^{(n-r)/2}} \exp\left(-\tfrac{1}{2}s^2 \mathbf{d}'_* R_*^{-1} \mathbf{d}_*\right) s^{n-r-1}\, ds$$

$$= \frac{|R_*|^{-1/2}}{(\mathbf{d}'_* R_*^{-1} \mathbf{d}_*)^{(n-r)/2}} \int_0^\infty \frac{1}{(2\pi)^{(n-r)/2}} \exp\left(-\tfrac{1}{2}s^2 c^2\right)(sc)^{n-r-1}\, dsc$$

$$= \frac{|R_*|^{-1/2}}{A_{n-r}(\mathbf{d}'_* R_*^{-1} \mathbf{d}_*)^{(n-r)/2}}$$

$$= \frac{\Gamma\left(\dfrac{n-r}{2}\right)|R_*|^{-1/2}}{2\pi^{(n-r)/2}\,(\mathbf{d}'_* R_*^{-1} \mathbf{d}_*)^{(n-r)/2}}$$

$$= \frac{|P'_2 R P_2|^{-1/2}}{A_{n-r}[\mathbf{d}' P_2 (P'_2 R P_2)^{-1} P'_2 \mathbf{d}]^{(n-r)/2}} \tag{6-50}$$

where we use the integration results in Sec. 6-2-5 and use $c = (\mathbf{d}'_* R_*^{-1} \mathbf{d}_*)^{1/2}$.

For the uncorrelated case with $R = I$ the density $h_I(\mathbf{d})$ reduces to the uniform distribution for **d** on the unit sphere in $\mathscr{L}^\perp(X)$ or for $\mathbf{d}_*$ on the unit sphere in $\mathbb{R}^{n-r}$:

$$h_I(\mathbf{d}) = \frac{1}{A_{n-r}} = \frac{\Gamma\left(\dfrac{n-r}{2}\right)}{2\pi^{(n-r)/2}} \tag{6-51}$$

Compare with Sec. 6-2-5.

For the correlated case the distribution is proportional to

$$(\mathbf{d}'_* R_*^{-1} \mathbf{d}_*)^{-(n-r)/2} = [\mathbf{d}' P_2 (P'_2 R P_2)^{-1} P'_2 \mathbf{d}]^{-(n-r)/2} \tag{6-52}$$

where $\mathbf{d}'_* R_*^{-1} \mathbf{d}_* = \mathbf{d}' P_2 (P'_2 R P_2)^{-1} P'_2 \mathbf{d}$ is, of course, a quadratic expression in terms of the $n - r$ coordinates of $\mathbf{d}_*$ or the n coordinates of **d**. This distribution is the distribution for the rays from the origin; it is recorded on the unit sphere—where the rays intersect the sphere—and is called the *projected normal distribution on the sphere*.

6-3-3 Inference for Serial Correlation

For response values that form a sequence in time or space there is often a concern for the possible presence of serial correlation. The common regression analysis based on uncorrelated variation was summarized in the results in Sec. 6-2-5. This common analysis can, of course, be altered in the presence of serial correlation, and we would follow the general pattern of Secs. 6-2-2, 6-2-3, and 6-2-5 but with the appropriate correlated model. As noted earlier, we do not examine these details here. Rather, we examine inference concerning serial correlation.

Certainly, in the pattern of Sec. 6-2-1 we can calculate the likelihood function for the correlation matrix R:

$$L(\mathbf{d}; R) = ch_R(\mathbf{d}) = \frac{c|P_2'RP_2|^{-1/2}}{[\mathbf{d}'P_2(P_2'RP_2)^{-1}P_2'\mathbf{d}]^{(n-r)/2}} \quad (6\text{-}53)$$

This can, of course, be easily calculated and plotted for the special correlation matrix R in (6-46), which has a single real parameter ρ.

For more specialized inference, we can construct, say, a test of significance for the hypothesis $H_0: R = I$ of zero correlation. The most powerful test of zero correlation against a specified alternative R is available in the form of the Neyman-Pearson likelihood ratio test:

$$u = \frac{h_R(\mathbf{d})}{h_I(\mathbf{d})} = \frac{|R_*|^{-1/2}/A_{n-r}(\mathbf{d}_*'R_*^{-1}\mathbf{d}_*)^{(n-r)/2}}{1/A_{n-r}} \quad (6\text{-}54)$$

Rejection for large values of u is equivalent to rejection for large values of

$$v = -(\mathbf{d}_*'R_*^{-1}\mathbf{d}_*) = -[\mathbf{d}'P_2(P_2'RP_2)^{-1}P_2'\mathbf{d}] \quad (6\text{-}55)$$

Note that v is a quadratic function of a point $\mathbf{d}$ on the unit sphere in $\mathscr{L}^\perp(X)$. The hypothesis-testing theory used here may be found in most standard texts; e.g., see Fraser (1976, p. 416).

Typically, it seems there will not exist a uniformly most powerful test, even for the relatively simple correlation matrix R in (6-46). However, for the special R in (6-46) we can determine the most powerful test for some specific and representative alternative, say $\rho = \frac{1}{2}$, and hope that the test behaves reasonably for the range of plausible correlation values $(0, 1)$.

The hypothesis distribution of a particular test statistic (6-55) can be obtained by computer integration from the uniform distribution (6-51) over the unit sphere in $\mathscr{L}^\perp(X)$. The corresponding power for a particular significance level can be obtained by computer integration using the general distribution (6-50) on the unit sphere in $\mathscr{L}^\perp(X)$.

6-3-4 An Example

The following example involves eighteen observations of y, which records wheat consumption, specifically domestic use of wheat and wheat products for food by civilians, for the years 1921 through 1938, in million bushels. As introduced by Hildreth and Lu (1960, p. 59), this uses six input variables:

$x_1 = 1$
$x_2 = $ average wholesale price at Kansas City of No. 2 hard red winter wheat per bushel in cents
$x_3 = $ average processing tax on wheat per bushel in cents
$x_4 = $ consumers' disposable income in billion dollars
$x_5 = $ wage rates of all factory workers, per hour, in cents
$x_6 = $ total population on January 1 in millions

124 INFERENCE AND LINEAR MODELS

The data are:

x_1	x_2	x_3	x_4	x_5	x_6	y
1	120.3	0	55.0	48.9	110.9	481
1	113.0	0	64.4	50.2	112.6	480
1	107.2	0	69.8	54.2	114.7	489
1	149.9	0	70.1	54.8	116.7	490
1	162.0	0	75.3	54.6	118.3	509
1	136.3	0	76.0	55.0	119.9	512
1	138.0	0	76.5	55.6	121.5	512
1	111.0	0	80.6	56.5	122.9	516
1	113.0	0	80.3	56.1	124.2	511
1	73.0	0	68.5	53.4	125.4	500
1	49.8	0	54.9	48.6	126.2	498
1	51.0	0	43.6	41.6	127.0	507
1	85.5	30	49.4	50.1	127.8	463
1	100.3	30	54.5	54.5	128.6	474
1	106.8	15	61.6	55.1	129.5	490
1	128.3	0	69.9	58.2	130.3	493
1	97.9	0	67.9	63.6	131.2	489
1	68.4	0	67.3	62.6	132.2	496

The likelihood function (6-53) for ρ using the special R matrix (6-46) is plotted in Fig. 6-1. The likelihood peaks at about 0.7 and has half-max likelihood in the range (0.05, 1.00).

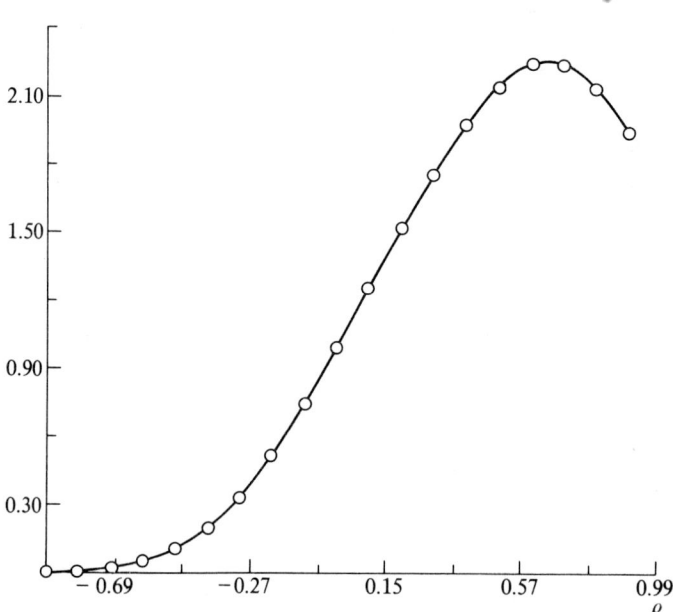

Figure 6-1 Observed likelihood function for ρ from wheat data.

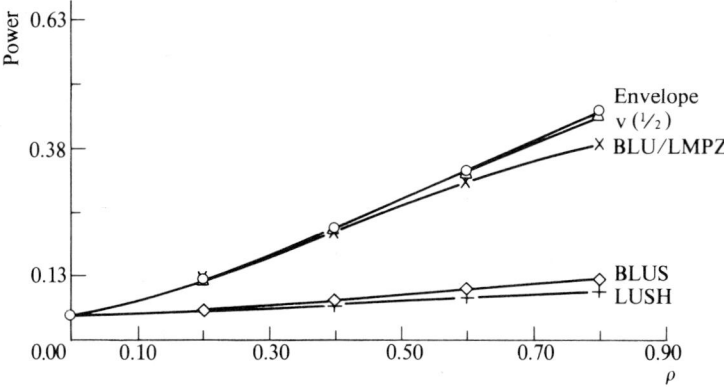

Figure 6-2 Power function for 5 percent test $v(\tfrac{1}{2})$. Envelope power function. Power function for four alternative test functions.

The test statistic (6-55) using the representative $\rho = \tfrac{1}{2}$ has observed the value

$$v = -1.58454$$

The 10, 5, and 1 percent values as obtained by computer integration are

$$v_{10\%} = -1.59198$$
$$v_{5\%} = -1.49042$$
$$v_{1\%} = -1.31270$$

The observed value is just significant at the 10 percent level.

The power function for the 5 percent test using $v(\rho = \tfrac{1}{2})$ was calculated by computer integration and is plotted in Fig. 6-2. For comparison, several other curves are also plotted: the curve just barely above that for $v(\tfrac{1}{2})$ is the envelope power function which records the power of $v(\rho)$ at the value ρ; the remaining curves record power functions for some alternative test statistics commonly used for serial correlation. For further details and background, see Fraser, Guttman, and Styan (1976). The data analysis and computer integrations were performed by Laurel L. Ward in association with George P. H. Styan on the McGill University IBM370(158).

6-4 REGRESSION WITH NONNORMAL VARIATION†

We now consider an example illustrating the regression methods with a nonnormal distribution for variation, specifically the Student (λ) family.

For purposes of calculation, it is useful to express a linear model in a canonical

† With Gordon Fick.

126 INFERENCE AND LINEAR MODELS

form. The natural presentation of the model may be

$$y = X\beta + \sigma z$$

where X represents the matrix of input vectors as recorded; an alternate presentation is

$$y = V\alpha + \sigma z$$

where V is an orthonormal matrix such that

$$\mathscr{L}(V) = \mathscr{L}(X)$$

Inferences can be made for the canonical parameter α and these usually have some direct interpretations for the applied problem. At a terminal stage of the analysis, inferences for β are available by observing that

$$X = VT$$

where T is upper triangular; from this, it follows that $\beta = T^{-1}\alpha$ and that $\mathbf{b}(\mathbf{y}) = T^{-1}\mathbf{a}(\mathbf{y})$ where, of course, $\mathbf{a}(\mathbf{y}) = V'\mathbf{y}$. Clearly $s^2(\mathbf{y})$ and $\mathbf{d}(\mathbf{y})$ do not depend on the basis for $\mathscr{L}(X)$.

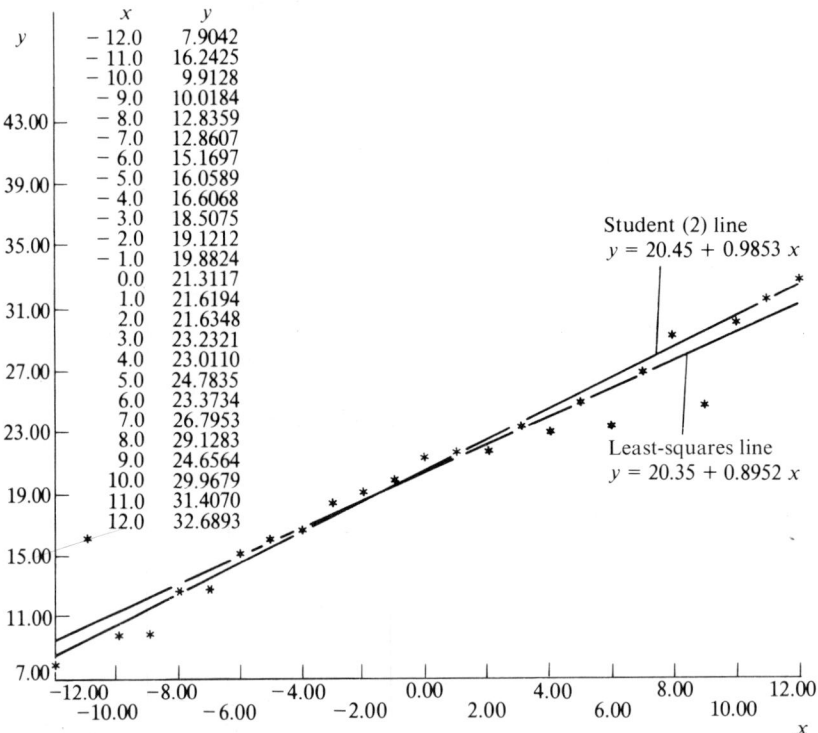

Figure 6-3 The 25 points (x, y); least-squares line; Student (2) line.

Consider the data set displayed in Fig. 6-3; we note that **1** and **x** are orthogonal. The context suggests the model

$$\mathbf{y} = \beta_1 \mathbf{1} + \beta_2 \mathbf{x} + \sigma \mathbf{z}$$
$$= \alpha_1 \mathbf{1}/\sqrt{25} + \alpha_2 \mathbf{x}/\sqrt{\Sigma x^2} + \sigma \mathbf{z}$$

where the distribution form for the variation **z** is the standardized Student (λ) family suggested in Chap. 2. For all the distributions except the normal, computer calculations are used.

For a given value for the shape parameter λ, the assessment of α and then β is based on the observed value of

$$\mathbf{t}(\mathbf{z}) = \mathbf{a}(\mathbf{z})/s_z = \sqrt{n-2}\,\mathbf{a}(\mathbf{z})/s(\mathbf{z})$$

together with the conditional distribution given $\mathbf{d}^o$ as computed using the selected λ value. Note that we are now using s_z as the standard deviation about regression.
Some preliminary calculations yield

$\mathbf{a}(\mathbf{y}) = (101.746, 32.278)'$

$\mathbf{b}(\mathbf{y}) = (20.3492, 0.8952)'$

$s(\mathbf{y}) = 8.6625$

$s_y = 1.80626$

$\mathbf{d} = (-0.1965 \quad\quad 0.6627 \quad -0.1713 \quad -0.2625 \quad -0.0406$
$\quad\quad -0.1411 \quad\quad 0.0221 \quad\quad 0.0215 \quad -0.0186 \quad\quad 0.0974$
$\quad\quad\;\; 0.0649 \quad\quad 0.0495 \quad\quad 0.1111 \quad\quad 0.0433 \quad -0.0583$
$\quad\quad\;\; 0.0228 \quad -0.1061 \quad -0.0048 \quad -0.2710 \quad\quad 0.0207$
$\quad\quad\;\; 0.1867 \quad -0.4329 \quad\quad 0.0769 \quad\quad 0.1397 \quad\quad 0.1844)'$

The largest positive deviation is 0.6627 (corresponding to $y_2 = 16.2425$ and $x_2 = -11.0$) and the largest negative deviation is -0.4329 (corresponding to $y_{22} = 24.6564$ and $x_{22} = 9.0$). These observations seem to be the most influential in determining distributions for $\mathbf{b}(\mathbf{y})$.

We now examine the data using the analysis in the beginning sections together with the Student (λ) family for variation. As with the location-scale analysis of Chap. 2, we begin by consulting the likelihood function for λ. In particular we examine

$$L^*(\mathbf{d}^o; \lambda) = A_{n-2}h_\lambda(\mathbf{d}^o) = A_{23}h_\lambda(\mathbf{d}^o)$$

Selected values of the likelihood function are

λ	1	2	3	4	5	6	9	∞
$L^*(\mathbf{d}^o; \lambda)$	138	302	166	82	45	27	10	1

This rather sharply discriminating likelihood suggests λ values between 1 and 6. For comparison, we have included the traditional normal analysis corresponding to $\lambda = \infty$.

Contour plots for the distribution of $\mathbf{t} = \mathbf{a}(\mathbf{z})/s_z$ are given in Fig. 6-4 for $\lambda = 1, 3, 6$, and ∞. Confidence regions for α or β are based on these distributions. For $\lambda = \infty$ the distribution of $\mathbf{t}$ is the bivariate Student (23) distribution with density

$$\frac{\Gamma\left(\dfrac{25}{2}\right)}{23\pi\,\Gamma\left(\dfrac{23}{2}\right)} \left(1 + \frac{t_1^2 + t_2^2}{23}\right)^{-25/2}$$

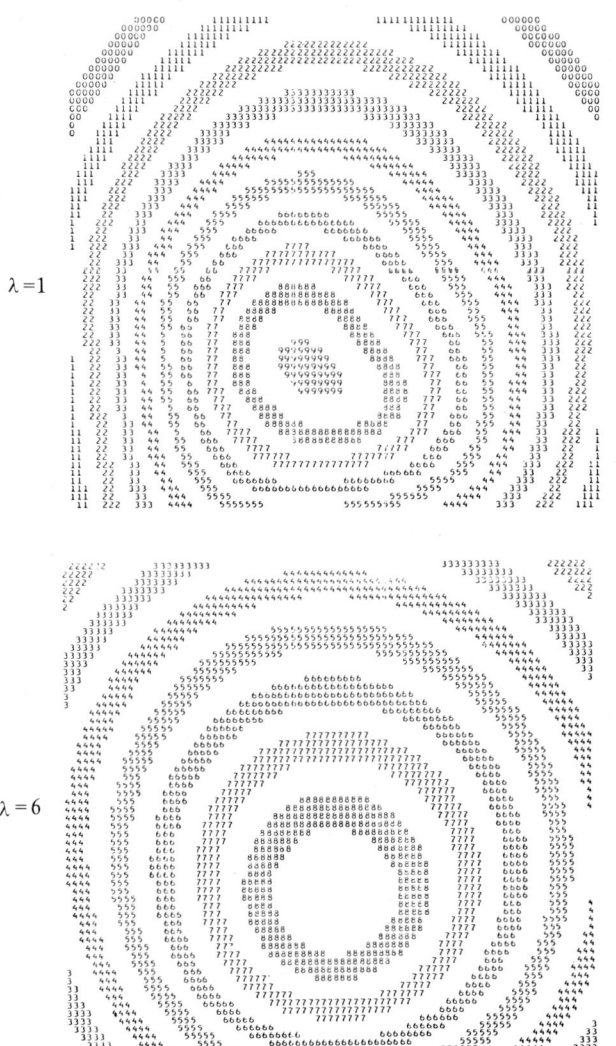

Figure 6-4 Density contour plots for (t_1, t_2) on $(-4, +4)^2$; $\lambda = 1, 3, 6, \infty$.

As λ tends to 1, the distribution shifts and becomes more concentrated. Note under normal analysis the distribution is independent of **d** and does not depend on V. However, under nonnormal analyses the distribution of **t** dramatically changes in shape, location, and concentration, depending on the value for **d** and V.

To assess β_1 and β_2 individually, we examine the component t statistics

$$t_1 = a_1(\mathbf{z})/s_z \qquad t_2 = a_2(\mathbf{z})/s_z$$

together with their distributions; the distributions are plotted in Figs. 6-5 and 6-6 for $\lambda = 1, 3, 6$, and ∞. Under normal analysis the distributions for both t_1 and t_2 are the ordinary Student (23). Note that the t_1 densities are more concentrated

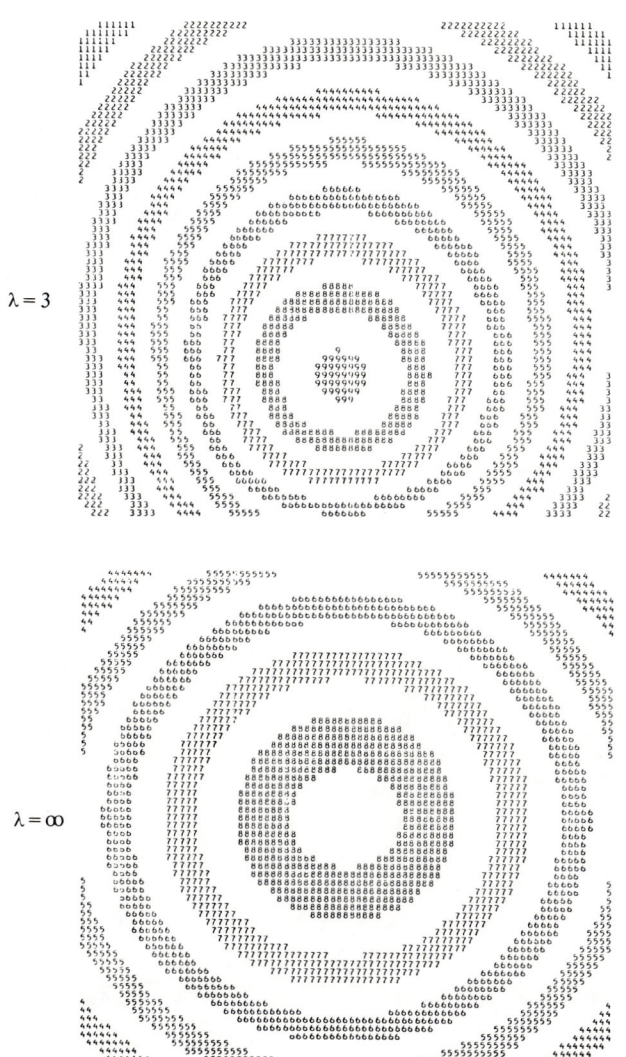

for $\lambda = 6, 3$, and 1 and do not shift substantially; also note that the t_2 densities are more concentrated and do shift substantially to the left.

Confidence intervals for β_1 and β_2 have the form

$$\beta_1: [b_1(\mathbf{y}) - t_{1U} s_y/5, b_1(\mathbf{y}) - t_{1L} s_y/5]$$

$$\beta_2: [b_2(\mathbf{y}) - t_{2U} s_y/\sqrt{1300}, b_2(\mathbf{y}) - t_{2L} s_y/\sqrt{1300}]$$

where (t_{1L}, t_{1U}) is a central interval for t_1 and (t_{2L}, t_{2U}) is a central interval for t_2. For example, 95 percent confidence intervals for β_2 are:

λ	Confidence interval
1	(0.92, 1.04)
3	(0.91, 1.04)
6	(0.87, 1.03)
∞	(0.79, 0.99)

Note that we have included the normal-theory least-squares line on the data plot.

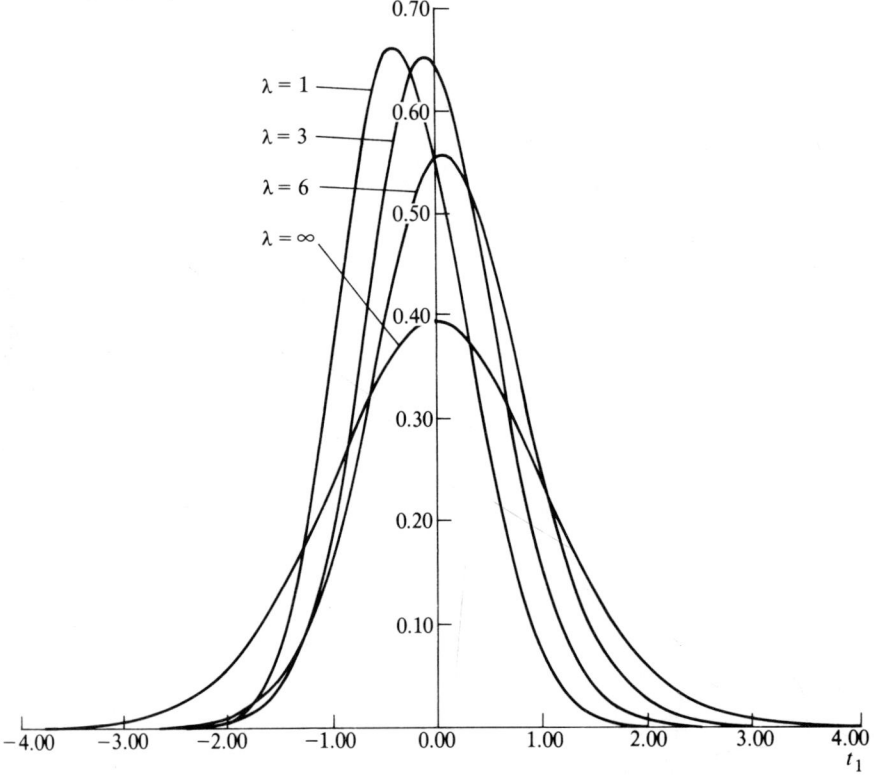

Figure 6-5 Density function for t_1; $\lambda = 1, 3, 6, \infty$.

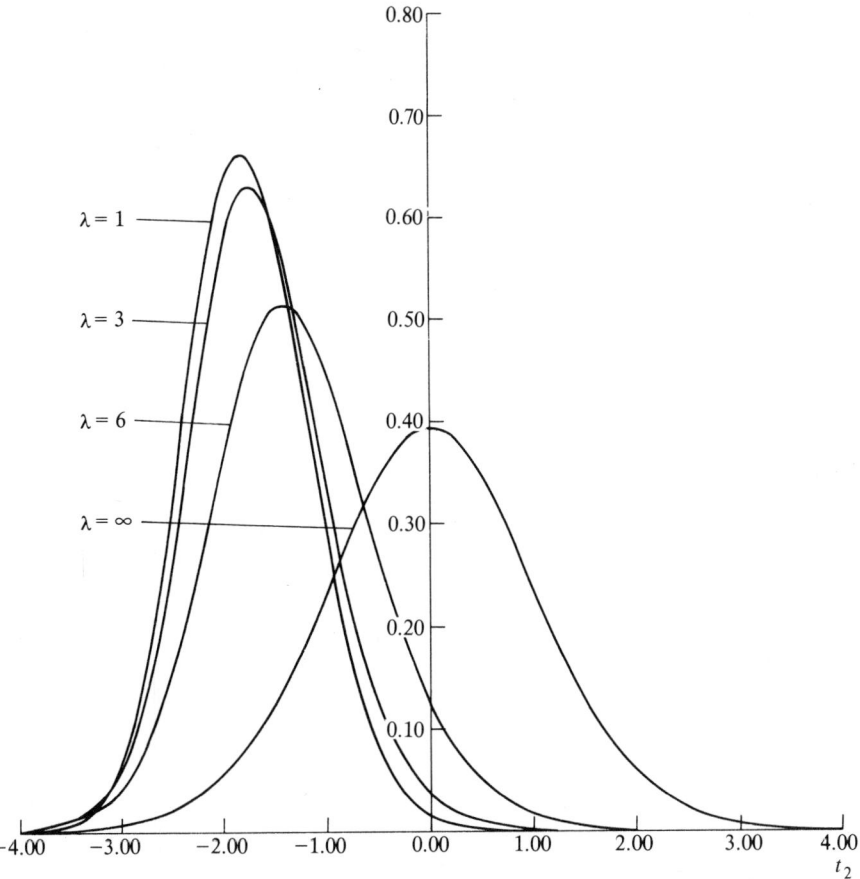

Figure 6-6 Density function for t_2; $\lambda = 1, 3, 6, \infty$.

The line based on the Student (2) analysis has the following values:

$$\text{INTERCEPT} = b_1(\mathbf{y}) - \text{median}(t_1)\, s_y/5 = 20.45$$

$$\text{SLOPE} = b_2(\mathbf{y}) - \text{median}(t_2)\, s_y/\sqrt{1300} = 0.9853$$

Note that the Student (2) line is steeper than the least-squares line and is resisting the effect of y_2 and y_{22}. The Student (λ) analysis provides a more robust and resistant fitting procedure than the usual least-squares procedure.

We find that with *normal* data the analyses are usually similar for various λ and with *nonnormal* data the Student analyses are usually quite different; recall the discussion in Sec. 2-5. The Student analyses with smaller λ values seem to have a very broad based reliability in producing the approximately correct analysis whatever the true value of λ; recall the discussion in Sec. 2-5-1.

The data set was generated using $\beta_1 = 20$, $\beta_2 = 1$, $\sigma = 1.1966$, and $\lambda = 3$.

The likelihood functions and distributions were obtained by three-dimensional integration procedures on the computer. For regression analyses with more than the two regression parameters various simulation and integration procedures are under development.

REFERENCES AND BIBLIOGRAPHY

Fraser, D. A. S.: On Local Inference and Information, *Jour. Roy. Stat. Soc.*, ser. B, vol. 26, pp. 253–260, 1964.
———: "Lecture Notes on Statistical Inference," Department of Mathematics, University of Toronto, 1965.
———: Data Transformations and the Linear Model, *Ann. Math. Stat.*, vol. 38, pp. 1456–1465, 1967.
———: "The Structure of Inference," John Wiley and Sons, New York, 1968.
———: "Probability and Statistics, Theory and Applications," Duxbury Press, North Scituate, Mass., 1976.
———, I. Guttman, and G. P. H. Styan: Serial Correlation and Distributions on the Sphere, *Commun. Statis.—Theory and Method*, vol. A5, pp. 97–118, 1976.
Hildreth, C., and J. Y. Lu: "Demand Relations with Autocorrelated Disturbances," Technical Bulletin no. 276, Agricultural Experiment Station, Michigan State University, East Lansing, Mich., 1960.

CHAPTER
SEVEN
COHERENT MODELS

Statistical models and inference bases were introduced in Chap. 1. As our primary illustration we considered a real-valued response, with location and scale unknown but with distribution form known or known up to a shape parameter λ. For this we discussed in Sec. 1-2-3 how the distribution form could be known—that the distribution form was by itself an objective element of the system under investigation and was amenable to direct sampling, or more informally, that it could be observed directly apart from where it was located and how it was scaled. We now use the term *coherence* for this property of the system being examined, and we call the corresponding model *coherent*.

In this chapter we investigate random systems that have this objective characteristic of distribution form—that have coherence. We then develop models that involve a description of this objective form, the structural models, and we discuss some basic methods of analysis for the corresponding inference bases.

7-1 THE STRUCTURAL MODEL

In this section we consider a random system for which the distribution form is objective and we develop the corresponding coherent model, a model that includes a description for this identifiable form. The model is called a *structural model*.

7-1-1 Identifying Distribution Form

In Sec. 1-2 we considered a real-valued response, with location and scaling unknown but with distribution form known or known up to a shape parameter λ.

Now more generally let Y be the response variable for a random system with sample space $\mathscr{S}$. Given the specification for the system we now consider how the distribution form can be known—how the distribution form can be identified as an objective element of the system, amenable to direct sampling.

In the pattern of Sec. 1-2-2 let Z be a variable for some standardized or nominal presentation of the distribution form, and let the one-one functions h from $\mathscr{S}$ onto $\mathscr{S}$ with h in a class Φ give the various presentations hZ of the distribution form. The shape or form of the distribution can then be given by the equivalence class

$$\{hZ : h \in \Phi\} \tag{7-1}$$

This records the various presentations hZ but does not formally keep a record of the manner of presentation; accordingly we can consider the equivalence class

$$\{(hZ, h) : h \in \Phi\} \tag{7-2}$$

This keeps a record of the manner of presentation h, but does, at the same time, single out the original nominal presentation. The formal way of not singling out the original nominal presentation is to reexpress (7-2) using an arbitrary initial presentation hZ:

$$\{(khZ, k) : k \in \Psi\} \tag{7-3}$$

where $\Psi = \Phi h^{-1} = \{h'h^{-1} : h' \in \Phi\}$, and then take the equivalence class over the various h in Φ. However, rather than being so formal, we follow the pattern in Sec. 1-2-2 and take the distribution form to be the equivalence class (7-1) of different presentations but with, implicitly, the interconnections provided by the various h in Φ.

We now investigate whether the distribution form by itself is objective and amenable to direct sampling. As part of this the class Φ cannot be too big—the possibilities represented by the elements of Φ cannot be more numerous than the possibilities for a sample, for some sample size. Accordingly, we make the following assumption:

Assumption 7-1 There exists a sample size n such that the images $hZ_1, \ldots, hZ_n$ from the initial points $Z_1, \ldots, Z_n$ completely determine the function h in Φ.

In other words, for a sample size n or larger there is at most one transformation carrying an initial vector point into a final vector point.

Now suppose that a sample $(Y_1, \ldots, Y_n)$ is available from the response Y but that the particular manner of presentation ϕ in Φ (as $Y = \phi Z$) is unknown. We let $(Z_1, \ldots, Z_n)$ designate the corresponding values for the nominal presentation mentioned earlier. The various samples corresponding to the distribution form (7-1) are then given by the equivalence class

$$\{(hZ_1, \ldots, hZ_n) : h \in \Phi\} = \{(h\phi^{-1}Y_1, \ldots, h\phi^{-1}Y_n) : h \in \Phi\} \tag{7-4}$$

If this class of samples is available from the observed sample $(Y_1, \ldots, Y_n)$ then the choice of trial ϕ for use in (7-4) cannot affect the expression (7-4). For a sample size n, larger than that specified by Assumption 7-1, the preceding gives that

$$\Phi \phi_1^{-1} = \Phi \phi_2^{-1} \tag{7-5}$$

for any ϕ_1, ϕ_2 in Φ. Conversely, we have of course that the condition (7-5) ensures that the choice of trial ϕ in (7-4) does not affect the expression (7-4).

Consider the implications of equation (7-5). The class $\Phi \phi_1^{-1}$ contains the identity $\phi_1 \phi_1^{-1}$. An element $\phi_2 \phi_1^{-1}$ has the inverse $\phi_1 \phi_2^{-1}$ which belongs to the right-hand side and hence to the left-hand side. For a pair of elements $\phi_3 \phi_1^{-1}$ and $\phi_2 \phi_1^{-1}$ we can reexpress $\phi_3 \phi_1^{-1}$ as $\phi_4 \phi_2^{-1}$ by (7-5) and thus verify that the product $\phi_4 \phi_2^{-1} \phi_2 \phi_1^{-1} = \phi_4 \phi_1^{-1}$ belongs to $\Phi \phi_1^{-1}$. It follows that for any ϕ_1, $\Phi \phi_1^{-1} = G$ is a group of transformations and that

$$\Phi = G \phi_1 \tag{7-6}$$

This says that we can represent each ϕ in Φ as $\theta \phi_1$ with θ in the group G. Conversely, it is clear that (7-6) gives (7-5).

We thus see that the distribution form is objective and can be sampled directly if and only if the class of presentations $\Phi = G\phi_1$ is essentially a group of transformations. By using $\phi_1 Z$ in place of Z for the initial nominal presentation we have that the distribution form is directly observable if and only if the new class of presentations is a group G of transformations on the sample space $\mathscr{S}$. This objectivity of distribution form is one of two slightly different ways of defining *coherence*.

7-1-2 The Structural Model

Consider a random system with sample space $\mathscr{S}$ and response variable Y. In the pattern suggested by Sec. 7-1-1, we suppose that the particular presentation θ of the response is unknown with respect to the transformations of a group G but that the distribution form is known or known up to some shape parameter λ in Λ.

Following the development in Sec. 1-2-2 we would typically choose some standardized presentation for the distribution form; accordingly, we suppose that some reasonable standardization has been applied. Let Z be the variable corresponding to this standardized presentation; then the response presentation is

$$Y = \theta Z \tag{7-7}$$

with θ in G. We call Z the *variation*, but, formally, the variation refers to all the equivalent presentations (7-1).

We now have the following variation-based model called a *structural model*:

$$\mathscr{M}_V = (\Omega; \mathscr{S}, \mathscr{A}, \mathscr{V}, G) \tag{7-8}$$

where $\Omega = \{(\theta, \lambda)\} = G \times \Lambda$, $\mathscr{A}$ is the appropriate σ algebra, $\mathscr{V}$ is the class of distributions for variation with parameter λ in Λ, and G is a group of transformations or presentations. We can abbreviate this as

$$Y = \theta Z \text{ with } \theta \text{ in } G$$
$$Z \text{ has distribution in the class } \mathscr{V} \qquad (7\text{-}9)$$

For completeness we record the form of the corresponding response-based model:

$$\mathscr{M}_R = (\Omega; \mathscr{S}, \mathscr{A}, \mathscr{F}) \qquad (7\text{-}10)$$

where $\mathscr{F}$ is the class of distributions obtained by applying the class G of presentations to the standardized distributions in $\mathscr{V}$ (compare with Sec. 1-2).

Earlier versions of the structural model may be found in Fraser (1965, 1966, 1968b).

7-1-3 Identifying Events for Variation

Consider an inference base using the structural model $\mathscr{M}_V$ and an observed response Y^0:

$$\mathscr{I} = (\mathscr{M}_V, Y^0) \qquad (7\text{-}11)$$

For the location-scale model in Sec. 2-2 and for the regression model in Sec. 6-1 we have seen how a certain function of the variation is observable. We now examine this for the general inference base (7-11).

Or more generally consider the situation examined in Sec. 7-1-1. For this we consider a value Z on a space $\mathscr{S}$ and suppose that the only information concerning Z is the value

$$Y = \phi Z \qquad (7\text{-}12)$$

obtained from some unknown function ϕ in a known class Φ of one-one transformations of $\mathscr{S}$ onto $\mathscr{S}$. We now investigate whether the information concerning Z is equivalent to having the value of a well-defined function on the space for Z. As part of this we obtain the results for the preceding paragraph and in addition obtain a modified version of coherence.

Clearly Z must be one of the values in the *antecedent set*:

$$A(Y) = \Phi^{-1}Y = \{\phi^{-1}Y : \phi \in \Phi\} \qquad (7\text{-}13)$$

The alternatives for Y based on an arbitrary element in Φ then give the *alternative set*:

$$T(Y) = \Phi\Phi^{-1}Y = \{\phi_2\phi_1^{-1}Y : \phi_i \in \Phi\} \qquad (7\text{-}14)$$

The information concerning Z from a response value Y can be displayed as

$$D(\Phi, Y) = \{(\phi^{-1}Y, \phi) : \phi \in \Phi\} \qquad (7\text{-}15)$$

which records possible preimage points $\phi^{-1}Y$, each with the corresponding function ϕ. We are familiar with how the functions $\bar{y}$ and Σy_i are equivalent in the information they provide concerning an antecedent y; the equivalence is based on a one-one onto transformation on the range of the functions. Accordingly, for our purposes here let $\mathscr{B}$ be the group of one-one onto transformations of $\mathscr{S}$ onto $\mathscr{S}$. We make the display of information (7-15) independent of reexpression on the response space by saying that $D(t\Phi, tY)$ is equivalent to $D(\Phi, Y)$ for any t in the group $\mathscr{B}$. Formally this gives the information

$$I(\Phi, Y) = \{D(t\Phi, tY): t \in \mathscr{B}\}$$

concerning Z from the response Y.

The information concerning Z is equivalent to having the value of a single function if each of the possible values for Z as recorded in (7-13) lead to the same information, that is, $I(\Phi, \cdot)$ is constant valued for points in the alternative set (7-14). This means that $D(\Phi, Y')$ is equivalent to $D(\Phi, Y)$ for Y' in $T(Y)$, or that $D(\Phi, Y') = D(t\Phi, tY)$ for some one-one function having $Y' = tY$, or that the functions t such that $\Phi = t\Phi$ are transformations that generate the set $T(Y)$ from Y, or that the class of transformations $\mathscr{G} = \{t: t\Phi = \Phi\}$ generates the set $T(Y)$ from Y.

The class of transformations $\mathscr{G} = \{t: t\Phi = \Phi\}$ is easily seen to be closed under the formation of products and inverses; it is called the *invariant group* of Φ. Thus we have that the information concerning Z is equivalent to having the value of a single function if, and only if, the orbits of $\mathscr{G}$ are the alternative sets $T(Y)$. Indeed, the information can then be written as $\mathscr{G}\phi_1 Z = \mathscr{G}Y$ for any choice of ϕ_1.

The condition $\mathscr{G}\Phi = \Phi$ shows that $\Phi = \mathscr{G}\phi_1$ as in formula (7-6) or more generally that

$$\Phi = \cup_\alpha \mathscr{G}\phi_\alpha \qquad (7\text{-}16)$$

where the various ϕ_α each map antecedent sets $A(Y)$ into corresponding alternative sets $T(Y)$. A class of functions satisfying (7-16) is called *event coherent*; we reserve the term *coherence* for the special case (7-6) in which distribution form is objective.

For an exploratory version of the methods here see Fraser (1971); for the formal details see Brenner and Fraser (1978).

7-1-4 Reduction of the Inference Base

We now consider reduction for the inference base $(\mathscr{M}_V, Y^0)$ using the structural model (7-8) and then comment briefly on reduction for $(\mathscr{M}_R, Y^0)$ using the response-based model (7-10). For the inference base $(\mathscr{M}_V, Y^0)$, let Z^0 designate the realized value of the standardized variation corresponding to the value Y^0 for the response presentation. We have then that $Y^0 = \theta Z^0$ for the true value of the parameter θ in G.

What information is available concerning the realized Z^0 without information concerning θ in G? We can write $Z^0 = \theta^{-1} Y^0$ for some θ in G. This identifies Z^0 as a point on the orbit GY^0 of the observed Y^0. In fact we can write

$$GZ^0 = GY^0 \qquad (7\text{-}17)$$

Thus we obtain the observed value of the function GZ of the variation Z. The results in Sec. 7-1-3 show that the data give no differential information concerning where Z^0 lies on the orbit GY^0. The information concerning Z^0 is equivalent to the value GY^0 for the function GZ.

By the necessary reduction method RM_3 in Sec. 3-3 we then have the factorization or separation of the inference base into

(a) The marginal model for GZ with observed value GY^0.
(b) The conditional distribution given $GZ = GY^0$ describing Z^0, together with the presentation $Y^0 = \theta Z^0$ for some θ in G.

The preceding suggests that the space $\mathscr{S}$ be examined in terms of the orbits GZ and the position of Z on an orbit. For this it is convenient to choose a *reference point* on each orbit; let $D(Z)$ be the reference point on the orbit GZ. We can then record the position of Z on the orbit by finding a transformation $[Z]$ in G which generates Z from $D(Z)$:

$$Z = [Z]D(Z) \tag{7-18}$$

In the cases of interest this transformation will be unique; this is covered by the following *exactness* assumption.

Assumption 7-2 The transformation group G is exact on the space $\mathscr{S}$. If $g_1 Y = g_2 Y$ for some g_1, g_2 in G and Y in $\mathscr{S}$, then $g_1 = g_2$.

This is Assumption 7-1 with no reference to sample size. For our present purposes this means that we are modeling a sample size large enough for the original Assumption 7-1 to hold.

Now let $Q = \{D(Z): Z \in \mathscr{S}\}$ be the set of reference points. Then by the preceding analysis we have a one-one correspondence

$$Z \leftrightarrow ([Z], D(Z))$$

between $\mathscr{S}$ and $G \times Q$. For notational simplicity we will often write

$$Z \leftrightarrow (g, D) \tag{7-19}$$

provided the context identifies $g = [Z]$ and $D = D(Z)$. Note that $Y = \theta Z = \theta[Z]D(Z) = [Y]D(Y)$ and thus that $Y = \theta Z$ is equivalent to $[Y] = \theta[Z]$, $D(Y) = D(Z)$.

We can now reexpress the factorization or separation of the inference base $(\mathscr{M}_Y, Y^0)$ using the coordinates just described. The necessary method RM_3 in Sec. 3-3 gives:

(a) $(\mathscr{M}_D, D^0)$, where $\mathscr{M}_D$ is the marginal model for $D(Z)$ with parameter λ in Λ and $D^0 = D(Y^0)$ is the observed value of $D(Z)$.
(b) $(\mathscr{M}_V^{D^0}, [Y^0])$, where $\mathscr{M}_V^{D^0}$ is the variation-based structural model recording the conditional distribution of $[Z]$ given $D(Z) = D(Y^0)$, together with the presentations $[Y] = \theta[Z]$ with θ in G.

Now consider briefly the inference base $(\mathcal{M}_R, Y^0)$ using the response-based model $\mathcal{M}_R$ in (7-10). The separation derived under $\mathcal{M}_V$ takes the following form for $\mathcal{M}_R$:

(a) $(\mathcal{M}_D, D^0)$, where $\mathcal{M}_D$ is the marginal model for $D(Y)$ with parameter λ in Λ and $D^0 = D(Y^0)$ is the observed value of $D(Y)$.
(b) $(\mathcal{M}^{D^0}, [Y^0])$ where $\mathcal{M}^{D^0}$ is the conditional model for $[Y]$ given $D^0 = D(Y^0)$ and $[Y^0]$ is the corresponding observed value.

Under the model $\mathcal{M}_R$ we can find the following grounds for examining these two components from the $\mathcal{M}_V$-based separation. The weak sufficiency principle in Sec. 4-4 gives some support for the use of the marginal distribution in (a) for inference concerning λ, and with attention confined to λ the necessary method RM_5 in Sec. 3-5 requires the use of the marginal distribution in (a) for inference concerning λ. For inference concerning θ given λ, the introduction of the ancillarity principle A in Sec. 4-2 can be used to support the conditional model (b); however, we recall from Sec. 4-2-4 that the ancillarity principle A is self-contradictory.

In this section we have introduced a general variation-based model, but have not entered into discussions involving density functions. In the next section, we develop methods appropriate for use with density functions.

We conclude this section with some comments on the notation for the position $[Z]$ on an orbit and the labeling $D(Z)$ for an orbit. Consider a first set of reference points $D_1(Z)$ on a cross section Q_1 with position $[Z]_1$ on the orbit, and a second set of reference points $D_2(Z)$ on a cross section Q_2 with position $[Z]_2$ on the orbit. Let $h(Z)$ be defined by

$$h(Z) = [D_2(Z)]_1$$

Note that $h(Z)$ is a function of the orbits GZ. We then obtain

$$[Z]_2 = [Z]_1 h^{-1}(Z) \qquad D_2(Z) = h(Z)D_1(Z)$$

This is a right multiplication adjustment for position and left multiplication adjustment for the reference point.

7-2 CHANGE OF VARIABLE

In Chap. 1 we noted that our emphasis would be on the wide range of problems involving continuous response variables, and that the discrete problems are far fewer and do have special simplicities. Now, in Sec. 7-1, we have developed the structural model, which covers almost all the linearity found in statistical applications, and all the interesting applications lie with continuous response variables. Accordingly, we now present the specialized notation for handling structural models with density functions—for making the change of variable for the needed marginal and conditional distributions discussed in the preceding section. For this there are substantial advantages in using the group properties of the class G of presentations. We do this by using densities with respect to measures that incorporate as much of the symmetry expressed by the group as possible.

7-2-1 Introduction

For most examples the sample space $\mathscr{S}$ for the response will be an open set in the Cartesian product of the reals $\mathbb{R}^N$. Accordingly we make the following assumption.

Assumption 7-3 The sample space $\mathscr{S}$ is an open set in the Cartesian product $\mathbb{R}^N$ of the reals.

A few examples, however, will deviate from this assumption and $\mathscr{S}$ will be a smooth manifold of dimension N embedded in a Cartesian product of the reals. For this we will use a Euclidean structure on the product of the reals and assume that local coordinates are derived from the Euclidean structure on the planes tangent to the manifold.

Now consider a group G acting on the sample space $\mathscr{S}$. We will cover most of the interesting cases by assuming that the group has continuous differentiability properties, both within itself and in its action on $\mathscr{S}$.

Assumption 7-4 G can be represented as an open set in $\mathbb{R}^L$ and the transformations

$$\tilde{g} = g^* g \qquad \tilde{Z} = g^* gZ$$

are continuously differentiable with respect to g^*, g in G, and Z in $\mathscr{S}$.

For a few examples we will deviate from this and also allow G to be a smooth manifold of dimension L in a Euclidean product of the reals. Thus differentiation, if not with respect to coordinates of $\mathbb{R}^L$, will be with respect to local coordinates, derived from the Euclidean properties of the embedding space.

We have noted in Sec. 7-1-4 that most cases are covered by having the group G exact on the space $\mathscr{S}$—at most one transformation connecting two sample points. Accordingly, we let $D(Z)$ give the reference point in a cross section Q of the orbits and we let $[Z]$ in G give position on an orbit:

$$Z = [Z]D(Z) \tag{7-20}$$

This gives a one-one correspondence between $\mathscr{S}$ and $G \times Q$. We add some continuity and summarize in the following assumption:

Assumption 7-5 The group G is exact on the space $\mathscr{S}$ and the functions $[Z]$ and $D(Z)$ are continuously differentiable on $\mathscr{S}$.

This assumption easily covers most of our examples. For any particular application, however, it suffices to have a local definition for the reference points $D(Z)$, local to the particular observed orbit; the needed continuity is then available.

7-2-2 Invariant Volume on $\mathscr{S}$

With the preceding assumptions we are now in a position to develop methods for obtaining the marginal and conditional distributions. Our first step toward this is to derive a support measure on $\mathscr{S}$ that incorporates the symmetries expressed by the group G.

On the space $\mathscr{S}$ in $\mathbb{R}^N$ the obvious first choice for support measure is Euclidean or Lebesgue volume measure:

$$V_N(A) = \int_A dZ \tag{7-21}$$

Now let us examine the effect of a transformation h in G on this measure. We have

$$J_N(h:Z) = \left|\frac{\partial hZ}{\partial Z}\right| \tag{7-22}$$

$$dhZ = J_N(h:Z)\,dZ$$

From this point onward we will take all Jacobians to be the absolute value of the Jacobian and thus positive; for example, $J_N(h:Z) = |\partial hZ/\partial Z|_+$.

Thus a neighborhood of the point Z has Euclidean volume dZ but the corresponding neighborhood of hZ has Euclidean volume that is increased by the factor $J_N(h:Z)$.

We have introduced a reference point $D(Z)$ in order to give the group position of a point Z. Accordingly, it seems reasonable to use Euclidean volume in the neighborhood of the reference point and then to measure "volume" elsewhere as the Euclidean volume for the corresponding neighborhood of the reference point.

For this let $J_N(Z)$ be the Jacobian from $D(Z)$ to Z:

$$J_N(Z) = J_N([Z]:D(Z)) \tag{7-23}$$

In passing note that

$$J_N(h:Z) = \frac{J_N(hZ)}{J_N(Z)} \tag{7-24}$$

We then define an adjusted measure:

$$M(A) = \int_A \frac{dZ}{J_N(Z)} \tag{7-25}$$

142 INFERENCE AND LINEAR MODELS

In terms of differentials we can write

$$dM(Z) = \frac{dZ}{J_N(Z)} \tag{7-26}$$

From this we see that the "volume" $dM(Z)$ in a neighborhood of Z is Euclidean volume as scaled back by the factor $J_N(Z)$—as scaled back by the volume change from $D(Z)$ to Z. Thus in effect $dM(Z)$ assigns to a neighborhood of Z the Euclidean volume of the corresponding neighborhood of the reference point.

First we show that the measure M is an *invariant* measure with respect to the group G:

$$M(hA) = \int_{hA} \frac{dZ}{J_N(Z)} = \int_A \frac{dhZ}{J_N(hZ)}$$

$$= \int_A \frac{J_N(h:Z)\,dZ}{J_N(h:Z)J_N(Z)}$$

$$= M(A) \tag{7-27}$$

Thus A and hA have the same "volume" as measured by M.

Second we show that M measures ordinary Euclidean volume in the neighborhood of reference points $D(Z)$. From our assumptions we have that $J_N(Z)$ is continuous in Z; also $J_N(Z) = J_N([Z]:D(Z))$, with $[Z] = i$, is the Jacobian for the identity transformation; thus $J_N(D(Z)) = 1$. For a neighborhood B of the reference point $D(Z)$ we then obtain

$$M(B) \doteq \int_B \frac{dZ}{1} = V_N(B)$$

Thus for the measure M we have *invariance* under the group and we have a correspondence with ordinary Euclidean volume along Q.

Example 7-1: Location-scale group on $\mathbb{R}^n$ Consider the location-scale group

$$G = \{[a, c] : a \in \mathbb{R}, c \in \mathbb{R}^+\} \tag{7-28}$$

with action $\tilde{\mathbf{z}} = [a, c]\mathbf{z} = a\mathbf{1} + c\mathbf{z}$ on $\mathbb{R}^n$. Following our notation from Chap. 2 we have

$$\mathbf{z} = [\bar{z}, s(\mathbf{z})]\mathbf{d}(\mathbf{z}) = \bar{z}\mathbf{1} + s(\mathbf{z})\mathbf{d}(\mathbf{z}) \tag{7-29}$$

where $s(\mathbf{z})$, $\mathbf{d}(\mathbf{z})$ are the residual length and unit residual. Then we obtain

$$J_n([a, c] : \mathbf{z}) = \left|\frac{\partial a\mathbf{1} + c\mathbf{z}}{\partial \mathbf{z}}\right| = c^n$$

$$J_n(\mathbf{z}) = s^n(\mathbf{z}) \tag{7-30}$$

$$M(A) = \int_A \frac{d\mathbf{z}}{s^n(\mathbf{z})}$$

$$dM(z) = \frac{d\mathbf{z}}{s^n(\mathbf{z})}$$

7-2-3 Invariant Volume on G

We now derive a measure on the group G itself, a measure that has symmetry properties paralleling those for the measure M.

For this consider the equation

$$\tilde{g} = hg \qquad (7\text{-}31)$$

where $\tilde{g}$, h, g are elements of the group G. We view this as a transformation h mapping g to $\tilde{g}$; a transformation by *left group multiplication*. Thus we are viewing G as a transformation group acting on the space G.

A natural reference point for the space G is the identity element i. Then by the results in Sec. 7-2-2 we define an invariant measure on G that agrees with Euclidean volume at the identity:

$$J_L(h:g) = \left| \frac{\partial hg}{\partial g} \right|$$

$$J_L(g) = J_L(g:i)$$

$$\mu(B) = \int_B \frac{dg}{J_L(g)} \qquad (7\text{-}32)$$

$$d\mu(g) = \frac{dg}{J_L(g)}$$

The letter L for the dimension of G is used as a label for the Jacobians and B is used for a measurable subset of G.

The measure μ is the *left invariant* measure that agrees with Euclidean volume at the identity; for we have

$$\mu(hB) = \mu(B) \qquad (7\text{-}33)$$

for left invariance, and we have

$$\mu(B) \div \int_B \frac{dg}{1} = V_L(B)$$

for a neighborhood B of the identity i where V_L gives Euclidean volume on G.

Perhaps the best way of viewing the measure μ, either intuitively or formally, is in terms of the differential $dg/J_L(g)$; this gives a clear picture of its formation in terms of a compensation for the effect of a transformation to g from the identity i. Any other left invariant measure on G can be examined in the same way and seen to be a propagation from the identity of a measure element at the identity; accordingly, any two left invariant measures differ only by a constant multiplicative factor, the ratio of the measure elements at the identity.

Now consider the equation

$$\tilde{g} = gh \qquad (7\text{-}34)$$

where h is viewed as a right transformation mapping g to $\tilde{g}$. As before we choose

144 INFERENCE AND LINEAR MODELS

the identity i as the reference point and use the results from Sec. 7-2-2 to obtain the *right invariant* measure v:

$$J_{\bar{L}}^*(h:g) = \left| \frac{\partial gh}{\partial g} \right|$$

$$J_{\bar{L}}^*(g) = J_{\bar{L}}^*(g:i)$$

$$v(B) = \int_B \frac{dg}{J_{\bar{L}}^*(g)} \tag{7-35}$$

$$dv(g) = \frac{dg}{J_{\bar{L}}^*(g)}$$

The measure v is the *right invariant* measure that agrees with Euclidean volume at the identity; for we have

$$v(Bh) = v(B) \tag{7-36}$$

for right invariance, and we have

$$v(B) \doteq \int_B \frac{dg}{1} = V_L(B)$$

for a neighborhood B of the identity i. Any two right invariant measures on G differ only by a constant multiplicative factor, the ratio of the measure elements at the identity.

Example 7-2: The location-scale group Consider further the location-scale group discussed in Sec. 7-2-2:

$$[\tilde{a}, \tilde{c}] = [a^*, c^*][a, c]$$
$$\tilde{a} = a^* + c^*a$$
$$\tilde{c} = c^*c$$

$$J_L(g^*:g) = c^{*2} \qquad\qquad J_{\bar{L}}^*(g:g^*) = c$$

$$J_L(g) = c^2 \qquad\qquad J_{\bar{L}}^*(g) = c$$

$$d\mu(g) = \frac{da\,dc}{c^2} \qquad\qquad dv(g) = \frac{da\,dc}{c} \tag{7-37}$$

$$\mu(B) = \int_B \frac{da\,dc}{c^2} \qquad\qquad v(B) = \int_B \frac{da\,dc}{c}$$

$$\mu([a,c]B) = \mu(B) \qquad\qquad v(B[a,c]) = v(B)$$

7-2-4 The Change of Variable

For use with the results in Sec. 7-1, we will start with a density function on $\mathscr{S}$ and derive the marginal distribution for the orbit expressed as $D(Z)$ or GZ, and the conditional distribution given the orbit as expressed by $[Z]$ relative to the reference point $D(Z)$. For this we need the change of variable,

$$Z \leftrightarrow ([Z], D(Z)) = (g, D) \tag{7-38}$$

where for simplicity we have written $[Z] = g$ in G and $D(Z) = D$ in Q. The corresponding Jacobian is

$$J(Z) = \left| \frac{\partial Z}{\partial (g, D)} \right| \tag{7-39}$$

where g in $G \subset \mathbb{R}^L$ has L coordinates and D in Q has effectively $N - L$ coordinates. As the Jacobian stands we would expect to use tangential Euclidean coordinates for D on Q, and, in fact, we follow this pattern for a few cases. In general, however, we will adopt an alternative and more convenient choice for the coordinates and volume measure for D; we discuss this briefly in Sec. 7-2-6.

We can rewrite (7-38) in three steps:

$$Z \leftrightarrow h^{-1}Z \leftrightarrow (h^{-1}g, D) \leftrightarrow (g, D) \tag{7-40}$$

The left-hand transformation involves a group element h applied on $\mathscr{S}$, and the right-hand transformation essentially involves just h^{-1} applied on the group G. The middle transformation is the original transformation (7-38) but at a different point. The left and right transformations are relatively easy to work with. We examine the three-step transformations for general h and then make a simplifying choice so that the middle transformation is special.

The Jacobian relation from the three-step transformation is

$$J(Z) = J_N(h : h^{-1}Z) J(h^{-1}Z) J_L(h^{-1} : g) \tag{7-41}$$

This holds for any h; as a special case we take $h = g$ and obtain

$$J(Z) = J_N(g : D) J(D) J_L(g^{-1} : g)$$
$$= J_N(Z) J(D) J_L^{-1}(g). \tag{7-42}$$

Thus the change of variable for the measure element can be expressed as

$$dZ = J_N(gD) J(D) J_L^{-1}(g) dg \, dD \tag{7-43}$$

This uses J_N and J_L which typically are easily obtained from transformations in which group elements operate on $\mathbb{R}^N$ or on $\mathbb{R}^L$. The factor $J(D)$ is the original Jacobian, but now needed only along the reference point contour Q. Often we can avoid calculating $J(D)$, for with the conditional for g given D the factor $J(D)$ enters as a constant which can be incorporated into the derivation of the normalizing constant. Note that by using formulas (7-26) and (7-32), formula (7-43) can be reexpressed as

$$dM(Z) = d\mu(g) J(D) \, dD \tag{7-44}$$

We will give some interpretations in Sec. 7-2-6.

Now consider a distribution for Z with density function $f_\lambda(Z)$. We make the change of variable $Z = gD$ and then determine the distributions described in

Sec. 7-1—the marginal for D and the conditional for g given D. The change of variable gives

$$f_\lambda(Z)\,dZ = f_\lambda(gD)J_N(gD)J(D)J_L^{-1}(g)\,dg\,dD$$
$$= f_\lambda(gD)J_N(gD)J(D)\,d\mu(g)\,dD \qquad (7\text{-}45)$$

First we obtain the *marginal distribution for D*:

$$h_\lambda(D)\,dD = \int_G f_\lambda(gD)J_N(gD)J(D)\,d\mu(g)\,dD \qquad (7\text{-}46)$$

This gives the model for the inference base (*a*) in Sec. 7-1-4. Next we obtain the *conditional distribution for* $g = [Z]$ *given* D:

$$g_\lambda(g:D)\,dg = h_\lambda^{-1}(D)f_\lambda(gD)J_N(gD)J(D)\,d\mu(g)$$
$$= h_\lambda^{-1}(D)f_\lambda(gD)J_N(gD)J(D)J_L^{-1}(g)\,dg \qquad (7\text{-}47)$$

This gives the distribution part of the model for the inference base (*b*) in Sec. 7-1-4.

Now consider the *response distribution for* $Y = \theta Z$. We obtain this by the substitution $Z = \theta^{-1}Y$:

$$f_\lambda(Z)\,dZ = f_\lambda(Z)J_N(Z)\,dM(Z)$$
$$= f_\lambda(\theta^{-1}Y)J_N(\theta^{-1}Y)\,dM(Y) \qquad (7\text{-}48)$$

where we have used the invariance of the support measure M. The *marginal distribution for* $D(Y) = D(\theta Z) = D(Z)$ is of course as given by formula (7-46). The *conditional distribution for* $[Y] = \theta g$ given D can be obtained from (7-48) or, more easily, from (7-47):

$$g_\lambda(g:D)\,dg = g_\lambda(\theta^{-1}[Y]:D)\,d\theta^{-1}[Y]$$
$$= g_\lambda(\theta^{-1}[Y]:D)J_L(\theta^{-1}[Y])\,d\mu\theta^{-1}[Y]$$
$$= h_\lambda^{-1}(D)f_\lambda(\theta^{-1}[Y]D)J_N(\theta^{-1}[Y]D)J(D)\,d\mu([Y])$$
$$= h_\lambda^{-1}(D)f_\lambda(\theta^{-1}[Y]D)J_N(\theta^{-1}[Y]D)J(D)J_L^{-1}([Y])\,d[Y] \qquad (7\text{-}49)$$

7-2-5 Left and Right Invariance

The two measures μ and ν on the group G have some useful interconnecting relations.

Consider a new measure μ_h defined by

$$\mu_h(B) = \mu(Bh) = \int_{g \in B} d\mu(gh) \qquad (7\text{-}50)$$

Clearly μ_h is left invariant, for we have

$$\mu_h(gB) = \mu(gBh) = \mu(Bh) = \mu_h(B)$$

Consequently we have that μ_h and μ differ by a constant factor, say $\Delta(h)$:

$$\mu_h(B) = \Delta(h)\mu(B) \tag{7-51}$$

or in differential form

$$d\mu(gh) = \Delta(h)\,d\mu(g) \tag{7-52}$$

with g as the variable.

The preceding constant of proportionality $\Delta(h)$ can be viewed as a function of h, the *modular function* $\Delta: G \to R^+$. We have the easily verified properties:

$$\Delta(i) = 1$$
$$\Delta(g_1 g_2) = \Delta(g_1)\Delta(g_2) \tag{7-53}$$
$$\Delta(g^{-1}) = \Delta^{-1}(g)$$

Thus Δ is a mapping that preserves group multiplication: it is a homomorphism from G to the positive reals.

Consider a new measure v_1 formed by using $\Delta(g)$ to adjust the left invariant measure $d\mu(g)$:

$$v_1(B) = \int_B \frac{d\mu(g)}{\Delta(g)} \tag{7-54}$$

We show that v_1 is right invariant:

$$v_1(Bh) = \int_{Bh} \frac{d\mu(g)}{\Delta(g)} = \int_{g \in B} \frac{d\mu(gh)}{\Delta(gh)}$$

$$= \int_B \frac{\Delta(h)\,d\mu(g)}{\Delta(h)\,\Delta(g)}$$

$$= \int_B \frac{d\mu(g)}{\Delta(g)} = v_1(B)$$

Also we see that $dv_1(g)$ agrees with $d\mu(g)$ near the identity, for we have $\Delta(i) = 1$; and thus it agrees with Euclidean volume near the identity. It then follows that $v_1 = v$, the unique right invariant measure that agrees with Euclidean volume near the identity. We thus obtain the relations:

$$v(A) = \int_A \frac{d\mu(g)}{\Delta(g)} \qquad \mu(A) = \int_A \Delta(g)\,dv(g)$$
$$dv(g) = \frac{d\mu(g)}{\Delta(g)} \qquad d\mu(g) = \Delta(g)\,dv(g) \tag{7-55}$$

Now consider a new measure v_2 formed by using μ on the inverse elements:

$$v_2(B) = \int_{g \in B} d\mu(g^{-1}) = \int_{B^{-1}} d\mu(g) = \mu(B^{-1}) \tag{7-56}$$

where $B^{-1} = \{g^{-1} : g \in B\}$. We see that v_2 is right invariant:
$$v_2(Bh) = \mu(h^{-1}B^{-1}) = \mu(B^{-1}) = v_2(B)$$

Also we see that it agrees with $d\mu(g)$ near the identity: take a symmetric neighbourhood B of the identity, $B = B^{-1}$, and then $v_2(B) = \mu(B^{-1}) = \mu(B)$; μ agrees with Euclidean volume near the identity. It follows that $v_2 = v$, the unique right invariant measure that agrees with Euclidean volume near the identity. We thus obtain the relations:

$$v(B) = \mu(B^{-1}) \qquad \mu(B) = v(B^{-1})$$
$$dv(g) = d\mu(g^{-1}) \qquad d\mu(g) = dv(g^{-1}) \qquad (7\text{-}57)$$

The following are a few bonus relations that can be very useful. From (7-55) we have

$$\Delta(g) = \left|\frac{d\mu(g)}{dv(g)}\right| = \left|\frac{dg/J_L(g)}{dg/J_L^*(g)}\right| = \frac{J_L^*(g)}{J_L(g)} \qquad (7\text{-}58)$$

Then from (7-57) we have

$$\Delta(g) = \left|\frac{d\mu(g)}{dv(g)}\right| = \left|\frac{dv(g^{-1})}{dv(g)}\right| = \left|\frac{dg^{-1}/J_L^*(g^{-1})}{dg/J_L^*(g)}\right| \qquad (7\text{-}59)$$

From these we obtain

$$\left|\frac{dg^{-1}}{dg}\right| = \frac{J_L^*(g^{-1})J_L^*(g)}{J_L^*(g)J_L(g)} = \frac{J_L^*(g^{-1})}{J_L(g)} = \frac{J_L(g^{-1})}{J_L^*(g)} \qquad (7\text{-}60)$$

The equations (7-58) and (7-60) can be summarized as

$$\Delta(g) = \frac{J_L^*(g)}{J_L(g)}$$
$$\left|\frac{dg^{-1}}{dg}\right| = \frac{J_L^*(g^{-1})}{J_L(g)} = \frac{J_L(g^{-1})}{J_L^*(g)} \qquad (7\text{-}61)$$

Example 7-3: The location-scale group Consider further the location-scale group from Sec. 7-2-3:

$$\Delta([a, c]) = \frac{J_2^*([a, c])}{J_2([a, c])} = \frac{c}{c^2} = \frac{1}{c}$$
$$= \frac{d\mu([a, c])}{dv([a, c])} \qquad (7\text{-}62)$$

and
$$\left|\frac{d[a, c]^{-1}}{d[a, c]}\right| = \frac{c^{-1}}{c^2} = \frac{1}{c^3} \qquad (7\text{-}63)$$

We can verify the Jacobian directly:

$$[a, c]^{-1} = [-c^{-1}a, c^{-1}]$$

$$\left| \frac{\partial(-c^{-1}a, c^{-1})}{\partial(a, c)} \right| = \left| \begin{array}{cc} -c^{-1} & 0 \\ c^{-2}a & -c^{-2} \end{array} \right| = \frac{1}{c^3}$$

7-2-6 Some Interpretations

Consider the change of variable $Z = gD \leftrightarrow (g, D)$ discussed in Sec. 7-2-4. From formula (7-43) we have

$$dZ = J_N(gD)J(D)J_L^{-1}(g)\, dg\, dD \qquad (7\text{-}64)$$

where dZ is Euclidean volume for Z on $\mathscr{S} \subset \mathbb{R}^N$, dg is Euclidean volume on $G \subset \mathbb{R}^L$, and dD is the temporarily chosen Euclidean volume with respect to tangential Euclidean coordinates on Q.

In the neighborhood of the cross section Q we have that J_N and J_L are approximately unity, and thus that

$$dZ \doteq J(D)\, dg\, dD \qquad (7\text{-}65)$$

If we remain with dD as volume along Q then the complementary factor $J(D)\, dg$ is Euclidean volume measured orthogonally to Q at the point D.

Now consider an alternative and generally more useful way of having coordinates and volume measures for D on the manifold Q. We now let dD be *Euclidean volume on the orthogonal complement of the orbit GD at the point D*. For this we have a neighborhood of D on the manifold Q; we project onto the orthogonal complement of the tangent to the orbit GD at D; we calculate the Euclidean volume of the projected neighborhood; and we use this as the measure for the neighborhood of D on Q.

From Eq. (7-65) we now use dD as volume orthogonal to the orbit GD and thus interpret the complementing factor $J(D)\, dg$ as Euclidean volume along the orbit GD in the neighborhood of D. As such

$$J(D) = \left| \frac{\partial gD}{\partial g} \right|$$

is the ratio of volume along the orbit to the corresponding volume on the group; or more directly it is volume change near the identity from the group G to its representation on the orbit GD.

7-3 INFERENCE, TESTS, AND CONFIDENCE REGIONS

The structural model $\mathscr{M}_V$ and the derived response-based model $\mathscr{M}_R$ were presented in Sec. 7-1-2. We now examine statistical inference for a model $\mathscr{M}_V$ or $\mathscr{M}_R$ together with data.

7-3-1 Inference for Shape λ

Consider the structural model

$$\mathcal{M}_V = (\Omega; \mathcal{S}, \mathcal{A}, \mathcal{V}, G) \tag{7-66}$$

recorded in formula (7-8). The parameter space $\Omega = \{(\theta, \lambda)\} = G \times \Lambda$ involves the parameter λ for shape and the parameter θ for the response presentation of the basic distribution for variation; $\mathcal{V}$ is the class of possible distributions for variation indexed by λ; and G is the transformation group recording possible response presentations. The corresponding response-based model is

$$\mathcal{M}_R = (\Omega; \mathcal{S}, \mathcal{A}, \mathcal{F}) \tag{7-67}$$

recorded in formula (7-10). The class $\mathcal{F}$ is obtained by applying the group G to the standardized distributions in $\mathcal{V}$. We will cover most cases of interest by having the exactness Assumption 7-2 hold.

In this section we examine statistical inference for the inference bases $(\mathcal{M}_V, Y^0)$ and $(\mathcal{M}_R, Y^0)$. Recall the reduction of these inference bases as discussed in Sec. 7-1-4.

For the inference base $(\mathcal{M}_V, Y^0)$, we noted in Sec. 7-1-4 that the basic distribution for variation involves only the parameter λ and that the only observation from this distribution is given by $GZ = GY^0$. Then in the special notation for the exactness case, we thus obtain the inference base $(\mathcal{M}_D, D^0)$ which has the parameter λ and has the observed value $D^0 = D(Y^0)$ for the function $D(Z)$ of the basic variation.

The complementary inference base $(\mathcal{M}_V^{D^0}, [Y^0])$ describes the unobservable part of the realized variation and relates entirely to the particular response presentation θ.

The basic distribution involves only λ, and only the function $D(Z)$ is observable with value $D^0 = D(Y^0)$. The inference base is

$$(\mathcal{M}_D, D^0) \tag{7-68}$$

For the location-scale analysis in Chap. 2 we investigated likelihood inference methods for λ but we were somewhat sceptical about the feasibility for the more detailed testing and confidence methods of inference. With one or two exceptions this seems to be the pattern generally.

For likelihood inference we need density functions and accordingly introduce the Assumptions 7-3, 7-4, and 7-5 from Sec. 7-2. The likelihood function for λ is available from (7-46):

$$L(D^0; \lambda) = ch_\lambda(D^0) \tag{7-69}$$

This can be assessed as discussed in Sec. 5-3 on likelihood; recall the discussion for the location-scale case in Sec. 2-2-1.

For the response-based model $(\mathcal{M}_R, Y^0)$ the situation is not as clear-cut and a variety of inference procedures may be found in the literature. However, with a formal decision to restrict attention to λ, the necessary method RM_5 produces the inference base $(\mathscr{M}_D, D^0)$ which we have just discussed. And as we have noted earlier, only the likelihood methods seem to be available in general.

7-3-2 Tests and Confidence Regions for θ

Consider further the variation-based model $(\mathscr{M}_V, Y^0)$ involving the structural model of Sec. 7-1.

From Sec. 7-1-4 we have the separation of the given inference base into two component inference bases: the component $(\mathscr{M}_D, D^0)$ just discussed in Sec. 7-3-1 and the component

$$(\mathscr{M}_V^{D^0}, [Y^0]) \tag{7-70}$$

to be discussed in the remainder of this section. The inference base (7-70) records the conditional structural model on the observed orbit $GZ = GY^0$, as well as the observed response position on the orbit. For notational ease we are using in (7-70) the special notation appropriate under the exactness Assumption 7-2.

Of the two component inference bases only the latter component (7-70) involves the parameter θ. This component (7-70) typically involves the parameter λ, but only in a noninformative way. Accordingly, the inferences concerning the parameter θ will be conditional on chosen values for λ. In practice the choice of λ values will be guided by the results obtained from the λ analysis discussed in Sec. 7-3-1; recall the particular location-scale analysis in Sec. 2-3-2.

First, consider a hypothesis concerning θ:

$$H_0 : \theta = \theta_0$$

On the assumption that $\theta = \theta_0$ the value of Z can be calculated as

$$Z = \theta_0^{-1} Y^0 \tag{7-71}$$

This observed value can be compared with the distribution for variation in $\mathscr{M}_V^{D^0}$ to see whether it is a plausible, reasonable value, or a questionable value, or an impossible value; recall the discussion in Secs. 6-2-2, 6-2-3, 2-2-2, and 2-2-3. We have here the direct calculation of the realized value and for comparison we have the alleged distribution describing that value. Thus we are not faced with the usual arbitrariness involved in finding a test statistic for a test of significance for a hypothesis; we *have* the test statistic necessarily.

With the specialized notation following the exactness Assumption 7-2, the solution (7-71) can be rewritten as

$$[Z] = \theta_0^{-1} [Y^0] \tag{7-72}$$

This observed value of $[Z]$ is then compared with the distribution for variation in $\mathscr{M}_V^{D^0}$, but expressed in terms of the group coordinates. For the model with density functions in Sec. 7-2, this distribution for $[Z] = g$ is recorded as $g_\lambda(g : D^0) \, dg$ in formula (7-47).

If the value in (7-72) is extreme, an explanation can be sought in terms of alternatives to the hypothesis H_0 being tested. Consider an alternative parameter value, say θ_1, together with a "reasonable" g value. The corresponding response value would be $\theta_1 g$ and the value in (7-72) to be tested would be $\theta_0^{-1}\theta_1 g = \delta g$. Thus an extreme value as tested can in part be explained by left repositioning of the basic distribution from (7-47); that is, instead of an extreme value under $g_\lambda(g:D^0)\,dg$ we would be considering a "reasonable" value under the displaced distribution $g_\lambda(\delta^{-1}g:D^0)J_L(\delta^{-1}g)\,d\mu(g)$ for some δ.

Second, consider confidence regions for the parameter θ. A $1-\alpha$ region can, of course, be constructed by determining a $1-\alpha$ acceptance region for each possible parameter value. Certain confidence regions, however, can be more useful and meaningful than others; recall the examples in Sec. 5-2. In particular, if the confidence level is directly interpretable in terms of properties of the basic variation then we have additional useful and meaningful attributes for the particular region.

Accordingly, we consider an acceptance region A for the variation Z that has $1-\alpha$ probability under the conditional model in (7-70). Then

$$P(\theta^{-1}Y \text{ in } A) = 1 - \alpha$$
$$P[\theta \text{ in } S(Y)] = 1 - \alpha \qquad (7\text{-}73)$$

where

$$S(Y) = \{\theta : \theta^{-1}Y \in A\} \qquad (7\text{-}74)$$

Thus $S(Y)$ is a $1-\alpha$ confidence region for θ and $S(Y^0)$ is the observed $1-\alpha$ region. Note, of course, that this region has $1-\alpha$ confidence using the conditional distribution in (7-70), but as such it will have, of course, $1-\alpha$ confidence unconditionally or marginally.

Now let us take advantage of the notation under the exactness assumption. For this let $\bar{A}$ now be the $1-\alpha$ acceptance region but expressed in terms of the group coordinates as chosen in Sec. 7-1-4. Then (7-73) can be rewritten as

$$P(\theta^{-1}[Y] \text{ in } \bar{A}) = 1 - \alpha$$
$$P[\theta \text{ in } S(Y)] = 1 - \alpha \qquad (7\text{-}75)$$

where

$$S(Y) = \{\theta : \theta^{-1}[Y] \in \bar{A}\}$$
$$= [Y]\bar{A}^{-1} \qquad (7\text{-}76)$$

with $\bar{A}^{-1} = \{[Z] : [Z]^{-1} \in \bar{A}\}$. Thus $S(Y)$ is the $1-\alpha$ confidence region and

$$S(Y^0) = [Y^0]\bar{A}^{-1}$$

is the observed $1-\alpha$ confidence region. Again we note that this region has $1-\alpha$ confidence conditionally given the identified orbit and, of course, $1-\alpha$ confidence marginally.

For models involving density functions we have the derived distributions available from Sec. 7-2. Accordingly, we introduce the Assumptions 7-3, 7-4, and 7-5. The distribution for $g = [Z]$ given the orbit $D(Z) = D(Y^0) = D^0$ is available from (7-47):

$$g_\lambda(g:D^0)\,dg = h_\lambda^{-1}(D^0)f_\lambda(gD^0)J_N(gD^0)J(D^0)\,d\mu(g) \qquad (7\text{-}77)$$

Accordingly, the acceptance region $\bar{A}$ satisfies

$$\int_{\bar{A}} g_\lambda(g:D^0)\,dg = 1 - \alpha \qquad (7\text{-}78)$$

The confidence region for θ then has the form $S(Y^0) = [Y^0]\bar{A}^{-1}$ as obtained from (7-76). For examples, see Secs. 6-2-2, 2-2-2, and 2-3-2.

7-3-3 Composite Hypotheses for θ

Consider a composite hypothesis:

$$\theta \in H_0 \qquad (7\text{-}79)$$

where H_0 is a subset of the parameter space G for θ. The difficulties of testing composite hypotheses for multivariate problems are well known. Often, attention is focused just on whether or not some plausible test can be found; questions then of reliability, merit, or any absolute criterion are ignored.

We investigate here those hypotheses $\theta \in H_0$ that are amenable under the structure and properties of the coherent models in this chapter. We first find the amenable hypotheses and then find the form of the appropriate test.

Consider further the inference base from (7-70):

$$(\mathscr{M}_Y^{D^0}, [Y^0]) \qquad (7\text{-}80)$$

We have the equation $Y = \theta Z$ which relates response values to values for the standardized variation. On the assumption that $\theta \in H_0$ we have that

$$Z \in \{\theta^{-1}Y^0 : \theta \in H_0\} = H_0^{-1}Y^0 \qquad (7\text{-}81)$$

This is, of course, a subset of the orbit $GZ = GY^0$ that was identified (Sec. 7-1-4) without the specific information under the present hypothesis.

For the information (7-81) to be useful with the distribution for variation we need the event identifiability properties investigated in Sec. 7-1-3. There we found that this requires a group that generates the various presentations in H_0 and produces orbits, one of which corresponds to the specific information (7-81). For our notation here this can be expressed most easily in terms of a subgroup G_1 of G: the group G_1 generates orbits and one of these corresponds to (7-81). It follows that the set H_0 has the form

$$H_0 = \psi_0 G_1 \qquad (7\text{-}82)$$

This is a left coset of the subgroup G_1 in G. The information (7-81) can then be

reexpressed as
$$Z \in G_1 \psi_0^{-1} Y^0 \tag{7-83}$$
which is, of course, an orbit of the subgroup G_1.

Now consider testing an amenable hypothesis
$$\theta \in \psi_0 G_1 \tag{7-84}$$
as just discussed. On the assumption that θ is in $\psi_0 G_1$ the G_1 orbit of Z can be calculated:
$$G_1 Z = G_1 \psi_0^{-1} Y^0 \tag{7-85}$$
This observed G_1 orbit can be compared with the distribution for such G_1 orbits as derived from $\mathcal{M}_Y^{D^0}$ and the hypothesis can be assessed accordingly; compare with Sec. 7-3-2. The observed G_1 orbit is an orbit on the previously identified G orbit of the model $\mathcal{M}_Y^{D^0}$, a suborbit of an orbit. Again note the lack of the usual arbitrariness in forming a test statistic for a test of significance.

We can rewrite the preceding with an extension of the exactness notation, following Assumption 7-2. For this let $D_1(Z)$ be a reference point on the $G_1 Z$ orbit and let $[Z]_1$ in G_1 be the position of Z on its orbit:
$$Z = [Z]_1 D_1(Z) \tag{7-86}$$
For the continuous case we will assume the same requirements as for the main group G. Also let $(Z)_1 = [D_1(Z)]$ in G be the position of $D_1(Z)$ relative to $D(Z)$:
$$Z = [Z]_1 D_1(Z) = [Z]_1 (Z)_1 D(Z) = [Z] D(Z) \tag{7-87}$$
Equation (7-85) can then be rewritten as
$$(Z)_1 = (\psi_0^{-1} Y^0)_1 \tag{7-88}$$
This observed value of $(Z)_1$ under the hypothesis (7-84) can then be compared with the distribution of $(Z)_1$ under the model $\mathcal{M}_Y^{D^0}$ and the hypothesis assessed accordingly. For the model with density functions in Sec. 7-2, the distribution for $(Z)_1$ can be obtained from the Z distribution (7-47) by integrating along the G_1 orbits; the details are recorded in the next section.

7-3-4 Distributions Relative to a Subgroup

The test of significance in Sec. 7-3-3 needed a distribution for the orbit of a subgroup G_1 of the main group G. We now derive such distributions under the density function Assumptions 7-3, 7-4, and 7-5.

For the main group G we have
$$Z = [Z] D(Z) = g D(Z) \tag{7-89}$$
where $[Z] = g$ gives the position of Z relative to the reference point $D(Z)$ in the collection Q of reference points. Then for the subgroup G_1 we have
$$Z = [Z]_1 D_1(Z) = g_1 D_1(Z) \tag{7-90}$$

where $[Z]_1 = g_1$ gives the position of Z relative to the reference point $D_1(Z)$ in the collection, say Q_1, of such reference points. The various suborbits $G_1 Z$ on an orbit GZ can be indexed by giving the positions of the points $D_1(Z)$ relative to the main reference point $D(Z)$:

$$D_1(Z) = (Z)_1 D(Z) = k_1 D(Z) \tag{7-91}$$

with $(Z)_1 = k_1$ in G recording the position. Thus the subset $K_1 = \{k_1\}$ of G indexes the points of Q_1 on the particular orbit GZ. Typically, we will use the same set K_1 on each GZ orbit.

The preceding gives the following representation for a point Z:

$$Z = g_1 k_1 D(Z) = g D(Z) \tag{7-92}$$

where $g_1 \in G_1$, $k_1 \in K_1$, and $D(Z) \in Q$; note that $k_1 D(Z) \in Q_1$.

In this section we start with the distribution for g as given by (7-47), and we derive the marginal distribution for $(Z)_1 = k_1$ as needed for Sec. 7-3-3. We also derive the conditional distribution for $[Z]_1 = g_1$ for use in Sec. 7-4.

For the group G_1 with dimension L_1 let μ_1 be the left invariant measure,

$$d\mu_1(g_1) = \frac{dg_1}{J_{L_1}(g_1)} \tag{7-93}$$

calculated as in Sec. 7-2-3. Now consider formula (7-44) with the measure M on $\mathscr{S}$ replaced by μ on G, and with μ on G replaced by μ_1 on G_1; we obtain

$$d\mu(g_1 k_1) = d\mu_1(g_1) H_1(k_1) \, dk_1 \tag{7-94}$$

The standardization of the measure μ does not directly correspond to that used for the measure M: the measure μ was standardized relative to the identity whereas M was standardized relative to the reference point which, here, is k_1. The factor $H_1(k_1)$ thus includes, of course, the factor corresponding to $J(D)$ in formula (7-44), but also an effect due to the present difference in standardization.

We now make the change of variable (7-92) in formula (7-47) using equation (7-94):

$$g_\lambda(g : D) \, dg = h_\lambda^{-1}(D) f_\lambda(g_1 k_1 D) J_N(g_1 k_1 D) J(D) \, d\mu_1(g_1) H_1(k_1) \, dk_1 \tag{7-95}$$

The *marginal distribution for* k_1 is obtained by integrating over the group G_1, giving

$$h_\lambda^{-1}(D) h_\lambda(k_1, D) H_1(k_1) \, dk_1 \tag{7-96}$$

where

$$h_\lambda(k_1, D) = \int_{G_1} f_\lambda(g_1 k_1 D) J_N(g_1 k_1 D) J(D) \, d\mu_1(g_1) \tag{7-97}$$

The particular choice of factors to include in the integral is largely a matter of convenience but does relate to the factors used in the resulting expression for the distribution in (7-96); note that (7-96) is conditional given D but is marginal for the suborbits corresponding to the given D.

The *conditional distribution for* g_1 *given* k_1 is then obtained by division:

$$h_\lambda^{-1}(k_1, D) f_\lambda(g_1 k_1 D) J_N(g_1 k_1 D) J(D) \, d\mu_1(g_1) \qquad (7\text{-}98)$$

Note that the notation here does not directly correspond to (7-47) with G replaced by G_1; this is because the Jacobian J_N relates to the reference point D rather than the reference point D_1 for the group G_1.

The distribution (7-96) with $D = D^0$ is the appropriate distribution for the test of significance in Sec. 7-3-3.

7-3-5 Confidence Regions: Component Parameters

In Sec. 7-3-3 we noted the difficulties of testing composite hypotheses for multivariate problems. The general difficulties are even greater for confidence regions. In this section we follow the earlier pattern and investigate confidence regions for those parameters that are amenable under the structure and properties of the coherent models in this chapter.

Consider a component parameter $\psi = \psi(\theta)$ for the inference base recorded in (7-70) and (7-80); we restrict ourselves here to the case satisfying the exactness Assumption 7-2. To form a $1 - \alpha$ confidence region for ψ we need a $1 - \alpha$ acceptance region for each value of the parameter ψ. For this we recall the results in Sec. 7-3-3 on tests of significance. To use the available structure we need a subgroup G_1 that generates orbits, and we use the marginal distribution for those orbits for the test of significance. Accordingly, the $1 - \alpha$ acceptance region for a value of ψ is a set of orbits of a subgroup G_1.

Also, as we noted in Sec. 7-3-2, certain confidence regions can be more meaningful than others; in particular, if the confidence level is directly interpretable in terms of the basic variation then we have additional and meaningful attributes for the region. Accordingly, we use the *same* set of G_1 orbits for each parameter value ψ. In a parallel way we have that the component parameter specifies a left coset of the subgroup G_1:

$$\theta = \psi \theta_1 \qquad (7\text{-}99)$$

with θ_1 in G_1 and ψ in the space Ψ for the component parameter. For a comparison recall that in Sec. 2-2-2 we used the same interval for the t distribution for each tested value of the location parameter μ.

We now investigate confidence regions for the component parameter ψ that indexes (7-99), the left cosets of the subgroup G_1 on the parameter space G. Let $\bar{A}$ on the space of possibilities for $[Z]$ be a set of G_1 orbits that has probability $1 - \alpha$ under the conditional model in (7-70) and (7-80). Then

$$P(\theta^{-1}[Y] \text{ in } \bar{A}) = 1 - \alpha$$
$$P(\theta_1^{-1} \psi^{-1}[Y] \text{ in } \bar{A}) = 1 - \alpha$$
$$P(G_1 \psi^{-1}[Y] \subset \bar{A}) = 1 - \alpha$$
$$P[\psi \text{ in } S(Y)] = 1 - \alpha$$

where
$$S(Y) = \{\psi : G_1\psi^{-1}[Y] \subset \bar{A}\}$$
$$= \{\psi : \psi^{-1}[Y] \in \bar{A}\}$$
$$= \{\psi : \psi \in [Y]\bar{A}^{-1}\} \tag{7-100}$$

Thus note that the confidence region is the Ψ section through the set $[Y]\bar{A}^{-1}$ on the group G. The set $S(Y^0)$ is the observed $1 - \alpha$ confidence region for the parameter ψ.

For models involving density functions we have the derived distributions available from Sec. 7-3-4. For this we need Assumptions 7-3, 7-4, and 7-5. The distribution for $[Z] = g_1 k_1$ is given by (7-47). The component variable k_1 indexes the G_1 orbits and has the marginal distribution (7-96):

$$h_\lambda^{-1}(D^0) h_\lambda(k_1, D^0) H_1(k_1) \, dk_1 \tag{7-101}$$

Let $\bar{A}$ as described above be a $1 - \alpha$ region for g and $\bar{A}_1$ be the corresponding $1 - \alpha$ region for the indexing variable k_1. Then

$$\int_{\bar{A}_1} h_\lambda^{-1}(D^0) h_\lambda(k_1, D^0) H_1(k_1) \, dk_1 = 1 - \alpha \tag{7-102}$$

The confidence region for ψ is then given by

$$S(Y) = \{\psi : \psi^{-1}[Y] \in \bar{A}\}$$
$$= \{\psi : (\psi^{-1}[Y])_1 \in \bar{A}_1\} \tag{7-103}$$

as obtained from (7-100).

7-3-6 Supplement

In this section we have derived confidence regions for θ and for the component $\psi(\theta)$. As noted earlier, a spectrum of confidence regions for a parameter can be viewed as a confidence distribution for the parameter. For the sake of completeness here we record the confidence distributions for θ. For structural models these have been called structural distributions; see, for example, Fraser (1968b).

For the parameter θ the equation $[Y] = \theta[Z]$ or $\theta^{-1}[Y] = [Z]$ is used to obtain the confidence region for θ from the acceptance region for the variation $[Z]$. Accordingly, we take the distribution (7-77) for $[Z] = g$,

$$h_\lambda^{-1}(D^0) f_\lambda(gD^0) J_N(gD^0) J(D^0) \, d\mu(g) \tag{7-104}$$

and transfer it to the parameter space by the substitution $g = \theta^{-1}[Y^0]$. We obtain the following *confidence distribution for* θ:

$$h_\lambda^{-1}(D^0) f_\lambda(\theta^{-1} Y^0) J_N(\theta^{-1} Y^0) J(D^0) \, d\mu(\theta^{-1}[Y^0])$$
$$= h_\lambda^{-1}(D_0) f_\lambda(\theta^{-1} Y^0) J_N(\theta^{-1} Y^0) J(D^0) \Delta([Y^0]) \, dv(\theta) \tag{7-105}$$

In this we have used (7-52) and (7-57).

For the component parameter ψ the preceding steps are notationally awkward unless we have certain simplifications. We defer this to Sec. 7-4-5.

The preceding confidence or structural distribution has been derived for a coherent model as defined in Sec. 7-1-3. Such distributions on the parameter space Ω are sometimes derived in a similar manner for the pivotal-type models defined in Sec. 1-2-5, but without the requirement of coherence that we have used here. Without coherence these "probabilities" on the parameter spaces need not be unique, and upper and lower probabilities for a parameter set have been defined by Dempster (1966, 1968); see also Beran (1971).

Whether these extended probabilities are in a reasonable way reliable seems in doubt; see, for example, McGilchrist (1973) and the comments thereon by Fraser (1973). The identification of events—the coherence in Sec. 7-1-3—does seem to be an essential ingredient for calculating meaningful extended probabilities.

7-4 MULTIPLE TESTS AND CONFIDENCE REGIONS

In the preceding section we examined tests and confidence intervals for a component parameter $\psi_1 = \psi_1(\theta)$. The complementary parameter $\theta_1 = \theta_1(\theta)$ might also be of interest, and we could want tests and confidence regions for it. We would then have the correspondence $\theta \leftrightarrow (\psi_1, \theta_1)$ and be interested in testing the components ψ_1 and θ_1. In a similar way we might have three or more parameter components and be interested in testing them individually. In this section we examine the case of three components. This covers the case of two components by making one component trivial, and it amply illustrates the procedures to follow for the more general case with more than three components.

As a preliminary, however, consider briefly an aspect of tests and confidence regions for parameter components—and for this the two-components case suffices: $\theta \leftrightarrow (\psi_1, \theta_1)$. Reasonably, we could want to have a test or confidence region for ψ_1 and to have a test or confidence region for θ_1. Having methods for both parameters may or may not be possible within the techniques discussed in the preceding section, but even if it is possible, there is one concern of relevance. The component levels of significance or the component confidence levels may well not combine to give an overall significance level or overall confidence level; the exceptional case would be where statistical independence effectively separated the parameters.

The separate examination of the component parameters may not be possible and, even if possible, the overall confidence level is typically not available. The methods in this section build on the available structure of the model and are concerned with tests and confidence regions for the parameter components taken in sequence. One attractive by-product of this sequential approach is the availability of the overall test or confidence level.

7-4-1 Model with Three Components

We consider the structural model (7-66) and suppose that the shape parameter λ has been investigated and a value (or values) of λ chosen for the subsequent analysis. We thus obtain the reduced inference base

$$(\mathcal{M}_V^{D^0}, [Y^0]) \tag{7-106}$$

as examined in Secs. 7-3-2 to 7-3-6. This is a structural model with parameter θ in a group G and with sample space the orbit GY^0 which, in the exactness case, is an effective copy of the group G.

As we have shown in Secs. 7-3-3 and 7-3-5 tests and confidence regions can be derived for a component $\psi_1(\theta)$ using the structural properties of the model for the case that

$$\theta = \psi_1 \theta_1 \tag{7-107}$$

where θ_1 is in a subgroup G_1 and ψ_1 is in a set Ψ_1 that indexes the left cosets of G_1.

The tests and confidence regions were then based on a reverse factorization of the group coordinates for the orbit:

$$g = g_1 k_1 \tag{7-108}$$

where g_1 is in the subgroup G_1 and k_1 is in a set K_1 that indexes, by means of reference points, the orbits of G_1. This factorization on the group has a corresponding factorization on the orbit:

$$Z = [Z]D(Z)$$
$$= [Z]_1(Z)_1 D(Z)$$
$$= g_1 k_1 D(Z)$$

with $g_1 = [Z]_1$ and $k_1 = (Z)_1$.

For a given value of ψ_1 and for the correspondingly determined reference point k_1 for the G_1 orbit, we can then contemplate a parallel splitting of the parameter θ_1 in G_1. Accordingly, we consider a component $\psi_2(\theta)$ where

$$\theta = \psi_1 \psi_2 \theta_2 \tag{7-109}$$

with θ_2 in a subgroup G_2 of G_1 and with ψ_2 in a set Ψ_2 that (with ψ_1) uniquely indexes the left cosets of G_2.

The tests and confidence regions for ψ_2 would then be based on a reverse factorization as in (7-108) giving

$$g = g_2 k_2 k_1 \tag{7-110}$$

where g_2 is in the subgroup G_2 and k_2 is in a set K_2 that (with k_1) uniquely indexes the orbits of the group G_2. This factorization on the group has a corresponding factorization on the orbit:

$$Z = g_2 k_2 k_1 D(Z)$$
$$= [Z]_2 (Z)_2 (Z)_1 D(Z) \tag{7-111}$$

with $g_2 = [Z]_2$, $k_2 = (Z)_2$, and $k_1 = (Z)_1$.

7-4-2 Sequential Tests

Consider a sequence of hypotheses concerning θ:

$$\psi_1 = \psi_{1,0} \qquad \psi_2 = \psi_{2,0} \qquad \theta_2 = \theta_{2,0} \tag{7-112}$$

We consider testing the first hypothesis $\psi_1 = \psi_{1,0}$ as in Sec. 7-3-3. Under this hypothesis we can calculate the G_1 orbit of Z:

$$k_1 = (Z)_1 = (\psi_{1,0}^{-1} Y^0)_1 \tag{7-113}$$

This "observed" value of $(Z)_1$ under the hypothesis can be compared with the distribution of $(Z)_1$ under the model $\mathcal{M}_V^{D^0}$; the hypothesis can then be assessed accordingly.

We now consider testing the second hypothesis $\psi_2 = \psi_{2,0}$ on the assumption that the first hypothesis holds, that is, $\psi_1 = \psi_{1,0}$. Under this hypothesis we can calculate the G_2 orbit of Z, obtaining

$$k_2 = (Z)_2 = (\psi_{2,0}^{-1} \psi_{1,0}^{-1} Y^0)_2 \tag{7-114}$$

together with the value in (7-113). The "observed" value of $(Z)_2$ under the hypothesis can be compared with the distribution of $(Z)_2$ conditional on $(Z)_1$ from (7-113) and using the model $\mathcal{M}_V^{D^0}$; the hypothesis can then be assessed accordingly.

We now consider testing the third hypothesis $\theta_2 = \theta_{2,0}$ on the assumption that the first and second hypotheses hold, that is, $\psi_1 = \psi_{1,0}$, $\psi_2 = \psi_{2,0}$. Under the hypothesis we can calculate the position of Z on the G_2 orbit:

$$g_2 = [Z]_2 = [\theta_{2,0}^{-1} \psi_{2,0}^{-1} \psi_{1,0}^{-1} Y^0]_2$$
$$= \theta_{2,0}^{-1} [\psi_{2,0}^{-1} \psi_{1,0}^{-1} Y^0]_2 \tag{7-115}$$

together with the values in (7-113) and (7-114). This "observed" value for $[Z]_2$ under the hypothesis can be compared with the distribution of $[Z]_2$ conditional on $(Z)_1$ from (7-113) and $(Z)_2$ from (7-114), and using the model $\mathcal{M}_V^{D^0}$; the hypothesis can then be assessed accordingly.

The sequential testing here may be found in a more general context in Fraser and MacKay (1975, 1976).

7-4-3 Component Distributions

We now derive the appropriate distributions for g_2, k_2, and k_1 based on the reduced model $\mathcal{M}_V^{D^0}$ and using the continuity Assumptions 7-3, 7-4, and 7-5 for the groups G, G_1, G_2.

The change of variable is given by

$$g = g_2 k_2 k_1 \tag{7-116}$$

From formula (7-94) we have

$$d\mu(g_1 k_1) = d\mu_1(g_1) H_1(k_1) \, dk_1 \tag{7-117}$$

where $H_1(k_1)$ has a derivation that parallels that for $J(D)$ in Sec. 7-2-4. We then have the further factorization $g_1 = g_2 k_2$ and can reapply (7-117) to the measure μ_1; we obtain

$$d\mu(g_2 k_2 k_1) = d\mu_2(g_2) H_2(k_2) \, dk_2 \, H_1(k_1) \, dk_1 \tag{7-118}$$

The change of variable (7-117) and (7-118) can be used in formula (7-47). Then in the pattern in Sec. 7-3-4 we obtain the *marginal distribution for k_1* as in (7-96):

$$h_\lambda^{-1}(D) h_\lambda(k_1, D) H_1(k_1) \, dk_1 \tag{7-119}$$

where

$$h_\lambda(k_1, D) = \int_{G_1} f_\lambda(g_1 k_1 D) J_N(g_1 k_1 D) J(D) \, d\mu_1(g_1) \tag{7-120}$$

The *conditional distribution for k_2 given k_1* is obtained from the joint for $k_2 k_1$ divided by the preceding marginal for k_1; we have

$$h_\lambda^{-1}(k_1, D) h_\lambda(k_2, k_1, D) H_2(k_2) \, dk_2 \tag{7-121}$$

where

$$h_\lambda(k_2, k_1, D) = \int_{G_2} f_\lambda(g_2 k_2 k_1 D) J_N(g_2 k_2 k_1 D) J(D) \, d\mu_2(g_2) \tag{7-122}$$

The *conditional distribution for g_2 given k_1 and k_2* is then

$$h_\lambda^{-1}(k_2, k_1, D) f_\lambda(g_2 k_2 k_1 D) J_N(g_2 k_2 k_1 D) J(D) \, d\mu_2(g_2) \tag{7-123}$$

The preceding distributions (7-119), (7-121), and (7-123) are those appropriate to the tests (7-113), (7-114), and (7-115) in Sec. 7-4-2.

Consider a special case that frequently arises with a group and its subgroups, $G \supset G_1 \supset G_2$. Specifically suppose that the sets K_1, K_2 can be chosen as subgroups of the full group G.

The subgroup K_1 complements G_1 within the full group G. Let μ_1^* and Δ_1^* be the left measure and modular functions for this complementary group K_1. From properties in Sec. 7-2-5 we have that

$$\frac{d\mu(g_1 k_1)}{\Delta(k_1)} \qquad d\mu_1(g_1) \frac{d\mu_1^*(k_1)}{\Delta_1^*(k_1)}$$

are invariant under both left multiplication by elements of G_1 and right multiplication by elements of K_1. Accordingly, they differ by a constant of proportionality:

$$\frac{d\mu(g_1 k_1)}{\Delta(k_1)} = H_1 \, d\mu_1(g_1) \frac{d\mu_1^*(k_1)}{\Delta_1^*(k_1)} \tag{7-124}$$

$$d\mu(g_1 k_1) = d\mu_1(g_1) H_1 \frac{\Delta(k_1)}{\Delta_1^*(k_1)} d\mu_1^*(k_1) \tag{7-125}$$

The modular functions at the identity are equal to 1, and accordingly the constant of proportionality arises from the local nonorthogonality of the subgroups K_1 and G_1. We have thus used group properties to determine the function $H_1(k_1)$:

$$H_1(k_1) \, dk_1 = H_1 \frac{\Delta(k_1)}{\Delta_1^*(k_1)} d\mu_1^*(k_1) \tag{7-126}$$

In a similar manner let μ_2^* and Δ_2^* be the left measure and modular functions for the group K_2 that complements G_2 within G_1. We obtain

$$d\mu_1(g_2 k_2) = d\mu_2(g_2) H_2 \frac{\Delta_1(k_2)}{\Delta_2^*(k_2)} d\mu_2^*(k_2) \tag{7-127}$$

$$H_2(k_2) \, dk_2 = H_2 \frac{\Delta_1(k_2)}{\Delta_2^*(k_2)} d\mu_2^*(k_2) \tag{7-128}$$

The preceding determinations of $H_1(k_1)$ and $H_2(k_2)$ can be used in the formulas (7-119), (7-121), and (7-123) for the appropriate distributions for k_1, k_2, and g_2.

7-4-4 Sequential Confidence Regions

Consider confidence regions for the parameter components in the order ψ_1, ψ_2, θ_2.

From Sec. 7-3-5 we can obtain the region for ψ_1. Let A_1 be a set of G_1 orbits on the space for Z with a total probability of β_1 and let $\bar{A}_1$ be the corresponding set for the indexing value k_1 for such orbits.

$$P(Z \text{ in } A_1) = P(G_1 Z \subset A_1)$$
$$= P(k_1 \text{ in } \bar{A}_1) = \beta_1 \tag{7-129}$$

Then

$$S_1(Y) = \{\psi_1 : (\psi_1^{-1} Y)_1 \text{ in } \bar{A}_1\} \tag{7-130}$$

is a β_1 confidence region for the parameter ψ_1. The region $\bar{A}_1$ can be determined from the distribution for k_1 in formula (7-119).

Now consider a confidence region for ψ_2 conditional on a value for the parameter ψ_1. The results in the preceding paragraph can be used with G_1 replaced by G_2 and the distribution for k_1 replaced by that for k_2 given $k_1 = (\psi_1^{-1} Y)_1$. Let $A_2(k_1)$ be a set of G_2 orbits on the space for Z conditioned by k_1 with a total probability of β_2 and let $\bar{A}_2(k_1)$ be the corresponding set for the indexing variable k_2 for such orbits.

$$P[Z \text{ in } A_2(k_1):(Z)_1 = k_1] = P[G_2 Z \subset A_2(k_1): k_1]$$
$$= P[k_2 \text{ in } \bar{A}_2(k_1): k_1] = \beta_2 \qquad (7\text{-}131)$$

Then

$$S_2(Y: \psi_1) = \{\psi_2 : (\psi_2^{-1}\psi_1^{-1} Y)_2 \text{ in } \bar{A}_2(k_1)\} \qquad (7\text{-}132)$$

is a β_2 confidence region for the parameter ψ_2 given ψ_1; note that $k_1 = (\psi_1^{-1} Y)_1$. The region $A_2(k_1)$ can be determined from the distribution for k_2 given k_1 in formula (7-121).

Finally, consider a confidence region for θ_2 given values for ψ_1 and ψ_2. For this we use the results from Sec. 7-3-2 with G replaced by G_2 and the distribution for g replaced by that for g_2 given $k_1 = (\psi_1^{-1} Y)_1$ and $k_2 = (\psi_2^{-1}\psi_1^{-1} Y)_2$. Let $A_3(k_2, k_1)$ be a set of Z values on the G_2 orbit formed with reference point $k_2 k_1 D(Y)$ and with a total probability of β_3 and let $\bar{A}_3(k_2, k_1)$ be the corresponding set for the group variable g_2 on the orbit.

$$P[Z \text{ in } A_3(k_2,k_1):(Z)_2 = k_2, (Z)_1 = k_1] = P[g_2 \text{ in } \bar{A}_3(k_2,k_1): k_2, k_1] = \beta_3 \qquad (7\text{-}133)$$

Then

$$S_3(Y: \psi_2, \psi_1) = \{\theta_2 : \theta_2^{-1}[\psi_2^{-1}\psi_1^{-1} Y]_2 \text{ in } \bar{A}_3(k_2, k_1)\} \qquad (7\text{-}134)$$

is a β_3 confidence region for the parameter θ_2 given ψ_2, ψ_1; note that $k_2 = (\psi_2^{-1}\psi_1^{-1} Y)_2$, $k_1 = (\psi_1^{-1} Y)_1$. The region $\bar{A}_3(k_2, k_1)$ can be determined from the distribution for g_2 given k_2, k_1 in formula (7-123).

The preceding gives us a sequence of confidence regions for ψ_1, ψ_2, θ_2, each conditional on values for preceding parameters. Can these confidence regions be compounded?

For this consider the following. Let $\bar{A}$ be the compound acceptance region

$$\bar{A} = \{g_2 k_2 k_1 : g_2 \in \bar{A}_3(k_2, k_1), k_2 \in \bar{A}_2(k_1), k_1 \in \bar{A}_1\} \qquad (7\text{-}135)$$

and let $S(Y)$ be formed as follows:

$$S(Y) = \{\psi_1\psi_2\theta_2 : \theta_2 \in S_3(Y: \psi_2, \psi_1), \psi_2 \in S_2(Y: \psi_1), \psi_1 \in S_1(Y)\} \qquad (7\text{-}136)$$

We now show that $S(Y)$ is a $\beta_1\beta_2\beta_3$ confidence region for $\theta = \psi_1\psi_2\theta_2$ based on the acceptance region $\bar{A}$.

Note, first, that $\bar{A}$ has probability $\beta_1\beta_2\beta_3$ on the space for the variation $[Z]$:

$$P(\bar{A}) = P[g_2 \text{ in } \bar{A}_3(k_2,k_1): k_2 \text{ in } \bar{A}_2(k_1), k_1 \text{ in } \bar{A}_1]$$
$$\times P[k_2 \text{ in } \bar{A}_2(k_1): k_1 \text{ in } \bar{A}_1]$$
$$\times P(k_1 \text{ in } \bar{A}_1)$$
$$= \beta_3\beta_2\beta_1 \qquad (7\text{-}137)$$

164 INFERENCE AND LINEAR MODELS

We now verify that $S(Y)$ is the confidence region corresponding to this acceptance region $\bar{A}$:

$$\{\psi_1\psi_2\theta_2 : \theta_2^{-1}\psi_2^{-1}\psi_1^{-1}[Y] \text{ in } \bar{A}\}$$
$$= \{\psi_1\psi_2\theta_2 : \theta_2^{-1}[\psi_2^{-1}\psi_1^{-1}Y]_2 \text{ in } A_3(k_2, k_1),$$
$$(\psi_2^{-1}\psi_1^{-1}Y)_2 \text{ in } A_2(k_1), (\psi_1^{-1}Y)_1 \text{ in } A_1\}$$
$$= \{\psi_1\psi_2\theta_2 : \theta_2 \text{ in } S_3(Y : \psi_2, \psi_1),$$
$$\psi_2 \text{ in } S_2(Y : \psi_1), \psi_1 \text{ in } S_1(Y)\}$$
$$= S(Y) \tag{7-138}$$

where throughout we have $k_2 = (\psi_2^{-1}\psi_1^{-1}Y)_2$ and $k_1 = (\psi_1^{-1}Y)_1$.

Thus the compound confidence region has the compound confidence level $\beta_3\beta_2\beta_1$.

7-4-5 Supplement

In Sec. 7-3-6 we calculated the confidence distribution for the parameter θ but postponed the calculations for the component parameter $\psi(\theta)$.

Now consider $\theta = \psi_1\psi_2\theta_2$ with the components ψ_1, ψ_2, θ_2 as discussed in the preceding sections. We examine the continuous case satisfying Assumptions 7-3, 7-4, and 7-5 for G and for G_1 and G_2. We also assume that the set Ψ_1 complementary to G_1 is a subgroup and that Ψ_2 complementary to G_2 is a subgroup. We can then take the index set for G_1 orbits to be $K_1 = \Psi_1$ and the index set for G_2 orbits to be $K_2 = \Psi_2$; compare with the end of Sec. 7-4-3.

The equation $[Y] = \theta[Z]$ or $\theta^{-1}[Y] = [Z]$ is used to obtain the confidence regions for θ and for its components from the acceptance region for the variation Z. For the components ψ_1 and ψ_2 in this Sec. 7-4 the analysis of Sec. 7-3-6 becomes awkward unless we fit our coordinates on the orbit G compatibly with the components on the group. For this we choose our reference point $D(Y^0)$ in a very special way; we choose the reference point *at the observed* Y^0. The equation $\theta^{-1}[Y^0] = [Z]$ then becomes $\theta^{-1} = [Z]$. In terms of the components we then have

$$\theta_2^{-1}\psi_2^{-1}\psi_1^{-1} = g_2 k_2 k_1 \tag{7-139}$$

with θ_2 and g_2 in G_2, ψ_2 and k_2 in the group Ψ_2, and ψ_1 and k_1 in the group Ψ_1. Equation (7-139) then splits into

$$\theta_2^{-1} = g_2 \qquad \psi_2^{-1} = k_2 \qquad \psi_1^{-1} = k_1 \tag{7-140}$$

The distribution for k_1 in (7-119) with (7-126) can be written as

$$h_\lambda^{-1}(Y^0)h_\lambda(k_1, Y^0)H_1 \frac{\Delta(k_1)}{\Delta_1^*(k_1)} d\mu_1^*(k_1) \tag{7-141}$$

Transferred to Ψ_1 by (7-140), this becomes the confidence distribution for ψ_1:

$$h_\lambda^{-1}(Y^0)h_\lambda(\psi_1^{-1}, Y^0)H_1 \frac{\Delta_1^*(\psi_1)}{\Delta(\psi_1)} dv_1^*(\psi_1) \tag{7-142}$$

The distribution for k_2 in (7-121) with (7-128) can be written as

$$h_\lambda^{-1}(\psi_1^{-1}, Y^0)h_\lambda(k_2, \psi_1^{-1}, Y^0)H_2 \frac{\Delta_1(k_2)}{\Delta_2^*(k_2)} d\mu_2^*(k_2) \tag{7-143}$$

Transferred to Ψ_2 by (7-140) this becomes the confidence distribution for ψ_2 given ψ_1:

$$h_\lambda^{-1}(\psi_1^{-1}, Y^0)h_\lambda(\psi_2^{-1}, \psi_1^{-1}, Y^0)H_2 \frac{\Delta_2^*(\psi_2)}{\Delta_1(\psi_2)} dv_2^*(\psi_2) \tag{7-144}$$

The distribution for g_2 in (7-123) is

$$h_\lambda^{-1}(\psi_2^{-1}, \psi_1^{-1}, Y^0)f_\lambda(g_2\psi_2^{-1}\psi_1^{-1}Y^0)J_N(g_2\psi_2^{-1}\psi_1^{-1}Y^0)J(Y^0) d\mu_2(g) \tag{7-145}$$

Transferred to G_2 by (7-140) this becomes the confidence distribution for θ_2 given ψ_2, ψ_1:

$$h_\lambda^{-1}(\psi_2^{-1}, \psi_1^{-1}, Y^0)f_\lambda(\theta_2^{-1}\psi_2^{-1}\psi_1^{-1}Y^0)J_N(\theta_2^{-1}\psi_2^{-1}\psi_1^{-1}Y^0)J(Y^0) dv_2(\theta_2) \tag{7-146}$$

The preceding distributions were obtained by taking the appropriate marginal and conditional distributions for the right-hand side of (7-139) and then inverting to the appropriate component on the left-hand side. They can, of course, be obtained directly from the distribution of θ in (7-105) by taking the appropriate marginal and conditional distributions. This route by parameter space integration does not immediately give the confidence distribution interpretation, whereas our present derivation does.

REFERENCES AND BIBLIOGRAPHY

Beran, R. J.: On Distribution-Free Statistical Inference with Upper and Lower Probabilities, *Ann. Math. Stat.*, vol. 42, pp. 157–168, 1971.

Bondar, J. V.: "Invariance and Structural Duality," Ph.D. thesis, University of Toronto, 1967.

———: Borel Cross-Sections and Maximal Invariants, *Ann. Stat.*, vol. 4, pp. 866–877, 1976.

Brenner, D.: "When is a Class of Functions a Function?" Ph.D. Thesis, University of Toronto, 1977.

——— and D. A. S. Fraser: "When is a Class of Functions a Function?," Department of Statistics, University of Toronto, 1978.

Brillinger, D. R.: Necessary and Sufficient Conditions for a Problem to be Invariant under a Lie Group, *Ann. Math. Stat.*, vol. 34, pp. 492–500, 1962.

Dawid, A. P., M. Stone, and J. W. Zidek: Marginalization Paradoxes in Bayesian and Structural Inference, *Jour. Roy. Stat. Soc.*, ser. B, vol. 35, pp. 189–233, 1973.

Dempster, A. P.: New Methods for Reasoning towards Posterior Distributions Based on Sample Data, *Ann. Math. Stat.*, vol. 37, pp. 355–374, 1966.

———: A Generalization of Bayesian Inference, *Jour. Roy. Stat. Soc.*, ser. B, vol. 30, pp. 205–247, 1968.

Fraser, D. A. S.: On Information in Statistics, *Ann. Math. Stat.*, vol. 36, pp. 890–896, 1965:

———: Structural Probability and a Generalization, *Biometrika*, vol. 53, pp. 1–9, 1966.

———: Data Transformations and the Linear Model, *Ann. Math. Stat.*, vol. 30, pp. 1456–1465, 1967.

———: Statistical Models and Invariance, *Ann. Math. Stat.*, vol. 30, pp. 1061–1067, 1967b.
———: A Black Box or a Comprehensive Model, *Technometrics*, vol. 10, no. 2, pp. 219–229, 1968a.
———: "The Structure of Inference," John Wiley and Sons, New York, and Krieger Publishing Company, Huntington, N.Y., 1968b.
———: Events, Information Processing, and the Structural Model, in V. P. Godambe and D. A. Sprott (eds.), "Proc. Symposium on the Foundations of Statistical Inference," Holt, Rinehart and Winston of Canada, Toronto and Montreal, pp. 32–55, 1971.
———: The Determination of Likelihood and the Transformed Regression Model, *Ann. Math. Stat.*, vol. 43, pp. 898–916, 1972.
———: Comments on McGilchrist's paper "Post-Data Two Sample Tests of Location," *Jour. Amer. Stat. Assoc.*, vol. 68, pp. 101–104, 1973.
——— and Jock MacKay: Parameter Factorization and Inference Based on Significance, Likelihood, and Objective Posterior, *Ann. Math. Stat.*, vol. 3, pp. 559–572, 1975.
——— and ———: On the Equivalence of Standard Inference Procedures (also Comments by Professor Lindley), in Harper and Hooker (eds.), "Foundations of Probability Theory, Statistical Inference, and Statistical Theories of Science," vol. II, D. Reidel Publishing Company, Dordrecht, Holland, pp. 47–62, 1976.
McGilchrist, C. A.: Post-Data Two Sample Tests of Location, *Jour. Amer. Stat. Assoc.*, vol. 68, pp. 97–101, 1973.
Ng, V. M.: "Generalized Structural Distribution, Prediction Distribution, and Related Topics," Ph.D. Thesis, University of Waterloo, 1976.
Wijsman, R. A.: Cross-Sections of Orbits and their Application to Densities of Maximal Invariants, in J. Neyman (ed.), *Proc. Fifth Berkeley Symposium*, vol. 1, pp. 389–400, 1967.

CHAPTER

EIGHT

SOME MULTIVARIATE MODELS

Consider a random system with p response variables $y_1, \ldots, y_p$. In this chapter we investigate several statistical models with various location-scale components that can be applicable to such a system. The choice among these models will depend on the degree of identification of the distribution form for the variation.

8-1 LOCATION-SCALE MULTIVARIATE MODEL

Consider a random system with p response variables $y_1, \ldots, y_p$. We suppose that the location and scaling for each variable is unknown but that otherwise the joint distribution form—the identifiable variation—is known or known up to a shape parameter λ.

8-1-1 The Model

Let $f_\lambda(z_1, \ldots, z_p)$ be the density function describing the objective variation. As part of this we assume that f_λ has been standardized with respect to the location-scale transformations for each coordinate. For example, the standardization for a particular axis could be such that the marginal distribution for the variation on that axis would satisfy the conditions recorded in formula (1-18) for the simple location-scale model.

Let $[\mu_1, \sigma_1]$ record the location and scale parameters for the response presentation of the first coordinate, ..., and $[\mu_p, \sigma_p]$ record the location and scale

168 INFERENCE AND LINEAR MODELS

for the pth coordinate. Then using the transformation notation from Sec. 1-2-2 we can write

$$\mathbf{y}_1 = [\mu_1, \sigma_1]\mathbf{z}_1, \ldots, \mathbf{y}_p = [\mu_p, \sigma_p]\mathbf{z}_p \tag{8-1}$$

Let G_i be the location-scale group as a transformation group for the ith coordinate:

$$G_i = \{[a_i, c_i]: a_i \in \mathbb{R}, c_i \in \mathbb{R}^+\}$$
$$[a_i, c_i]z_i = a_i + c_i z_i \tag{8-2}$$

And let G record the transformation group for the full response vector:

$$G = \{([a_1, c_1], \ldots, [a_p, c_p]): [a_i, c_i] \in G_i\}$$
$$= G_1 \times \cdots \times G_p \tag{8-3}$$

$$([a_1, c_1], \ldots, [a_p, c_p])(z_1, \ldots, z_p) = (a_1 + c_1 z_1, \ldots, a_p + c_p z_p)$$

The class G is a group of transformations, the *direct product* of the groups G_1, ..., G_p; the multiplication is coordinate by coordinate

$$([A_1, C_1], \ldots, [A_p, C_p])([a_1, c_1], \ldots, [a_p, c_p])$$
$$= ([A_1 + C_1 a_1, C_1 c_1], \ldots, [A_p + C_p a_p, C_p c_p]) \tag{8-4}$$

and similarly for the inverse.

For multiple performances of the system we have an n-vector for each of the basic variables of the system. The response presentation then has the form

$$\mathbf{y}_1 = [\mu_1, \sigma_1]\mathbf{z}_1, \ldots, \mathbf{y}_p = [\mu_p, \sigma_p]\mathbf{z}_p \tag{8-5}$$

and the distribution for the variation has the form

$$\Pi f_\lambda(z_{1i}, \ldots, z_{pi}) \tag{8-6}$$

For compact notation we will write

$$Y = (\mathbf{y}_1, \ldots, \mathbf{y}_p)$$
$$Z = (\mathbf{z}_1, \ldots, \mathbf{z}_p) \tag{8-7}$$
$$\theta = ([\mu_1, \sigma_1], \ldots, [\mu_p, \sigma_p])$$
$$f_\lambda(Z) = \Pi f_\lambda(z_{1i}, \ldots, z_{pi}) \tag{8-8}$$

We thus obtain the following structural model.

$$\mathcal{M}_V = (\Omega; \mathbb{R}^{pn}, \mathscr{B}^{pn}, \mathscr{V}, G) \tag{8-9}$$

where the parameter space is $\Omega = G \times \Lambda$, $\mathscr{B}^{pn}$ is the Borel class on $\mathbb{R}^{pn}$, $\mathscr{V}$ is class of densities $f_\lambda(\cdot)$ in (8-8), and G is the transformation group (8-3). This can be abbreviated informally as

$$Y = \theta Z \text{ with } \theta \text{ in } G$$
$$Z \text{ has distribution in the class } \mathscr{V} \tag{8-10}$$

The exactness Assumption 7-2 holds provided $n \geq 2$ and we exclude a set of measure zero consisting of points for which some coordinate vector y_j lies on the extended one-vector $\mathscr{L}(1)$ in $\mathbb{R}^n$.

8-1-2 The Analysis

We freely use the notation developed in Sec. 2-1:

$$\bar{z}_j = n^{-1} \sum_i z_{ji}$$

$$s^2(\mathbf{z}_j) = \sum_i (z_{ji} - \bar{z}_j)^2$$

$$\mathbf{d}(\mathbf{z}_j) = s^{-1}(\mathbf{z}_j)(\mathbf{z}_j - \bar{z}_j \mathbf{1}) \tag{8-11}$$

giving the location-scale position $[\bar{z}_j, s(\mathbf{z}_j)]$ relative to $\mathbf{d}(\mathbf{z}_j)$ for the jth coordinate:

$$\mathbf{z}_j = [\bar{z}_j, s(\mathbf{z}_j)]\mathbf{d}(\mathbf{z}_j)$$

The corresponding position for the full Z is

$$g = [Z] = ([\bar{z}_1, s(\mathbf{z}_1)], \ldots, [\bar{z}_p, s(\mathbf{z}_p)]) \tag{8-12}$$

relative to the reference point

$$D(Z) = (\mathbf{d}(\mathbf{z}_1), \ldots, \mathbf{d}(\mathbf{z}_p)) \tag{8-13}$$

Note that the points excluded by the exactness assumption are those for which one of the $s(\mathbf{z}_j)$ would be equal to zero with the resulting lack of definition for the corresponding $\mathbf{d}(\mathbf{z}_j)$.

The Jacobians from Secs. 7-2-2 and 7-2-3 are easily calculated:

$$J_{pn}(g:Z) = (c_1 \ldots c_p)^n$$

$$J_{pn}(Z) = (s(\mathbf{z}_1) \ldots s(\mathbf{z}_p))^n = (s_1 \ldots s_p)^n \tag{8-14}$$

$$dM(Z) = \frac{dZ}{(s_1 \ldots s_p)^n}$$

$$J_{2p}(g) = (s_1 \ldots s_p)^2$$

$$d\mu(g) = \frac{d\bar{z}_1 \, ds_1 \ldots d\bar{z}_p \, ds_p}{(s_1 \ldots s_p)^2} \tag{8-15}$$

The special Jacobian $J(D)$ refers to volume change at the point D; we have $\sqrt{n}$ for each of the vectors $\mathbf{z}_j$ at $\mathbf{d}(\mathbf{z}_j)$ (because the $\mathbf{1}$ vector has length $\sqrt{n}$). Thus

$$J(D) = n^{p/2} \tag{8-16}$$

Now consider the inference base $(\mathscr{M}_V, Y^0)$. We follow the analysis in Secs. 7-1 and 7-2 and make references to the earlier discussion of the simple case in Chap. 2.

The observed orbit gives the observed value for the variation:

$$D(Z) = D(Y^o) = D^o \tag{8-17}$$

or

$$\mathbf{d}(\mathbf{z}_j) = \mathbf{d}(\mathbf{y}_j^o) = \mathbf{d}_j^o \qquad j = 1, \ldots, p \tag{8-18}$$

This gives the inference base (a) from Sec. 7-1-4:

$$(\mathscr{M}_D, D^o) \tag{8-19}$$

The distributions in the model $\mathscr{M}_D$ are available from (7-46):

$$h_\lambda(D) = \int_G \Pi f_\lambda(\bar{z}_1 + s_1 d_{1i}, \ldots, \bar{z}_p + s_p d_{pi})(s_1 \ldots s_p)^{n-2} \, \Pi \sqrt{n} \, d\bar{z}_j \, ds_j \tag{8-20}$$

This, of course, leads to the observed likelihood function for λ from (7-69):

$$L(D^o; \lambda) = ch_\lambda(D^o) \tag{8-21}$$

The integrations (8-20) for this would typically require computer quadrature and simulation methods, indeed some sophisticated techniques even for medium to large values of p. Again we note that the model for possible likelihood functions seems generally inaccessible.

The unobservable characteristics of the variation lead to the inference base (b) from Sec. 7-1-4:

$$(\mathscr{M}_V^{D^o}, [Y^o]) \tag{8-22}$$

The distributions for the structural model $\mathscr{M}_V^{D^o}$ are available from (7-47):

$$h_\lambda^{-1}(D^o) \, \Pi f_\lambda(\bar{z}_1 + s_1 d_{1i}^o, \ldots, \bar{z}_p + s_p d_{pi}^o) \quad (s_1 \ldots s_p)^{n-2} \, \Pi \sqrt{n} \, d\bar{z}_j \, ds_j \tag{8-23}$$

on the group G. This distribution describing the unidentified variation is used with the transformation

$$\begin{aligned}
\bar{y}_1 &= \mu_1 + \sigma_1 \bar{z}_1 \\
s(\mathbf{y}_1) &= \quad \sigma_1 s_1 \\
&\cdots\cdots\cdots\cdots\cdots \\
\bar{y}_p &= \mu_p + \sigma_p \bar{z}_p \\
s(\mathbf{y}_p) &= \quad \sigma_p s_p
\end{aligned} \tag{8-24}$$

The corresponding observed values for the response characteristics are given by

$$[Y^o] = \{[\bar{y}_1^o, s(\mathbf{y}_1^o)], \ldots, [\bar{y}_p^o, s(\mathbf{y}_p^o)]\} \tag{8-25}$$

The methods from Secs. 7-3 and 7-4 are then available for inference concerning the parameters $\mu_1, \sigma_1, \ldots, \mu_p, \sigma_p$.

The response distribution corresponding to the identified $D(Y) = D(Y^o)$ is available from (7-49):

$$h_\lambda^{-1}(D^0)\Pi f_\lambda(\sigma_1^{-1}[\bar{y}_1 - \mu_1 + s(\mathbf{y}_1)d_{1i}], \ldots, \sigma_p^{-1}[\bar{y}_p - \mu_p + s(\mathbf{y}_p)d_{pi}])$$

$$\times \frac{[s(\mathbf{y}_1)\ldots s(\mathbf{y}_p)]^{n-2}}{(\sigma_1 \ldots \sigma_p)^n} \Pi \sqrt{n}\, d\bar{y}_j\, ds(\mathbf{y}_j) \tag{8-26}$$

To illustrate the method in Secs. 7-3-3 and 7-3-5 suppose we are interested in the parameter component $[\mu_p, \sigma_p]$. The corresponding group is the direct product of the location-scale groups for the remaining coordinates and the corresponding indexing set on the sample space for $[Z]$ is conveniently given by the location-scale element $[\bar{z}_p, s(\mathbf{z}_p)]$ for the pth variable. The marginal distribution for this is obtained by integrating over $G_1 \times \cdots \times G_{p-1}$:

$$h_\lambda^{-1}(D^0) \int_{G_1 \times \cdots \times G_{p-1}} \Pi f_\lambda(\bar{z}_1 + s_1 d_{1i}, \ldots, \bar{z}_p + s_p d_{pi})$$

$$\times (s_1 \ldots s_{p-1})^{n-2} \Pi_1^{p-1} d\sqrt{n}\, \bar{z}_j\, ds_j$$

$$\cdot s_p^{n-2} d\sqrt{n}\, \bar{z}_p\, ds_p \tag{8-27}$$

The corresponding equation is

$$\bar{y}_p = \mu_p + \sigma_p \bar{z}_p$$
$$s(\mathbf{y}_p) = \sigma_p s_p \tag{8-28}$$

This is a simple location-scale reduced model as in (2-16) and can be analyzed exactly as in Secs. 2-1 and 2-2.

8-1-3 A Normal Example

For a simple example to illustrate the mathematics of the model preceding consider the following normal distribution for the variation $(z_1, \ldots, z_p)$:

$$f_p(z_1, \ldots, z_p) = (2\pi)^{-p/2} |P|^{-1/2} \exp(-\tfrac{1}{2}\mathbf{z}'P^{-1}\mathbf{z}) \tag{8-29}$$

where P is a correlation matrix for the variation; note that the shape parameter λ is the correlation matrix P. We examine this for the bivariate case $p = 2$, with $\text{cov}(z_1, z_2) = \rho$.

The marginal distribution for the identified orbit for the variation can be obtained by integration from (8-20). For this let $r = r(\mathbf{z}_1, \mathbf{z}_2) = \Sigma d_{1i} d_{2i}$ be the correlation coefficient for the bivariate sample for variation:

$$h_\lambda(\mathbf{d}_1, \mathbf{d}_2) = (2\pi)^{-n}(1-\rho^2)^{-n/2}$$

$$\times \int \exp\left[-\Sigma \frac{(\bar{z}_1 + s_1 d_{1i})^2 - 2\rho(\bar{z}_1 + s_1 d_{1i})(\bar{z}_2 + s_2 d_{2i}) + (\bar{z}_2 + s_2 d_{2i})^2}{2(1-\rho^2)}\right]$$

$$\times (s_1 s_2)^{n-2} d\sqrt{n}\, \bar{z}_1\, d\sqrt{n}\, \bar{z}_2\, ds_1\, ds_2$$

172 INFERENCE AND LINEAR MODELS

$$= \int \frac{1}{(2\pi)(1-\rho^2)^{1/2}} \exp\left[-n\frac{\bar{z}_1^2 - 2\rho\bar{z}_1\bar{z}_2 + \bar{z}_2^2}{2(1-\rho^2)}\right] d\sqrt{n}\,\bar{z}_1\, d\sqrt{n}\,\bar{z}_2$$

$$\times \frac{1}{(2\pi)^{n-1}(1-\rho^2)^{(n-1)/2}} \int \exp\left[-\frac{s_1^2 - 2\rho r s_1 s_2 + s_2^2}{2(1-\rho^2)}\right]$$

$$\times (s_1 s_2)^{n-2}\, ds_1 ds_2$$

$$= \frac{2^{n-1}(1-\rho^2)^{(n-1)/2}}{(2\pi)^{n-1}} \int_0^\infty\!\!\int \exp(-t_1^2 - t_2^2 + 2\rho r t_1 t_2)(t_1 t_2)^{n-2}\, dt_1\, dt_2$$

$$= \frac{2^{n-1}(1-\rho^2)^{(n-1)/2}}{(2\pi)^{n-1}} \sum_{\alpha=0}^\infty \frac{(2\rho r)^\alpha}{\alpha!} \int_0^\infty\!\!\int \exp(-t_1^2 - t_2^2)(t_1 t_2)^{n-2+\alpha}$$

$$\times dt_1\, dt_2$$

$$= \frac{2^{n-1}(1-\rho^2)^{(n-1)/2}}{(2\pi)^{n-1}} \sum_{\alpha=0}^\infty \frac{(2\rho r)^\alpha}{\alpha!} \frac{1}{2^2} \int_0^\infty\!\!\int$$

$$\times \exp(-t_1^2 - t_2^2)(t_1^2 t_2^2)^{(n-1+\alpha)/2 - 1}\, dt_1^2\, dt_2^2$$

$$= \frac{2^{n-3}(1-\rho^2)^{(n-1)/2}}{(2\pi)^{n-1}} \sum_{\alpha=0}^\infty \frac{(2\rho r)^\alpha}{\alpha!} \Gamma^2\!\left(\frac{n-1+\alpha}{2}\right)$$

$$= \frac{2^{n-3}(1-\rho^2)^{(n-1)/2}}{(2\pi)^{n-1}} H_{n-1}(\rho r) \tag{8-30}$$

where the function $H_n(t)$ is an abbreviation for the following power series:

$$H_n(t) = \sum_{\alpha=0}^\infty \frac{(2t)^\alpha}{\alpha!} \Gamma^2\!\left(\frac{n+\alpha}{2}\right) \tag{8-31}$$

The conditional distribution for the unidentified variation $\bar{z}_1, s_1, \bar{z}_2, s_2$ is then available from (8-23):

$$\frac{h_\lambda^{-1}(\mathbf{d}_1, \mathbf{d}_2)}{(2\pi)^n(1-\rho^2)^{n/2}}$$

$$\times \exp\left[-\sum \frac{(\bar{z}_1 + s_1 d_{1i})^2 - 2\rho(\bar{z}_1 + s_1 d_{1i})(\bar{z}_2 + s_2 d_{2i}) + (\bar{z}_2 + s_2 d_{2i})^2}{2(1-\rho^2)}\right]$$

$$\times (s_1 s_2)^{n-2}\, d\sqrt{n}\,\bar{z}_1\, d\sqrt{n}\,\bar{z}_2\, ds_1\, ds_2$$

$$= \frac{1}{2\pi(1-\rho^2)^{1/2}} \exp\left[-n\frac{\bar{z}_1^2 - 2\rho\bar{z}_1\bar{z}_2 + \bar{z}_2^2}{2(1-\rho^2)}\right] d\sqrt{n}\,\bar{z}_1\, d\sqrt{n}\,\bar{z}_2$$

$$\times \frac{1}{2^{n-3}(1-\rho^2)^{n-1}H_{n-1}(\rho r)} \exp\left[-\frac{s_1^2 - 2\rho r s_1 s_2 + s_2^2}{2(1-\rho^2)}\right]$$

$$\times (s_1 s_2)^{n-2}\, ds_1 ds_2 \tag{8-32}$$

Thus $(\sqrt{n}\,\bar{z}_1, \sqrt{n}\,\bar{z}_2)$ is bivariate normal with means 0, variances 1, and correlation ρ. The distribution of (s_1^2, s_2^2) is a correlated bivariate gamma independent of the preceding normal.

The likelihood function for ρ from $\mathbf{d}_1, \mathbf{d}_2$ has the form

$$L(\mathbf{d}_1, \mathbf{d}_2; \rho) = c\,\frac{2^{n-3}(1-\rho^2)^{(n-1)/2}}{(2\pi)^{n-1}}\,H_{n-1}(\rho r) \tag{8-33}$$

This can be taken in ratio to the value of the function at $\rho = 0$, giving the representative likelihood

$$L(\mathbf{d}_1, \mathbf{d}_2; \rho) = \frac{(1-\rho^2)^{(n-1)/2} H_{n-1}(\rho r)}{\Gamma^2[(n-1)/2]} \tag{8-34}$$

From (8-33) it is seen that r is a sufficient statistic for the distribution of $(\mathbf{d}_1, \mathbf{d}_2)$. Accordingly, (8-34) is the likelihood function for the marginal distribution of r. The $\rho = 0$ distribution for the correlation coefficient r is easily derived from simple normal regression theory: the function

$$t = \frac{\sqrt{n-2}\, r}{\sqrt{1-r^2}}$$

has the Student $(n-2)$ distribution; thus the $\rho = 0$ density for r is

$$\frac{\Gamma[(n-1)/2]}{\Gamma(\tfrac{1}{2})\,\Gamma[(n-2)/2]}(1-r^2)^{(n-4)/2} = \frac{2^{n-3}\Gamma^2[(n-1)/2]}{\pi\Gamma(n-2)}(1-r^2)^{(n-4)/2}$$

on $(-1, +1)$; for the preceding we have used $\Gamma(2p) = 2^{2p-1}\Gamma(p)\Gamma(p+\tfrac{1}{2})/\Gamma(\tfrac{1}{2})$. The likelihood ratio (8-34) then gives the general noncentral distribution for the correlation r:

$$\frac{2^{n-3}}{\pi\Gamma(n-2)}(1-\rho^2)^{(n-1)/2}\,H_{n-1}(\rho r)(1-r^2)^{(n-4)/2}$$

on the interval $(-1, +1)$. This distribution was obtained here as a simple by-product of our analysis of the bivariate location-scale model.

Note that we have used a rather sophisticated tool called *likelihood modulation* for the derivation of the nonnull distribution of the correlation coefficient r. The technique was used in Watson (1956) and developed in Fraser (1968). For the present example we were able to derive the marginal density for $(\mathbf{d}_1, \ldots, \mathbf{d}_p)$ on the Cartesian product of p spheres in $\mathbb{R}^{n-1}$; we found that the likelihood depended only on the variable r and thus that we had the likelihood function from the marginal density for r (Fraser 1976, p. 338, for example); we then used the likelihood from r to modulate the $\rho = 0$ density for r to give the general density for r.

The general likelihood modulation technique can be summarized as follows:

(a) Derive the marginal density for some general variable D.
(b) Discover that the density for D involves only a parameter ϕ and some simple variable r; this then provides the likelihood function from the simple variable r.
(c) Use the preceding likelihood to modulate some special $\phi = \phi_0$ density for r.

8-2 MULTIVARIATE MODEL: PROGRESSION

Consider a random system with p response variables $y_1, \ldots, y_p$. We suppose that the distribution form has been identified or identified up to a parameter λ, but not identified to the degree considered in the preceding section. The model we consider here has some applications, but our interest in the model is primarily to develop methods and techniques for the model to be considered in Sec. 8-4.

For the progression model in this section we suppose that for the first response the location and scale are unknown, that for the second response the location and scale and also the regression on the first variable are unknown, and so on, so that for each response the location, scale, and regression on preceding variables are unknown. Clearly the applications for this model have a very special property in which the variables are ordered and have possible regression dependence on preceding variables.

8-2-1 On Notation

The response for the random system has p variables, $y_1, \ldots, y_p$. With n performances of the system we then have, in effect, np variables.

For notation in the preceding section we formed an n-vector for each of the basic system variables and then combined these to form an $n \times p$ matrix. This conforms with the usual notation for the regression model where the vectors are in $\mathbb{R}^n$ and the design matrix is an $n \times r$ matrix recording an n-vector for each of the r input variables.

For multivariate analysis there are two conventions: the use of the $n \times p$ matrix and the use of its transpose, the $p \times n$ matrix. If we stay with the use of the $n \times p$ matrix then our transformations will be matrix transformations from the right; if we reverse and use the $p \times n$ matrix then our transformations are matrix transformations from the left and thus conform to the general notation that we have been developing. We choose the advantages of this latter approach and now reverse the notation available from the preceding section.

Let $\mathbf{y} = (y_1, \ldots, y_p)'$ designate the p variate response of the system and then let

$$Y = (\mathbf{y}_1, \ldots, \mathbf{y}_n) = \begin{bmatrix} y_{11} & \cdots & y_{1n} \\ \vdots & & \vdots \\ y_{p1} & \cdots & y_{pn} \end{bmatrix} \tag{8-35}$$

be the compound response for n performances of the basic system. On the occasions when we want to refer to an n-vector for a response variable we will write, for example:

$$\begin{aligned} Y_1 &= (y_{11}, \ldots, y_{1n}) \\ Y_2 &= (y_{21}, \ldots, y_{2n}) \end{aligned} \tag{8-36}$$

for the first and second variables, these being the appropriate row vectors from the matrix Y.

8-2-2 The Model

Consider the random system with the p response variables $y_1, \ldots, y_p$. We suppose that the location and scaling of y_1 are unknown but that the distribution form is known or known up to a shape parameter λ, that the location and scaling of y_2 and the linear regression on y_1 are unknown but that the distribution form is known or known up to a shape parameter λ, and so on. Accordingly, we write

$$y_1 = \mu_1 + \sigma_{(1)} z_1$$
$$y_2 = \mu_2 + \tau_{21} z_1 + \sigma_{(2)} z_2$$
$$\vdots \qquad (8\text{-}37)$$
$$y_p = \mu_p + \tau_{p1} z_1 + \cdots + \tau_{p,p-1} z_{p-1} + \sigma_{(p)} z_p$$

or

$$\mathbf{y} = \boldsymbol{\mu} + \Upsilon \mathbf{z} = [\boldsymbol{\mu}, \Upsilon]\mathbf{z} \qquad (8\text{-}38)$$

where the residual (after regression) scalings $\sigma_{(1)}, \ldots, \sigma_{(p)}$ are greater than 0 and the μ and τ values are arbitrary; note that the matrix Υ is a *positive lower triangular matrix*, a matrix for which the diagonal elements are positive, the above diagonal elements are zero, and the below diagonal elements are real. Also let

$$f_\lambda(\mathbf{z}) = f_\lambda(z_1, \ldots, z_p) \qquad (8\text{-}39)$$

be the density function for the objective variation; we suppose that f_λ has been suitably standardized using, say, (1-18) on each axis and some reasonable procedure for the regressions.

For n independent performances we have $\mathbf{y}_i = \boldsymbol{\mu} + \Upsilon \mathbf{z}_i$ for the ith performance and

$$Y = \boldsymbol{\mu} \mathbf{1}' + \Upsilon Z = [\boldsymbol{\mu}, \Upsilon]Z = \theta Z \qquad (8\text{-}40)$$

for the compound response Y and variation Z. Note that $\mathbf{1}$ is an n-vector of ones and that $[\boldsymbol{\mu}, \Upsilon]$ operates on Z column by column as in (8-38). The distribution for the compound variation is

$$f_\lambda(Z) = \Pi_1^n f_\lambda(\mathbf{z}_i) \qquad (8\text{-}41)$$

The transformations $[\boldsymbol{\mu}, \Upsilon]$ form a group:

$$[\mathbf{a}_2, T_2][\mathbf{a}_1, T_1] = [\mathbf{a}_2 + T_2 \mathbf{a}_1, T_2 T_1]$$
$$[\mathbf{a}, T]^{-1} = [-T^{-1}\mathbf{a}, T^{-1}] \qquad (8\text{-}42)$$
$$[\mathbf{0}, I] = i$$

where the key item to check is that positive lower triangular (PLT) times positive lower triangular is positive lower triangular. We then have that

$$G = \{[\mathbf{a}, T] : \mathbf{a} \in \mathbb{R}^p, T \text{ is PLT}\} \qquad (8\text{-}43)$$

is a group. We will see that G satisfies the exactness Assumption 7-2 on the sample space, provided $n \geq p + 1$ and a certain set of measure zero is excluded.

176 INFERENCE AND LINEAR MODELS

We thus obtain the following structural model:

$$\mathcal{M}_V = (\Omega; \mathbb{R}^{pn}, \mathcal{B}^{pn}, \mathcal{V}, G) \tag{8-44}$$

where the parameter space is $\Omega = (G \times \Lambda)$, $\mathcal{V}$ is the class of densities f_λ in (8-41), and G is the transformation group (8-43) with action (8-38). We abbreviate this as

$$\begin{aligned} Y &= \theta Z \text{ with } \theta \text{ in } G \\ Z &\text{ has distribution in the class } \mathcal{V} \end{aligned} \tag{8-45}$$

and assume that $n \geq p + 1$.

8-2-3 The Analysis

For the analysis we can use notation available from Chaps. 2 and 6.

To begin with, consider the first response variable for a single performance in (8-37) and for the n performances in (8-40). For the n performances we have

$$Y_1 = \mu_1 \mathbf{1}' + \sigma_{(1)} Z_1$$

This is a simple location-scale transformation as in Chap. 2. Accordingly, we let $\bar{z}_1$ be the average for Z_1 and $s_{(1)}(Z)$ be the residual length after regression on the one-vector $\mathbf{1}$; and we let $D_1(Z)$ be the unit residual [see (2-5) and (2-6)]. Thus

$$Z_1 = \bar{z}_1 \mathbf{1}' + s_{(1)}(Z) D_1(Z) \tag{8-46}$$

Now consider the second response variable for a single performance in (8-37) and for the n performances in (8-40). Then for the n performances we have

$$Y_2 = \mu_2 \mathbf{1}' + \tau_{21} Z_1 + \sigma_{(2)} Z_2$$

This has the regression form in Chap. 6. Accordingly, we let $\bar{z}_2$ be the average for Z_2, $t_{21}(Z)$ be the regression coefficient on $D_1(Z)$, and $s_{(2)}(Z)$ be the residual length after the preceding regressions; and we let $D_2(Z)$ be the unit residual. Note that we are using the descending factorial notation "(2)" in the subscript to suggest that the "2" also covers the preceding "1," in this case as a regression residual. Thus

$$Z_2 = \bar{z}_2 \mathbf{1}' + t_{21}(Z) D_1(Z) + s_{(2)}(Z) D_2(Z) \tag{8-47}$$

Continuing in this pattern we have the following for Z_p:

$$Z_p = \bar{z}_p \mathbf{1}' + t_{p1}(Z) D_1(Z) + \cdots + t_{p,p-1}(Z) D_{p-1}(Z) + s_{(p)}(Z) D_p(Z) \tag{8-48}$$

where $\bar{z}_p, t_{p1}(Z), \ldots, t_{p,p-1}(Z)$ are the regression coefficients of Z_p on $\mathbf{1}, D_1(Z), \ldots, D_{p-1}(Z)$; $s_{(p)}(Z)$ is the residual length; and $D_p(Z)$ is the unit residual. The regression coefficients are particularly easy to calculate because the relevant vectors are orthogonal.

The preceding can be collected in the following equation:

$$\begin{bmatrix} Z_1 \\ \vdots \\ Z_p \end{bmatrix} = \begin{bmatrix} \bar{z}_1 \\ \vdots \\ \bar{z}_p \end{bmatrix} \mathbf{1}' + \begin{bmatrix} s_{(1)} & & 0 \\ \vdots & \ddots & \\ t_{p1} & & s_{(p)} \end{bmatrix} \begin{bmatrix} D_1 \\ \vdots \\ D_p \end{bmatrix}$$

$$Z = [Z]D(Z) = [\bar{z}, T(Z)]D(Z) \tag{8-49}$$

where

$$\bar{z} = \begin{bmatrix} \bar{z}_1 \\ \vdots \\ \bar{z}_p \end{bmatrix}$$

$$T(Z) = \begin{bmatrix} s_{(1)}(Z) & & & 0 \\ t_{21}(Z) & s_{(2)}(Z) & & \\ \vdots & & \ddots & \\ t_{p1}(Z) & \cdots & t_{p,p-1}(Z) & s_{(p)}(Z) \end{bmatrix} \tag{8-50}$$

Thus we have expressed a general point Z on the orbit GZ by means of the transformation $[\bar{z}, T(Z)]$ in the group G relative to the reference point $D(Z)$ obtained by successive orthonormalization. The preceding calculations can fail if one of the residual lengths $s_{(1)}(Z), \ldots, s_{(p)}(Z)$ is zero. In this case there is linear dependence among the vectors $\mathbf{1}', Z_1, \ldots, Z_p$; we exclude the corresponding set of measure zero from the effective sample space.

The Jacobians from Secs. 7-2-2 and 7-2-3 are readily calculated. In the sample space we have

$$J_{pn}(g : Z) = \left| \frac{\partial [\mathbf{a}, T]Z}{\partial Z} \right| = |T|^n \tag{8-51}$$

The transformation $[\mathbf{a}, T]$ operates column by column and $|T|$ is the determinant for any particular column. Thus

$$J_{pn}(Z) = |T(Z)|^n$$

$$dM(Z) = \frac{dZ}{|T(Z)|^n} = \frac{dZ}{s_{(1)}^n(Z) \cdots s_{(p)}^n(Z)} \tag{8-52}$$

On the group we have

$$[\tilde{\mathbf{a}}, \tilde{T}] = [\mathbf{a}, T][\mathbf{a}^*, T^*]$$
$$= [\mathbf{a} + T\mathbf{a}^*, TT^*]$$
$$\tilde{\mathbf{a}} = \mathbf{a} + T\mathbf{a}^* \tag{8-53}$$
$$\tilde{T} = TT^*$$

The left transformation operates column by column on $\mathbf{a}^*, T^*$; a determinant for a column involves the *relevant* part of the matrix T—specifically the diagonal elements that correspond to *real* coordinates for a column of T^*. Thus

$$J(g:g^*) = [s_{(1)} \ldots s_{(p)}] [s_{(1)} \ldots s_{(p)}] [s_{(2)} \ldots s_{(p)}] \ldots [s_{(p)}]$$
$$= s_{(1)}^2 \ldots s_{(p)}^{p+1} = |T| |T|_\Delta$$

where we introduce some convenient notation for *increasing* and *decreasing* determinants:

$$|T|_\Delta = \begin{vmatrix} s_{(1)} & & 0 \\ & \ddots & \\ t_{p1} & & s_{(p)} \end{vmatrix}_\Delta = s_{(1)}^1 \ldots s_{(p)}^p$$

$$|T|_\nabla = \begin{vmatrix} s_{(1)} & & 0 \\ & \ddots & \\ t_{p1} & & s_{(p)} \end{vmatrix}_\nabla = s_{(1)}^p \ldots s_{(p)}^1 \tag{8-54}$$

Thus

$$J(g) = |T| |T|_\Delta$$
$$d\mu(g) = \frac{d\mathbf{a}\, dT}{|T||T|_\Delta} = \frac{d\mathbf{a}\, dT}{s_{(1)}^2 \ldots s_{(p)}^{p+1}} \tag{8-55}$$

The right transformation carries T to $\tilde{T}$ row by row, and for given T carries **a** to **ã** by location change; a determinant for a row of T involves the relevant part of the matrix T^*—specifically the diagonal elements that correspond to real coordinates for a row of T. Thus

$$J^*(g^*:g) = s_{(1)}^*[s_{(1)}^* s_{(2)}^*] [s_{(1)}^* s_{(2)}^* s_{(3)}^*] \ldots [s_{(1)}^* \ldots s_{(p)}^*]$$
$$= s_{(1)}^{*p} \ldots s_{(p)}^{*1} = |T^*|_\nabla$$
$$J^*(g) = |T|_\nabla$$
$$dv(g) = \frac{d\mathbf{a}\, dT}{|T|_\nabla} = \frac{d\mathbf{a}\, dT}{s_{(1)}^p \ldots s_{(p)}^1} \tag{8-56}$$
$$\Delta(g) = \frac{s_{(1)}^p \ldots s_{(p)}^1}{s_{(1)}^2 \ldots s_{(p)}^{p+1}} = \frac{|T|_\nabla}{|T||T|_\Delta}$$

Consider the coordinates $[\bar{z}, T(Z)]$ as given by (8-49). Rows of Z come from rows of $[\bar{z}, T(Z)]$; for any row the corresponding basis vectors $\mathbf{1}', D_1(Z), \ldots, D_p(Z)$ are orthogonal and of length 1, except for the one-vector which has length $\sqrt{n}$. Accordingly, we have

$$J(D) = n^{p/2} \tag{8-57}$$

This uses the alternate interpretation for dD: Euclidean volume orthogonal to the orbit GD at D (recall Sec. 7-2-6).

Now consider the analysis of an inference base $(\mathcal{M}_V, Y^0)$; we follow the pattern in Secs. 7-1 and 7-2 and the first application in the preceding Sec. 8-1.

The observed orbit gives the observed value for the variation

$$D(Z) = D(Y^0) = D^0 \tag{8-58}$$

This gives the inference base (a) from Sec. 7-1-4:

$$(\mathcal{M}_D, D^0) \tag{8-59}$$

The distributions in $\mathcal{M}_D$ are available from (7-46):

$$h_\lambda(D) = \int_G f_\lambda(\bar{z}\mathbf{1}' + TD)s_{(1)}^{n-2} \ldots s_{(p)}^{n-p-1} n^{p/2} \, d\bar{z} \, dT \tag{8-60}$$

This, of course, leads to the observed likelihood for λ:

$$L(D^0; \lambda) = ch_\lambda(D^0) \tag{8-61}$$

The integration (8-60) is a $[p + p(p + 1)/2]$-dimensional integration, as opposed to the $2p$-dimensional integration in Sec. 8-1, and would need rather special computer techniques except for very small p.

The unobserved characteristics of the variation lead to the inference base (b) from Sec. 7-1-4:

$$(\mathcal{M}_V^{D^0}, [Y^0]) \tag{8-62}$$

The distributions for the structural model $\mathcal{M}_V^{D^0}$ are available from (7-47):

$$h_\lambda^{-1}(D^0) f_\lambda(\bar{z}\mathbf{1}' + TD)s_{(1)}^{n-2} \ldots s_{(p)}^{n-p-1} n^{p/2} \, d\bar{z} \, dT \tag{8-63}$$

a distribution on the group G. This distribution for the unidentified variation is used with the transformation

$$\begin{aligned} \bar{y} &= \mu + \Upsilon \bar{z} \\ T(Y) &= \quad \Upsilon T \end{aligned} \tag{8-64}$$

The observed values for the response coordinates are given by

$$[Y^0] = [\bar{y}^0, T(Y^0)] \tag{8-65}$$

The response distribution corresponding to the identified $D(Y) = D(Y^0) = D^0$ is available from (7-49):

$$h_\lambda^{-1}(D^0) f_\lambda(\Upsilon^{-1}[(\bar{y} - \mu)\mathbf{1}' + T(Y)D^0]) \left(\frac{s_{(1)}(Y) \ldots s_{(p)}(Y)}{\sigma_{(1)} \ldots \sigma_{(p)}} \right)^n \frac{n^{p/2} \, d\bar{y} \, dT(Y)}{s_{(1)}^2(Y) \ldots s_{(p)}^{p+1}(Y)} \tag{8-66}$$

8-2-4 Inference for Component Parameters

As we have noted in Secs. 7-3-3 and 7-3-5, some parameters may not index left cosets on G and thus may not be amenable to the strong inference methods investigated in Chap. 7. In this section we examine two important parameter components for the model, components that do have the special left coset property.

We are able to, and in fact will, examine these parameter components separately. The sequential methods of Sec. 7-4 are also available, however, for the two components, in either order.

For the location parameter μ we have the separation of the equation $Y = \theta Z$ giving a μ-specific component:

$$T^{-1}(Y)(\bar{y} - \mu) = T^{-1}\bar{z} = \mathbf{t} \tag{8-67}$$

The full parameter can be separated as

$$[\mu, \Upsilon] = [\mu, I][0, \Upsilon] \tag{8-68}$$

Note that μ indexes the left cosets of the scale group $G_S = \{[0, T] : T \text{ is PLT}\}$:

$$[\mu, \Upsilon]G_S = [\mu, I]G_S \tag{8-69}$$

The group coordinates can be separated in reverse order:

$$[\bar{z}, T] = [0, T][\mathbf{t}, I] \tag{8-70}$$

Note that $\mathbf{t}$ indexes the orbits (right cosets) of the scale group

$$G_S[\bar{z}, T] = G_S[\mathbf{t}, I] \tag{8-71}$$

The marginal distribution for $\mathbf{t}$ is easily derived from (8-63) using $d\bar{z} = |T| d\mathbf{t}$ for fixed T; the probability differential is

$$h_\lambda^{-1}(D^0) \int_{G_S} f_\lambda[T(\mathbf{t1}' + D^0)] s_{(1)}^{n-1} \ldots s_{(p)}^{n-p} \, dT \, n^{p/2} \, d\mathbf{t} \tag{8-72}$$

This distribution with equation (8-67) provides for tests and confidence regions for the location parameter μ.

For the scale parameter Υ we have the separation of the equation giving the Υ-specific component

$$\Upsilon^{-1}T(Y) = T \tag{8-73}$$

The full parameter can now be separated as

$$[\mu, \Upsilon] = [0, \Upsilon][\delta, I] \tag{8-74}$$

where $\delta = \Upsilon^{-1}\mu$ is the coefficient of variation. Note that Υ indexes the left cosets of the location group $G_L = \{[\mathbf{a}, I] : \mathbf{a} \in \mathbb{R}^p\}$:

$$[\mu, \Upsilon]G_L = [0, \Upsilon]G_L \tag{8-75}$$

The group coordinates can be separated in the reverse order, giving

$$[\bar{z}, T] = [\bar{z}, I][0, T] \tag{8-76}$$

Note that T indexes the orbits (right cosets) of the location group

$$G_L[\bar{z}, T] = G_L[0, T] \tag{8-77}$$

The marginal distribution for T can be obtained by directly integrating $\bar{z}$ from the distribution (8-63); the probability differential is

$$h_\lambda^{-1}(D^0) \int_{G_L} f_\lambda(\bar{z}1' + TD) s_{(1)}^{n-2} \ldots s_{(p)}^{n-p-1} n^{p/2} \, d\bar{z} \, dT \qquad (8\text{-}78)$$

This distribution with the equation (8-73) provides for tests and confidence regions for Υ.

For location the subcomponents $\mu_1, (\mu_1, \mu_2), (\mu_1, \mu_2, \mu_3), \ldots$ are amenable to the strong confidence procedures we have been examining. In a parallel way for scale, the subcomponents

$$\sigma_{(1)}, \begin{bmatrix} \sigma_{(1)} & 0 \\ \tau_{21} & \sigma_{(2)} \end{bmatrix}, \ldots$$

also are amenable to the method.

8-2-5 Supplement

Confidence distributions are not part of the theme of this book, but they are rather easily derived; the mathematics is attractive and they have importance. For this particular development, we choose to avoid them for special reasons. However, we record several here for their mathematical interest. We also record some useful relations for measure differentials.

For the full parameter θ we obtain the following confidence distribution from formula (7-105) with (8-63):

$$h_\lambda^{-1}(D^0) f_\lambda(\Upsilon^{-1}(Y^0 - \mu 1')) \left[\frac{s_{(1)}(Y^0) \ldots s_{(p)}(Y^0)}{\sigma_{(1)} \ldots \sigma_{(p)}} \right]^n$$

$$\times \frac{s_{(1)}^p(Y^0) \ldots s_{(p)}^1(Y^0) \; n^{p/2} \, d\mu \, d\Upsilon}{s_{(1)}^2(Y^0) \ldots s_{(p)}^{p+1}(Y^0) \; \sigma_{(1)}^p \ldots \sigma_{(p)}^1} \qquad (8\text{-}79)$$

For the location parameter μ we can proceed as in Sec. 7-4-5 by factoring the invariant measure in accordance with the component groups and then using (8-70). The resulting *formula*, however, is available directly by integrating Υ out of (8-79); the differential for μ is

$$h_\lambda^{-1}(D^0) \int_{G_S} f_\lambda(\Upsilon^{-1}(Y^0 - \mu 1')) \frac{d\Upsilon}{\sigma_{(1)}^{n+p} \ldots \sigma_{(p)}^{n+1}}$$

$$\times s_{(1)}^{n+p-2}(Y^0) s_{(2)}^{n+p-4}(Y^0) \ldots s_{(p)}^{n-p}(Y^0) n^{p/2} \, d\mu \qquad (8\text{-}80)$$

For the scale parameter Υ we can also use (8-70). Again, however, we can obtain the *formula* directly by integrating μ out of (8-79); the differential for Υ is

$$h_\lambda^{-1}(D^0) \int_{G_L} f_\lambda(\Upsilon^{-1}(Y^0 - \mu 1')) n^{p/2} \, d\mu \, s_{(1)}^{n+p-2}(Y^0) \ldots s_{(p)}^{n-p}(Y^0)$$

$$\times \frac{d\Upsilon}{\sigma_{(1)}^{n+p} \ldots \sigma_{(p)}^{n+1}} \qquad (8\text{-}81)$$

182 INFERENCE AND LINEAR MODELS

For completeness, we record the factorizations of the left invariant measure needed for the direct use of the results in Sec. 7-4-5. For the location group G_L the invariant measures are

$$d\mu_L([\mathbf{a}, I]) = d\mathbf{a} \qquad dv_L([\mathbf{a}, I]) = d\mathbf{a} \tag{8-82}$$

For the scale group G_S the measures are available by a simplification of the steps used with (8-55) and (8-56):

$$d\mu_S([\mathbf{0}, T]) = \frac{dT}{|T|_\Delta}$$

$$dv_S([\mathbf{0}, T]) = \frac{dT}{|T|_\nabla} \tag{8-83}$$

$$\Delta_S([\mathbf{0}, T]) = \frac{|T|_\nabla}{|T|_\Delta}$$

For the factorization

$$[\mathbf{a}, T] = [\mathbf{0}, T][\mathbf{t}, I] \tag{8-84}$$

with $\mathbf{t} = T^{-1}\mathbf{a}$ in (8-70), we have, from (7-125),

$$\frac{d\mathbf{a}\, dT}{|T||T|_\Delta} = \frac{dT}{|T|_\Delta} H_1\, d\mathbf{t}$$

$$= \frac{dT}{|T|_\Delta} d\mathbf{t} \tag{8-85}$$

where $H_1 = 1$ can be verified easily.

For the factorization

$$[\mathbf{a}, T] = [\mathbf{a}, I][\mathbf{0}, T] \tag{8-86}$$

in (8-76) we have, from (7-125),

$$\frac{d\mathbf{a}\, dT}{|T||T|_\Delta} = d\mathbf{a}\, H_1 \frac{|T|_\nabla/|T||T|_\Delta}{|T|_\nabla/|T|_\Delta} \frac{dT}{|T|_\Delta}$$

$$= d\mathbf{a}\, \frac{dT}{|T||T|_\Delta} \tag{8-87}$$

where $H_1 = 1$ trivially.

The formulas (8-85) and (8-87) are rather routine and hardly need the results in Sec. 7-4-3. The steps, however, contain some other useful material and do illustrate the factorization in Sec. 7-4-5.

8-3 MULTIVARIATE MODEL : NORMAL PROGRESSION

Now consider the multivariate progression model but with a normal pattern for the variation. As we have noted before, the multivariate normal has so many symmetries that many distinctions, useful and essential in general, vanish rather trivially. From this viewpoint the normal does not provide a good illustration for the results in the preceding section. The results from the preceding section, however, give us in a simple routine mechanical way all the basic distribution theory for the multivariate normal. Accordingly, we examine this normal case in detail.

8-3-1 The Normal Model

Consider a random system with a p-variate response **y**. From Sec. 8-2 the progression model $\mathcal{M}_Y$ for n performances has the form

$$Y = \mu\mathbf{1}' + \Upsilon Z \tag{8-88}$$

where Y and Z are $p \times n$, μ is the location for **y**, and Υ is the positive lower triangular $p \times p$ scale matrix.

In this section we examine a normal distribution for the variation. The standardization (1-18) together with a reasonable elimination of regressions gives the standard multivariate normal,

$$\begin{aligned} f(\mathbf{z}) &= (2\pi)^{-p/2} \exp\left(-\tfrac{1}{2}\Sigma z_i^2\right) \\ &= (2\pi)^{-p/2} \exp\left(-\tfrac{1}{2}\mathbf{z}'\mathbf{z}\right) \end{aligned} \tag{8-89}$$

The corresponding distribution for the sample matrix Z is

$$\begin{aligned} f(Z) &= (2\pi)^{-np/2} \exp\left(-\tfrac{1}{2}\operatorname{tr} Z'Z\right) \\ &= (2\pi)^{-np/2} \operatorname{etr}\left(-\tfrac{1}{2}Z'Z\right) \end{aligned} \tag{8-90}$$

where tr is the trace function and etr is an abbreviation for exp(tr —). We can write the preceding in the invariant form

$$f(Z)dZ = (2\pi)^{-np/2} \operatorname{etr}\left(-\tfrac{1}{2}Z'Z\right)|T|^n \, dM(Z) \tag{8-91}$$

The corresponding response distribution can be obtained by direct substitution and the use of the group transformation $Z = \Upsilon^{-1}(Y - \mu\mathbf{1}')$. For this we need the following simplification of the quadratic exponent:

$$\begin{aligned} \Sigma z_{ji}^2 &= \operatorname{tr} Z'Z \\ &= \operatorname{tr}(Y - \mu\mathbf{1}')'\Upsilon'^{-1}\Upsilon^{-1}(Y - \mu\mathbf{1}') \\ &= \operatorname{tr}(Y - \mu\mathbf{1}')'\Sigma^{-1}(Y - \mu\mathbf{1}') \\ &= \Sigma_i(\mathbf{y}_i - \mu)'\Sigma^{-1}(\mathbf{y}_i - \mu) \end{aligned} \tag{8-92}$$

where $\Sigma = \Upsilon\Upsilon'$ is the covariance matrix for $\mathbf{y} = \mu + \Upsilon\mathbf{z}$ and we use $|\Upsilon|^n = |\Sigma|^{n/2}$.

184 INFERENCE AND LINEAR MODELS

The response distribution is

$$(2\pi)^{-np/2}|\Sigma|^{-n/2} \text{ etr}\left[-\tfrac{1}{2}(Y-\mu\mathbf{1}')'\Sigma^{-1}(Y-\mu\mathbf{1}')\right]|T(Y)|^n \, dM(Y)$$
$$= (2\pi)^{-np/2}|\Sigma|^{-n/2} \text{ etr}\left[-\tfrac{1}{2}(Y-\mu\mathbf{1}')'\Sigma^{-1}(Y-\mu\mathbf{1}')\right]dY \quad (8\text{-}93)$$

Recall that the progression model describes the response variables in a prescribed order taken to be $y_1, \ldots, y_p$ and note that the relation $\Sigma = \Upsilon\Upsilon'$ represents Υ as a PLT square root taken with respect to that same ordering. A different ordering of coordinates would give a different square root of a variance matrix Σ. For an application of the progression model the particular order for the coordinates is part of the specification of the system being investigated.

8-3-2 The Analysis

Consider the analysis of the inference base $(\mathcal{M}_V, Y^0)$ using the normal model for variation in Sec. 8-3-1. Following Sec. 8-2-3 we determine the marginal distribution for $D(Z) = D(Y) = D$ and the conditional distribution for $[Z]$ given D.

For this we need the simplification of the normal exponent

$$\text{tr } Z'Z = \text{tr } (ZZ')$$
$$= \text{tr } (\bar{z}\mathbf{1}' + TD)(\bar{z}\mathbf{1}' + TD)'$$
$$= \text{tr } (\bar{z}n\bar{z}') + \text{tr } TDD'T' \quad (8\text{-}94)$$
$$= \text{tr } n\bar{z}'\bar{z} + \text{tr } TT'$$
$$= \Sigma(\sqrt{n}\,\bar{z}_j)^2 + \sum_{j>j'} t_{jj'}^2 + \Sigma s_{(j)}^2$$

where we have used the orthogonality of $\mathbf{1}'$ with the rows of D and the orthonormality of D, that is, $DD' = I$.

First we obtain the conditional distribution of $[Z]$ given D:

$$h^{-1}(D) f(\bar{z}\mathbf{1}' + TD) s_{(1)}^{n-2} \ldots s_{(p)}^{n-p-1} n^{p/2} \, d\bar{z} \, dT$$
$$= \frac{A_{n-1} \ldots A_{n-p}}{(2\pi)^{np/2}} \exp\left(-\tfrac{1}{2}\Sigma n\bar{z}_j^2 - \tfrac{1}{2}\Sigma t_{jj'}^2 - \tfrac{1}{2}\Sigma s_{(j)}^2\right) \quad (8\text{-}95)$$
$$\times s_{(1)}^{n-2} \ldots s_{(p)}^{n-p-1} \, \Pi d\sqrt{n}\,\bar{z}_j \, \Pi \, ds_{(j)} \, \Pi \, dt_{jj'}$$

For this we have used the normalizing constant $1/(2\pi)^{1/2}$ for the standard normal density form found for each of the $\sqrt{n}\,\bar{z}_j$ and $t_{jj'}$. Also we have used the normalizing constant for the chi density form found for each of the $s_{(j)}$:

$$\Gamma^{-1}\left(\frac{f}{2}\right)\exp\left(-\frac{s^2}{2}\right)\left(\frac{s^2}{2}\right)^{f/2-1}\frac{ds^2}{2} = \frac{1}{2^{f/2-1}\Gamma(f/2)}\exp\left(-\frac{s^2}{2}\right)s^{f-1}\,ds$$
$$= \frac{A_f}{(2\pi)^{f/2}}\exp\left(-\frac{s^2}{2}\right)s^{f-1}\,ds \quad (8\text{-}96)$$

where $A_f = 2\pi^{f/2}/\Gamma(f/2)$ is the surface volume of the unit sphere in $\mathbb{R}^f$. In the preceding distribution (8-95) we have used the volume constants A_f rather than the usual gamma functions. This allows the associated denominator involving powers of $(2\pi)^{1/2}$ to correspond directly to the dimension of the space from which the variable derives, and leaves the constant A_f as the representative of the unit sphere integration that is appropriate for the particular variable. We can write the distribution (8-95) more compactly; in particular, we let

$$A_f^{(p)} = A_f A_{f-1} \cdots A_{f-p+1} \qquad (8\text{-}97)$$

Note the suggestive use of the descending factorial notation. The distribution for $[Z]$ given D is

$$\frac{A_{n-1}^{(p)}}{(2\pi)^{np/2}} \operatorname{etr}(-\tfrac{1}{2}n\bar{z}\bar{z}' - \tfrac{1}{2}TT')|T|^n \frac{d\sqrt{n}\,\bar{z}\,dT}{|T||T|_\Delta}$$

$$= \frac{1}{(2\pi)^{p/2}} \operatorname{etr}(-\tfrac{1}{2}n\bar{z}\bar{z}')\, d\sqrt{n}\,\bar{z}\, \frac{A_{n-1}^{(p)}}{(2\pi)^{(n-1)p/2}} \operatorname{etr}(-\tfrac{1}{2}TT')|T|^{n-1} \frac{dT}{|T|_\Delta} \qquad (8\text{-}98)$$

This distribution for $\bar{z}$ and T can be described very simply. The components are statistically independent and

$$\sqrt{n}\,\bar{z}_j \text{ is normal } (0, 1)$$

$$T = \begin{bmatrix} s_{(1)} & & & 0 \\ t_{21} & s_{(2)} & & \\ \vdots & & \ddots & \\ t_{p1} & \cdots & t_{p,p-1} & s_{(p)} \end{bmatrix} = \begin{bmatrix} \chi_{n-1} & & & 0 \\ z_{21} & \chi_{n-2} & & \\ \vdots & & \ddots & \\ z_{p1} & \cdots & z_{p,p-1} & \chi_{n-p} \end{bmatrix} \qquad (8\text{-}99)$$

where the z variables here represent independent normal $(0, 1)$ variables and the χ variables represent independent chi variables with degrees of freedom as subscribed. The distribution of T is called the *triangular chi distribution* $\Delta\chi_p(n-1)$.

The conditional distribution (8-98) gives us the distribution component for the model $\mathscr{M}_V^{D^o}$. Thus as in Sec. 8-2-3 we obtain the component inference base (8-62):

$$(\mathscr{M}_V^{D^o}, [Y^o])$$

where the model $\mathscr{M}_V^{D^o}$ has the distribution for $[Z]$ given by (8-98) and the transformation

$$\bar{y} = \mu + \Upsilon\bar{z}$$
$$T(Y) = \Upsilon T$$

relating the response $[Y] = [\bar{y}, T(Y)]$ to the variation $[Z] = [\bar{z}, T]$. The observed values for the response coordinates are

$$[Y^o] = [\bar{y}^o, T(Y^o)]$$

186 INFERENCE AND LINEAR MODELS

Now we obtain the marginal distribution of D. This involves effectively the integration of the normal and chi components we found in (8-95). That integration was trivial as we happened to know the norming constants for the normal and chi density forms. The resulting constant $h(D)$ can be read from (8-95):

$$h(D)\, dD = \frac{dD}{A_{n-1}^{(p)}} \tag{8-100}$$

For this, as in Sec. 8-2, we have used the alternative and typically more useful interpretation of dD from Sec. 7-2-6. Formula (8-100) shows that D has a uniform distribution on Q relative to our particular choice of measure dD.

Note that D consists of p orthonormal row vectors in $\mathscr{L}^\perp(\mathbf{1}')$. Thus D is a $p \times n$ semiorthogonal matrix with rows orthogonal to $\mathbf{1}'$. The space Q of such matrices forms a manifold in $\mathbb{R}^{pn}$ of dimension $p(n-1) - p(p+1)/2$. It is an example of a Stiefel manifold. Our calculations here have determined that this Stiefel manifold has volume $A_{n-1}^{(p)}$ as calculated orthogonal to the orbits GD at each point D on Q. In Sec. 8-3-5 we will obtain the actual volume of the Stiefel manifold.

The marginal distribution (8-100) gives us the distribution component for the model $\mathscr{M}_D$. Thus, as in Sec. 8-2-3, we obtain the component inference base (8-59):

$$(\mathscr{M}_D, D^0)$$

where $\mathscr{M}_D$ has the single distribution (8-100). The model $\mathscr{M}_D$, however, has no parameter, and records a fixed probability distribution; the data value D^0 is an observed value for that distribution. Accordingly, the inference base collapses by necessary method RM_2 in Sec. 3-2.

The conditional response distribution for Y given $D(Y) = D(Y^0) = D^0$ is available from (8-66). For this we first simplify further the exponent from (8-92):

$$\begin{aligned}
\operatorname{tr} Z'Z &= \operatorname{tr}(Y - \mu\mathbf{1}')'\Sigma^{-1}(Y - \mu\mathbf{1}') \\
&= \operatorname{tr}[(\bar{y} - \mu)\mathbf{1}' + T(Y)D]'\Sigma^{-1}[(\bar{y} - \mu)\mathbf{1}' + T(Y)D] \\
&= \operatorname{tr} n(\bar{y} - \mu)'\Sigma^{-1}(\bar{y} - \mu) + \operatorname{tr} T(Y)T'(Y)\Sigma^{-1} \\
&= \operatorname{tr} n(\bar{y} - \mu)'\Sigma^{-1}(\bar{y} - \mu) + \operatorname{tr} S(Y)\Sigma^{-1}
\end{aligned} \tag{8-101}$$

where

$$\begin{aligned}
S(Y) &= T(Y)T'(Y) = T(Y)D(Y)D'(Y)T'(Y) \\
&= (Y - \bar{y}\mathbf{1}')(Y - \bar{y}\mathbf{1}')'
\end{aligned} \tag{8-102}$$

is the inner product matrix for the row vector residuals relative to the one vector; we have, of course, used the property $\operatorname{tr}(AB) = \operatorname{tr}(BA)$. The response density function in terms of $[Y] = [\bar{y}, T(Y)]$ is

$$\frac{A_{n-1}^{(p)}}{(2\pi)^{np/2}} \exp\left[-\frac{n}{2}(\bar{y} - \mu)'\Sigma^{-1}(\bar{y} - \mu)\right] \operatorname{etr}\left[-\tfrac{1}{2}S(Y)\Sigma^{-1}\right] \left[\frac{s_{(1)}(Y) \ldots s_{(p)}(Y)}{\sigma_{(1)} \ldots \sigma_{(p)}}\right]^n$$

$$\times \frac{d\sqrt{n}\,\bar{y}\,dT(Y)}{s^2_{(1)}(Y)\cdots s^{p+1}_{(p)}(Y)}$$

$$= \frac{|\Sigma|^{-1/2}}{(2\pi)^{p/2}} \exp\left[-\frac{n}{2}(\bar{y}-\mu)'\Sigma^{-1}(\bar{y}-\mu)\right] d\sqrt{n}\,\bar{y}\, \frac{A^{(p)}_{n-1}}{(2\pi)^{(n-1)p/2}}\, \text{etr}\left[-\tfrac{1}{2}S(Y)\Sigma^{-1}\right]$$

$$\times \frac{|S(Y)|^{(n-1)/2}}{|\Sigma|^{(n-1)/2}} \frac{dT(Y)}{|T(Y)|_\Delta} \tag{8-103}$$

Thus $\bar{y}$ is $N(\mu; n^{-1}\Sigma)$ and independently $T(Y)$ is a scaled triangular chi distribution designated $\Delta\chi_p(n-1:\Upsilon)$, where Υ is the PLT square root of Σ; for this note the relation

$$\text{tr}\, S(Y)\Sigma^{-1} = \text{tr}\, \Upsilon^{-1}T(Y)\,T'(Y)\Upsilon'^{-1} \tag{8-104}$$

8-3-3 The Scale Component

Consider the scale parameter Υ. As noted in Sec. 8-2-4 this indexes left cosets on the parameter space G and we have

$$[\mu, \Upsilon] = [0, \Upsilon][\delta, I] \tag{8-105}$$

where δ is the coefficient of variation. The presentation equation $[Y] = \theta[Z]$ has the scale component

$$T(Y) = \Upsilon T$$

or

$$\Upsilon^{-1}T(Y) = T \tag{8-106}$$

The marginal distribution for T is available from (8-78) and trivially from (8-95) and (8-98) by noting the independence; we have the probability differential

$$\frac{A^{(p)}_{n-1}}{(2\pi)^{(n-1)p/2}}\, \text{etr}\,\{-\tfrac{1}{2}TT'\}\,|T|^{n-1}\,\frac{dT}{|T|_\Delta} \tag{8-107}$$

This is the triangular chi distribution $\Delta\chi_p(n-1)$ of formula (8-99). This distribution with equation (8-106) gives tests and confidence regions for the parameter Υ.

The analysis of scale for the multivariate normal is usually in terms of the covariance matrix $\Sigma = \Upsilon\Upsilon'$ and the inner product matrix $S(Y) = T(Y)T'(Y)$. We have one-one correspondences:

$$\Upsilon \leftrightarrow \Sigma = \Upsilon\Upsilon'$$
$$T(Y) \leftrightarrow S(Y) = T(Y)T'(Y) \tag{8-108}$$
$$T \leftrightarrow S = TT'$$

where a left element is the PLT square root of the corresponding right element. We have distributions for T and $T(Y)$; we now obtain the corresponding distributions for S and $S(Y)$.

The transformation from T to S given by $S = TT'$ has a triangular Jacobian matrix provided we take coordinates of T row by row from left to right within rows.

Component transformation	Conditional Jacobian	
$s_{11} = s_{(1)}^2$	$2s_{(1)}$	
$s_{21} = t_{21}s_{(1)}$	$s_{(1)}$	
$s_{22} = t_{21}^2 + s_{(2)}^2$	$2s_{(2)}$	(8-109)
$s_{31} = t_{31}s_{(1)}$	$s_{(1)}$	
$s_{32} = t_{31}t_{21} + t_{32}s_{(2)}$	$s_{(2)}$	
$s_{33} = t_{31}^2 + t_{32}^2 + s_{(3)}^2$	$2s_{(3)}$	
...		

Accordingly, we obtain

$$\left|\frac{\partial S}{\partial T}\right| = 2s_{(1)}(s_{(1)}2s_{(2)})\cdots = 2^p |T|_\nabla$$
$$= 2^p |S|_\nabla^{1/2} \tag{8-110}$$

The alternative form uses the notation

$$|S|_\nabla = |s_{11}|\begin{vmatrix} s_{11} & s_{12} \\ s_{21} & s_{22} \end{vmatrix} \begin{vmatrix} s_{11} & s_{12} & s_{13} \\ s_{21} & s_{22} & s_{23} \\ s_{31} & s_{32} & s_{33} \end{vmatrix} \cdots$$
$$= |T|_\nabla^2 \tag{8-111}$$

which will be useful later. Then by substitution in (8-107) we obtain

$$\frac{A_{n-1}^{(p)}}{(2\pi)^{(n-1)p/2}} \operatorname{etr}(-\tfrac{1}{2}S)|S|^{(n-1)/2} \frac{dS}{|T|_\Delta \, 2^p |T|_\nabla}$$
$$= \frac{A_{n-1}^{(p)}}{(2\pi)^{(n-1)p/2}} \operatorname{etr}(-\tfrac{1}{2}S)|S|^{[n-1-(p+1)]/2} \, 2^{-p} \, dS \tag{8-112}$$

This is the standard p-variate Wishart distribution with $(n-1)$ degrees of freedom, $W_p(n-1)$.

The distribution of $S(Y)$ is obtained in the same manner from the distribution of $T(Y)$ implicit in (8-103):

$$\frac{A_{n-1}^{(p)}}{(2\pi)^{(n-1)p/2}} \operatorname{etr}[-\tfrac{1}{2}S(Y)\Sigma^{-1}] \frac{|S(Y)|^{(n-1)/2}}{|\Sigma|^{(n-1)/2} \, 2^p |S(Y)|^{(p+1)/2}} dS(Y) \tag{8-113}$$

This is the general $W_p(n-1, \Sigma)$ distribution, the p-variate Wishart with $(n-1)$ degrees of freedom and variance matrix Σ.

8-3-4 The Location Component

Now consider the location component μ. We examined the scale component in the preceding section so as to have available some integration results for the present section. The parameter μ indexes left cosets and we have
$$[\mu, \Upsilon] = [\mu, I][0, \Upsilon]$$
The presentation $[Y] = \theta[Z]$ has the location component
$$T^{-1}(Y)(\bar{y} - \mu) = T^{-1}\bar{z} = \mathbf{t} \qquad (8\text{-}114)$$
We now derive the marginal distribution for $\mathbf{t} = T^{-1}\bar{z}$ using (8-72), together with the normal case distribution recorded in (8-98); we record this in terms of the density for $\sqrt{n}\,\mathbf{t}$. Thus

$$\int_{G_S} \frac{A_{n-1}^{(p)}}{(2\pi)^{np/2}}\,\text{etr}\left[-\tfrac{1}{2}T(n\mathbf{tt}' + I)T'\right]|T|^n \frac{dT}{|T|_\Delta}$$

$$= \int_{G_S} \frac{A_{n-1}^{(p)}}{(2\pi)^{np/2}}\,\text{etr}\left(-\tfrac{1}{2}TEE'T'\right)|T|^n \frac{dT}{|T|_\Delta} \qquad (8\text{-}115)$$

where $E = (n\mathbf{tt}' + I)^L$ and we have written M^L for the PLT square root of an inner product matrix M.

For the integration in (8-115) we use the seemingly natural variable TE together with the relation
$$dTE = |E|_V\,dT$$
implicitly available from (8-83) with (7-35); we also use $|TE|_\Delta = |T|_\Delta |E|_\Delta$. Thus

$$\frac{A_{n-1}^{(p)}}{(2\pi)^{np/2}} \int_{G_S} \text{etr}\left(-\tfrac{1}{2}TEE'T'\right)|TE|^n \frac{dTE}{|TE|_\Delta} \frac{|E|_\Delta}{|E|^n|E|_V}$$

$$= \frac{A_{n-1}^{(p)}}{(2\pi)^{np/2}} \frac{(2\pi)^{np/2}}{A_n^{(p)}} \frac{|E|_\Delta}{|E|^n|E|_V} \qquad (8\text{-}116)$$

where we have now used the integration properties available from (8-107).

The density for $\sqrt{n}\,\mathbf{t}$ in (8-116) can be simplified using the notation in (8-111):

$$\frac{A_{n-1}^{(p)}}{A_n^{(p)}} \frac{1}{|E|^n} \frac{|E|_\Delta |E|_V}{|E|_V^2} = \frac{A_{n-p}}{A_n} \frac{1}{|I + n\mathbf{tt}'|^{n/2}} \frac{|I + n\mathbf{tt}'|^{(p+1)/2}}{|I + n\mathbf{tt}'|_V} \qquad (8\text{-}117)$$

$$= \frac{A_{n-p}}{A_n} \frac{1}{|I + n\mathbf{tt}'|^{[(n-1)+1]/2}} \frac{|I + n\mathbf{tt}'|^{(p+1)/2}}{|I + n\mathbf{tt}'|_V}$$

The distribution of $\sqrt{n}\,\mathbf{t}$ is called a *disguised Student* or *triangular Student* distribution; in the final expression it is written as a special case of a more general matrix distribution and is called the *triangular Student* $\Delta t_{p\times 1}(n-1)$ with $(n-1)$ *degrees of freedom*.

This distribution with equation (8-114) gives tests and confidence regions for the location parameter μ.

8-3-5 Supplement: Volume on the Cross Section

The reference points D lie on a manifold Q of semiorthogonal matrices. From formula (8-100) we saw that the "surface volume" of Q was $A^{(p)}_{n-1}$: the matrices D were $p \times n$ semiorthogonal matrices of p row vectors orthogonal to the $\mathbf{1}'$ vector and the volume was calculated at each point D *orthogonal* to the orbit GD at the point D. Now consider the volume of Q using the tangential coordinates on Q as initially suggested in Sec. 7-2-4.

For this we have the expression

$$TD \tag{8-118}$$

where T belongs to the scale group G_S of $p \times p$ PLT matrices and D is a semiorthogonal matrix of p row vectors in $\mathscr{L}^\perp(\mathbf{1})$ of $\mathbb{R}^n$. The matrix TD consists of p row vectors in $\mathscr{L}^\perp(\mathbf{1})$ of $\mathbb{R}^n$. We are considering measures and volumes based on Euclidean distance. Accordingly, we can make an orthogonal transformation of $\mathbb{R}^n$ to isolate $\mathscr{L}^\perp(\mathbf{1})$ and reconsider (8-118) with D now an arbitrary semiorthogonal matrix of row vectors in $\mathbb{R}^{n-1}$.

We consider the relationship between dD calculated orthogonal to GD at D and dD calculated tangential to the manifold Q. For this we again make an orthogonal transformation of $\mathbb{R}^{n-1}$ and represent the point D as

$$D = \begin{bmatrix} 1 & 0 & & & \cdots & & 0 \\ 0 & 1 & \ddots & & & & \vdots \\ \vdots & & \ddots & \ddots & 0 & \cdots & \\ 0 & \cdots & & 0 & 1 & 0 & \cdots & 0 \end{bmatrix} \tag{8-119}$$

The change from D to an adjacent point of Q (actually on the tangent plane) has the following representation:

$$\delta D = \begin{bmatrix} 0 & \delta_{12} & & \cdots & & & \delta_{1n-1} \\ -\delta_{12} & 0 & \ddots & & & & \vdots \\ \vdots & & \ddots & & \delta_{p-1,p} & & \\ -\delta_{1p} & & -\delta_{p-1,p} & 0 & \delta_{p,p+1} & \delta_{p,n-1} \end{bmatrix} \tag{8-120}$$

for clearly to a first derivative approximation $D + \delta D$ is semiorthogonal. In a similar way a first derivative change in T from the identity I has representation

$$\Delta T \cdot D = \begin{bmatrix} \Delta_{11} & 0 & & \cdots & & 0 \\ \vdots & \ddots & \ddots & & & \vdots \\ \Delta_{p1} & \cdots & \Delta_{pp} & 0 & \cdots & 0 \end{bmatrix} \tag{8-121}$$

If we use the coordinates of δD *above the diagonal*, then we have orthogonality of (8-120) and (8-121). Thus, locally, the above-diagonal coordinates in (8-120) represent Q projected in the orthogonal complement of GD at D, and the full array of coordinates represents the tangent plane to the manifold. How do these two representations differ?

Consider a special case. Suppose we record x for length on $\mathbb{R}^1$ and record (x, x) on $\mathbb{R}^2$. Unit length for the original is represented as length $\sqrt{2}$ on the diagonal line.

In the more general case we have $p(p-1)/2$ coordinates that are represented (locally) in this double manner. Accordingly, the volume of Q tangentially is $2^{p(p-1)/4}$ times the volume orthogonally.

It follows that the volume of the particular Stiefel manifold in Sec. 8-3-2 is $A_{n-1}^{(p)} 2^{p(p-1)/4}$. For further discussion see Bishop (1977).

8-4 MULTIVARIATE MODEL: LINEAR

Consider a random system with a p-variate response $\mathbf{y} = (y_1, \ldots, y_p)'$. For the linear model in this section we suppose that the location of the response vector is unknown and that the scale of the response in the form of a positive linear transformation is unknown but that otherwise the distribution form is known or known up to a shape parameter λ.

8-4-1 The Model

We use the modified notation introduced in Sec. 8-2. For n performances of the system we have $\mathbf{y}_1, \ldots, \mathbf{y}_n$ where each is the appropriate p-variate response vector; thus

$$Y = (\mathbf{y}_1, \ldots, \mathbf{y}_n) = \begin{bmatrix} Y_1 \\ \vdots \\ Y_p \end{bmatrix} = \begin{bmatrix} y_{11} & \cdots & y_{1n} \\ \cdots & \cdots & \cdots \\ y_{p1} & \cdots & y_{pn} \end{bmatrix}$$

and Y_j is the row vector recording the n performances for the jth response.

For the response $\mathbf{y}$ we suppose the location $\boldsymbol{\mu}$ is unknown; that the scale Γ is an unknown $p \times p$ transformation matrix with positive determinant; but that the distribution form is known or known up to a parameter λ. Accordingly, we have

$$\mathbf{y} = \boldsymbol{\mu} + \Gamma \mathbf{z} \tag{8-122}$$

or

$$\begin{bmatrix} y_1 \\ \vdots \\ y_p \end{bmatrix} = \begin{bmatrix} \mu_1 \\ \vdots \\ \mu_p \end{bmatrix} + \begin{bmatrix} \gamma_{11} & \cdots & \gamma_{1p} \\ \cdots & \cdots & \cdots \\ \gamma_{p1} & \cdots & \gamma_{pp} \end{bmatrix} \begin{bmatrix} z_1 \\ \vdots \\ z_p \end{bmatrix} \tag{8-123}$$

where $|\Gamma| > 0$ and $\boldsymbol{\mu} \in \mathbb{R}^p$. Also we let

$$f_\lambda(\mathbf{z}) = f_\lambda(z_1, \ldots, z_p) \tag{8-124}$$

be the density function for the objective variation. We suppose that f_λ has been suitably standardized using, say, (1-18) on each axis and some reasonable standardization for the linear transformation.

For n independent performances of the system we then have $\mathbf{y}_i = \boldsymbol{\mu} + \Gamma \mathbf{z}_i$

192 INFERENCE AND LINEAR MODELS

for the ith performance and

$$Y = \mu 1' + \Gamma Z = [\mu, \Gamma] Z = \theta Z \qquad (8\text{-}125)$$

for the compound response Y and compound variation Z. Note that $[\mu, \Gamma]$ operates on Z column by column as in (8-122). The distribution for the compound variation is

$$f_\lambda(Z) = \prod_1^n f_\lambda(z_i) \qquad (8\text{-}126)$$

The transformations $[\mu, \Gamma]$ form a group:

$$[\mathbf{a}_2, C_2][\mathbf{a}_1, C_1] = [\mathbf{a}_2 + C_2 \mathbf{a}_1, C_2 C_1]$$
$$[\mathbf{a}, C]^{-1} = [-C^{-1}\mathbf{a}, C^{-1}] \qquad (8\text{-}127)$$
$$[\mathbf{0}, I] = i$$

Note, of course, that the $p \times p$ positive matrices are closed under multiplication and inverse. We thus have the *positive affine* group

$$G = \{[\mathbf{a}, C] : \mathbf{a} \in \mathbb{R}^p, |C| > 0\} \qquad (8\text{-}128)$$

We will see that G is exact on the sample space provided $n \geq p + 1$ and a certain set of measure zero is excluded.

We thus obtain the following structural model:

$$\mathcal{M}_V = (\Omega; \mathbb{R}^{pn}, \mathcal{B}^{pn}, \mathcal{V}, G) \qquad (8\text{-}129)$$

where the parameter space is $\Omega = (G \times \Lambda)$, $\mathcal{V}$ is the class of densities f_λ in (8-126), and G is the positive affine group (8-128) with action (8-125). We abbreviate this as

$$\begin{aligned} & Y = \theta Z \text{ with } \theta \text{ in } G \\ & Z \text{ has distribution in the class } \mathcal{V} \end{aligned} \qquad (8\text{-}130)$$

and assume that $n \geq p + 1$.

8-4-2 The Analysis

For the analysis we use the notation from Chaps. 2 and 6 together with that from Sec. 8-2.

Consider a transformation $[\mathbf{a}, C]$ in the group G:

$$\begin{bmatrix} \tilde{Z}_1 \\ \vdots \\ \tilde{Z}_p \end{bmatrix} = \begin{bmatrix} a_1 \\ \vdots \\ a_p \end{bmatrix} 1' + \begin{bmatrix} c_{11} & \cdots & c_{1p} \\ \vdots & & \vdots \\ c_{p1} & \cdots & c_{pp} \end{bmatrix} \begin{bmatrix} Z_1 \\ \vdots \\ Z_p \end{bmatrix} \qquad (8\text{-}131)$$

We find it convenient to view Z as a sequence $Z_1, \ldots, Z_p$ of p points in $\mathbb{R}^n$ and to observe the effect of the transformation on this sequence of p points.

Let $\mathscr{L}^+(1', Z_1, \ldots, Z_p)$ be the $(p+1)$-dimensional subspace of $\mathbb{R}^n$:

$$\mathscr{L}^+(1', Z_1, \ldots, Z_p) = \{c_0 1' + c_1 Z_1 + \cdots + c_p Z_p : c_j \in \mathbb{R}\} \qquad (8\text{-}132)$$

together with the orientation of the sequence $\mathbf{1}', Z_1, \ldots, Z_p$ treated as the *positive* orientation for $p+1$ vectors in (8-132). A group element $g = [\mathbf{a}, C]$ carries the sequence $Z_1, \ldots, Z_p$ into the sequence $\tilde{Z}_1, \ldots, \tilde{Z}_p$ also in $\mathscr{L}^+(\mathbf{1}', Z_1, \ldots, Z_p)$, with the same positive orientation. In fact, from (8-131) we see that we can get any sequence in $\mathscr{L}^+(\mathbf{1}, Z_1, \ldots, Z_p)$ provided it has the positive orientation.

We wish to choose a basis for the space $\mathscr{L}^+(\mathbf{1}', Z_1, \ldots, Z_p)$. Let $\mathbf{1}'$ be one of the basis vectors and let

$$D(Z) = \begin{bmatrix} D_1(Z) \\ \vdots \\ D_p(Z) \end{bmatrix} \tag{8-133}$$

consist of p orthonormal vectors, orthogonal to $\mathbf{1}'$, and with the positive orientation. We use $\mathbf{1}'$ with $D(Z)$ as a basis for the space. The choice of basis must not depend on $Z_1, \ldots, Z_p$ directly, only on the *space* $\mathscr{L}^+(\mathbf{1}, Z_1, \ldots, Z_p)$. Such a basis can be formed, for example, in the following way: take a sequence of p linearly independent vectors, say the first p coordinate vectors; project them into the subspace; and orthonormalize them in sequence as done to obtain $D_1, \ldots, D_p$ in Sec. 8-2-3. Such a procedure gives a basis except, of course, for a set of measure zero for which the projections have linear dependence.

Now let $\bar{z} = (\bar{z}_1, \ldots, \bar{z}_p)'$, $C(Z)$ together record the regression coefficients for $Z_1, \ldots, Z_p$ on $\mathbf{1}', D_1(Z), \ldots, D_p(Z)$:

$$\begin{aligned} Z &= \bar{z}\mathbf{1}' + C(Z)D(Z) \\ &= [\bar{z}, C(Z)]D(Z) \end{aligned} \tag{8-134}$$

where $[Z] = [\bar{z}, C(Z)]$ is an element of the group G that gives the position of Z relative to $D(Z)$ as the reference point.

The Jacobians from Secs. 7-2-2 and 7-2-3 are readily available with less complication than in Sec. 8-2-3. A transformation on the sample space operates column by column. Thus, as with (8-51), we have

$$\begin{aligned} J_{pn}(Z) &= |C(Z)|^n \\ dM(Z) &= \frac{dZ}{|C(Z)|^n} \end{aligned} \tag{8-135}$$

On the group we have

$$\begin{aligned} [\tilde{\mathbf{a}}, \tilde{C}] &= [\mathbf{a}, C][\mathbf{a}^*, C^*] \\ &= [\mathbf{a} + C\mathbf{a}^*, CC^*] \\ \tilde{\mathbf{a}} &= \mathbf{a} + C\mathbf{a}^* \\ \tilde{C} &= CC^* \end{aligned} \tag{8-136}$$

The left transformation operates column by column on $\mathbf{a}^*, C^*$. Accordingly,

$$\begin{aligned} J(g) &= |C|^{p+1} \\ d\mu(g) &= \frac{d\mathbf{a}\, dC}{|C|^{p+1}} \end{aligned} \tag{8-137}$$

The right transformation operates row by row on C and for given C operates only by location on $\mathbf{a}$. Accordingly,

$$J^*(g) = |C|^p \qquad (8\text{-}138)$$

$$dv(g) = \frac{d\mathbf{a}\, dC}{|C|^p}$$

and

$$\Delta(g) = |C|^{-1} \qquad (8\text{-}139)$$

Consider the transformation $[\bar{\mathbf{z}}, C(Z)]$ and its application giving $Z = [\bar{\mathbf{z}}, C(Z)]D(Z)$. Rows of Z come from rows of $[\bar{\mathbf{z}}, C(Z)]$; for any row the corresponding basis vectors $\mathbf{1}', D_1(Z), \ldots, D_p(Z)$ are orthogonal and of length 1, except for the one-vector which has length $\sqrt{n}$; accordingly, we have

$$J(D) = n^{p/2} \qquad (8\text{-}140)$$

This uses the alternate interpretation for dD: Euclidean volume orthogonal to the orbit GD at D (recall Sec. 7-2-6).

Now consider the analysis of an inference base $(\mathcal{M}_Y, Y^0)$; we follow the pattern in Secs. 7-1, 7-2, and 8-2.

The observed orbit gives the observed value for the variation

$$D(Z) = D(Y^0) = D^0 \qquad (8\text{-}141)$$

This gives the inference base (a) from Sec. 7-1-4:

$$(\mathcal{M}_D, D^0) \qquad (8\text{-}142)$$

The distributions in $\mathcal{M}_D$ are available from (7-46):

$$h_\lambda(D) = \int_G f_\lambda(\bar{z}\mathbf{1}' + CD)\,|C|^{n-p-1}\, n^{p/2}\, d\bar{\mathbf{z}}\, dC \qquad (8\text{-}143)$$

At a minimum this gives us the observed likelihood for λ:

$$L(D^0; \lambda) = ch_\lambda(D^0) \qquad (8\text{-}144)$$

The unobserved characteristics of the variation lead to the inference base (b) from Sec. 7-1-4:

$$(\mathcal{M}_V^{D^0}, [Y^0]) \qquad (8\text{-}145)$$

The distributions for the structural model $\mathcal{M}_V^{D^0}$ are available from (7-47):

$$h_\lambda^{-1}(D^0) f_\lambda(\bar{z}\mathbf{1}' + CD^0)\,|C|^{n-p-1}\, n^{p/2}\, d\bar{\mathbf{z}}\, dC \qquad (8\text{-}146)$$

This distribution for the unobserved variation is used with the transformation

$$\bar{y} = \mu + \Gamma\bar{z}$$
$$C(Y) = \Gamma C \qquad (8\text{-}147)$$

together with the observed response position

$$[Y^0] = [\bar{z}^0, C(Y^0)] \qquad (8\text{-}148)$$

The response distribution corresponding to the identified $D(Y) = D(Y^0) = D^0$ is available from (7-49):

$$h_\lambda^{-1}(D^0) f_\lambda(\Gamma^{-1}[(\bar{y} - \mu)\mathbf{1}' + C(Y)D^0]) \frac{|C(Y)|^n}{|\Gamma|^n} \frac{n^{p/2} \, d\bar{y} \, dC(Y)}{|C(Y)|^{p+1}} \qquad (8\text{-}149)$$

8-4-3 Component Parameters

We have noted in Secs. 7-3-3 and 7-3-5 that some parameters index left cosets on the parameter space G and accordingly give unequivocal tests and confidence regions expressible directly in terms of the variation. As in Sec. 8-2-4 we derive the appropriate distributions for use with the location and scale parameters; the tests and confidence regions are then available from the formulas in Sec. 7-3.

For the location parameter μ we have the separation of $Y = \theta Z$ giving the μ-specific component

$$C^{-1}(Y)(\bar{y} - \mu) = C^{-1}\bar{z} = \mathbf{t} \qquad (8\text{-}150)$$

The full parameter can be separated as

$$[\mu, \Gamma] = [\mu, I][0, \Gamma] \qquad (8\text{-}151)$$

showing the left coset form

$$[\mu, \Gamma]G_S = [\mu, I]G_S \qquad (8\text{-}152)$$

where

$$G_S = \{[0, C] : |C| > 0\} \qquad (8\text{-}153)$$

is the scale group. The group coordinates can be separated in the reverse order

$$[\bar{z}, C] = [0, C][\mathbf{t}, I] \qquad (8\text{-}154)$$

where $\mathbf{t}$ indexes the orbits

$$G_S[\bar{z}, C] = G_S[\mathbf{t}, I] \qquad (8\text{-}155)$$

of the scale group.

The marginal distribution of $\mathbf{t}$ is easily obtained from (8-146) using $d\bar{z} = |C| d\mathbf{t}$ for fixed C; the probability differential is

$$h_\lambda^{-1}(D^0) \int_{G_S} f_\lambda(C(\mathbf{t}\mathbf{1}' + D^0)) |C|^{n-p} \, dC \, n^{p/2} \, d\mathbf{t} \qquad (8\text{-}156)$$

This distribution with equation (8-150) provides for tests and confidence regions for the location parameter μ.

For the scale parameter Γ we have the separation of $Y = \theta Z$ giving the Γ-specific component

$$\Gamma^{-1}C(Y) = C \qquad (8\text{-}157)$$

The full parameter can be separated as

$$[\mu, \Gamma] = [0, \Gamma][\delta, I] \qquad (8\text{-}158)$$

where $\delta = \Gamma^{-1}\mu$ is the coefficient of variation, and we have the left coset form

$$[\mu, \Gamma]G_L = [0, \Gamma]G_L \tag{8-159}$$

where

$$G_L = \{[\mathbf{a}, I] : \mathbf{a} \in \mathbb{R}^p\} \tag{8-160}$$

is the location group. The group coordinates can be separated in the reverse order

$$[\bar{z}, C] = [\bar{z}, I][0, C] \tag{8-161}$$

where C indexes the orbits

$$G_L[\bar{z}, C] = G_L[0, C] \tag{8-162}$$

of the scale group.

The marginal distribution for C is easily obtained by directly integrating $\bar{z}$ from the distribution (8-146); the probability differential is

$$h_\lambda^{-1}(D^0) \int_{G_L} f_\lambda(\bar{z}\mathbf{1}' + CD^0)\, n^{p/2}\, d\bar{z}\, |C|^{n-p-1}\, dC \tag{8-163}$$

This distribution with the equation (8-157) provides for tests and confidence regions for Γ.

The location subcomponents such as $\mu_1, (\mu_1, \mu_2)$ are not amenable to the strong procedures that are directly variation-based: some analysis shows they do not have left coset form. In a similar way specific component entries in the matrix Γ do not seem to be amenable to these variation-based methods.

8-4-4 Supplement

As noted in Sec. 8-2-5 confidence distributions are not part of the theme of this book. However, they are certainly of mathematical interest in relation to what we have done, and we record several for this reason. We also record some useful relations for measure differentials.

For the full parameter θ we obtain the following confidence distribution from formula (7-105) with (8-146):

$$h_\lambda^{-1}(D^0) f_\lambda(\Gamma^{-1}(Y^0 - \mu\mathbf{1}')) \frac{|C(Y^0)|^n}{|\Gamma|^n} \frac{1}{|C(Y^0)|} \frac{n^{p/2}\, d\mu\, d\Gamma}{|\Gamma|^p} \tag{8-164}$$

For the location and scale parameters we can proceed as in Sec. 7-4-5, or, as noted in Sec. 8-2-5, equivalently, we can integrate directly on the parameter space. Accordingly for μ we obtain

$$h_\lambda^{-1}(D^0) \int_{G_S} f_\lambda(\Gamma^{-1}(Y^0 - \mu\mathbf{1}')) \frac{d\Gamma}{|\Gamma|^{n+p}} |C(Y^0)|^{n-1}\, n^{p/2}\, d\mu \tag{8-165}$$

and for Γ we obtain

$$h_\lambda^{-1}(D^0) \int_{G_L} f_\lambda(\Gamma^{-1}(Y^0 - \mu\mathbf{1}'))\, n^{p/2}\, d\mu \frac{|C(Y^0)|^{n-1}}{|\Gamma|^n} \frac{d\Gamma}{|\Gamma|^p} \tag{8-166}$$

For completeness we record the left and right invariant measures for the location and scale groups; these can be used for the Sec. 7-4-5 analysis and will be of use later. For the location group we have

$$d\mu_L([\mathbf{a}, I]) = d\mathbf{a} \qquad dv_L([\mathbf{a}, I]) = d\mathbf{a}$$
$$\Delta_L([\mathbf{a}, I]) = 1 \tag{8-167}$$

And for the scale group we have

$$d\mu_S([\mathbf{0}, C]) = \frac{dC}{|C|^p} \qquad dv_S([\mathbf{0}, C]) = \frac{dC}{|C|^p} \tag{8-168}$$

$$\Delta_S([\mathbf{0}, C]) = 1$$

8-5 MULTIVARIATE MODEL: NORMAL LINEAR

Now consider the multivariate linear model but with a normal pattern for the variation.

8-5-1 The Normal Model

Consider a random system with a p-variate response $\mathbf{y}$. From Sec. 8-4 the linear model $\mathcal{M}_V$ for n performances has the form

$$Y = \mu \mathbf{1}' + \Gamma Z \tag{8-169}$$

where Y and Z are $p \times n$, μ is the response location, and Γ is a positive $p \times p$ scale matrix.

In this section we examine the model using the standard normal distribution for the variation:

$$f(Z) = (2\pi)^{-pn/2} \operatorname{etr}(-\tfrac{1}{2}Z'Z) \tag{8-170}$$

The corresponding response model for Y is available following the pattern of calculation in Sec. 8-3-1:

$$(2\pi)^{-np/2} |\Sigma|^{-n/2} \operatorname{etr}\left[-\tfrac{1}{2}(Y - \mu\mathbf{1}')'\Sigma^{-1}(Y - \mu\mathbf{1}')\right] dY \tag{8-171}$$

where $\Sigma = \Gamma\Gamma'$ is the variance matrix for Y.

Note that Γ is a square root of the inner product matrix Σ. There are, of course, many square roots for a given matrix Σ, including the PLT square root Υ of Sec. 8-2 and the obvious permutation variants on that PLT square root. This suggests that Γ is not fully identifiable from sample values—the arbitrariness or unidentifiability corresponds to the rotational symmetry of the standard normal. We return to this point later.

In summary we have the following structural model:

$$\mathcal{M}_V = (\Omega; \mathbb{R}^{pn}, \mathcal{B}^{pn}, \mathcal{V}, G) \tag{8-172}$$

where $\Omega = G$ is the parameter space, $\mathscr{V}$ has a single distribution (8-171), and G is the positive affine group. We abbreviate this as

$$Y = \theta Z \text{ with } \theta \text{ in } G$$

$$Z \text{ is standard normal on } \mathbb{R}^{pn} \tag{8-173}$$

8-5-2 The Analysis

We use the results from Sec. 8-4 together with some details from Sec. 8-3.
The change of variables to get coordinates on the orbit is given by

$$Z = \bar{z}\mathbf{1}' + C(Z)D(Z) = [\bar{z}, C(Z)]D(Z) \tag{8-174}$$

where the rows of $D(Z)$ form an orthonormal basis for the complement of $\mathbf{1}'$ in $\mathscr{L}^+(\mathbf{1}', Z_1, \ldots, Z_p)$. The exponent of the normal density involves the following quadratic expression:

$$\operatorname{tr} ZZ' = \operatorname{tr}(\bar{z}\mathbf{1}' + CD)(\bar{z}\mathbf{1}' + CD)'$$
$$= \operatorname{tr}(n\bar{z}\bar{z}') + \operatorname{tr}(CC') \tag{8-175}$$

For the integration on the group and the use of results from Sec. 8-2-3 we will find it convenient to express C as

$$C = TO \tag{8-176}$$

where T is positive lower triangular and O is positive orthogonal. The above factoring of C into positive lower triangular and semiorthogonal is precisely the factoring of $Z - \bar{z}\mathbf{1}'$ used for the group coordinates in Sec. 8-2-3. Note that the $p \times p$ size of O and the positive determinants for C and T then gives that O is a positive orthogonal matrix or a *rotation* matrix. This change of variable has the following effect on the quadratic expression:

$$\operatorname{tr} ZZ' = \operatorname{tr}(n\bar{z}\bar{z}') + \operatorname{tr}(TOO'T')$$
$$= \operatorname{tr}(n\bar{z}\bar{z}') + \operatorname{tr}(TT')$$
$$= \Sigma(\sqrt{n}\,\bar{z}_j)^2 + \sum_{j>j'} t_{jj'}^2 + \Sigma s_{(j)}^2 \tag{8-177}$$

To use the change of coordinates $C = TO$ to facilitate the integration we need the corresponding effect on the differentials. For this we use results from Sec. 7-4-3 together with the invariant measures from Secs. 8-2-5 and 8-4-4. We also need properties of the positive orthogonal matrices O; these matrices form a group

$$G_R = \{O : OO' = I, |O| = 1\} \tag{8-178}$$

Euclidean volume is invariant under orthogonal transformations; accordingly, we have

$$d\mu_R(O) = dO \qquad dv_R(O) = dO \tag{8-179}$$

where dO is Euclidean volume tangential to the group G_R as embedded in $\mathbb{R}^{p^2}$. Then from formula (7-127) together with (8-83) and (8-168) we obtain

$$d\mu(C) = \frac{dC}{|C|^p}$$

$$= \frac{dT}{|T|_\Delta} H \frac{1}{1} dO \qquad (8\text{-}180)$$

$$= \frac{dT}{|T|_\Delta} H dO$$

From Sec. 8-3-5 we can rewrite this as

$$\frac{dC}{|C|^p} = \frac{dT}{|T|_\Delta} dO \qquad (8\text{-}181)$$

provided we now interpret dO as volume orthogonal to the orbits of the PLT scale group. Otherwise we would need to write

$$\frac{dC}{|C|^p} = \frac{dT}{|T|_\Delta} 2^{-p(p-1)/4} dO \qquad (8\text{-}182)$$

where we use the original interpretation of dO as volume tangent to G_R in $\mathbb{R}^{p^2}$. For tidiness of interpretation we use (8-181) with the orthogonal interpretation for dO (recall Sec. 7-2-6).

We can now determine the conditional and marginal distributions. For $[Z] = [\bar{z}, C(Z)]$ given D we have

$$h^{-1}(D) f(\bar{z}\mathbf{1}' + CD) |C|^n n^{p/2} \frac{d\bar{z}}{|C|} \frac{dC}{|C|^p}$$

$$= h^{-1}(D)(2\pi)^{-np/2} \operatorname{etr}(-\tfrac{1}{2}n\bar{z}\bar{z}' - \tfrac{1}{2}TT')|T|^{n-1} n^{p/2} \, d\bar{z} \, \frac{dT}{|T|_\Delta} dO$$

$$= \frac{A_{n-1}^{(p)}}{(2\pi)^{np/2}} \exp\left[-\tfrac{1}{2}\Sigma(\sqrt{n}\,\bar{z}_j)^2 - \tfrac{1}{2}\sum_{j>j'} t_{jj'}^2 - \tfrac{1}{2}\Sigma s_{(j)}^2\right]$$

$$\times s_{(1)}^{n-2} \ldots s_{(p)}^{n-p-1} \Pi d\sqrt{n}\,\bar{z}_j \, \Pi \, ds_{(j)} \, \Pi \, dt_{jj'} \, \frac{dO}{A_p \ldots A_2} \qquad (8\text{-}183)$$

$$= \frac{1}{(2\pi)^{p/2}} \operatorname{etr}\left(-\frac{n}{2}\bar{z}\bar{z}'\right) d\sqrt{n}\,\bar{z} \, \frac{A_{n-1}^{(p)}}{(2\pi)^{(n-1)p/2}} \operatorname{etr}(-\tfrac{1}{2}TT')|T|^{n-1} \frac{dT}{|T|_\Delta} \frac{dO}{A_p^{(p-1)}}$$

$$= \frac{1}{(2\pi)^{p/2}} \operatorname{etr}\left(-\frac{n}{2}\bar{z}\bar{z}'\right) d\sqrt{n}\,\bar{z} \, \frac{A_{n-1}^{(p)}}{A_p^{(p-1)}(2\pi)^{(n-1)p/2}} \operatorname{etr}(-\tfrac{1}{2}CC')|C|^{n-1} \frac{dC}{|C|^p}$$

This uses (8-95) for the normal and gamma components and it uses (8-100) with Sec. 8-3-5 for the positive orthogonal component. Note the presence of $A_p^{(p-1)}$

rather than $A_p^{(p)}$, thus omitting $A_1 = 2$. This is due to the restriction to *positive* semiorthogonal matrices O: with $p - 1$ rows of O given there is only 1 rather than 2 possibilities for the last row, because of the *positive* determinant. Note for (8-183) the particular interpretation for dO mentioned after formula (8-181).

The conditional distribution (8-183) gives us the distribution component for the model $\mathcal{M}_V^{D^o}$. Thus as in Sec. 8-4-2 we obtain the component inference base (8-145):

$$(\mathcal{M}_V^{D^o}, [Y^o]) \tag{8-184}$$

where the model $\mathcal{M}_V^{D^o}$ has the distribution for $[Z]$ given by (8-183) and the transformation

$$\begin{aligned} \bar{y} &= \mu + \Gamma \bar{z} \\ C(Y) &= \quad \Gamma C \end{aligned} \tag{8-185}$$

relating the response $[Y] = [\bar{y}, C(Y)]$ to the variation $[Z] = [\bar{z}, C]$. The observed values for the response coordinates are

$$[Y^o] = [\bar{y}^o, C(Y^o)] \tag{8-186}$$

The marginal distribution for D can be read from (8-183):

$$h(D)\, dD = \frac{A_p^{(p-1)}}{A_{n-1}^{(p)}}\, dD \tag{8-187}$$

Note that D has a uniform distribution on Q relative to our particular choice of Euclidean volume orthogonal to the orbit GD at each D.

The marginal distribution (8-187) gives us the distribution component for the model $\mathcal{M}_D$. Thus, as in Sec. 8-4-2, we obtain the component inference base

$$(\mathcal{M}_D, D^o) \tag{8-188}$$

where $\mathcal{M}_D$ has the single distribution (8-187). This model, however, has no parameter; accordingly, the inference base collapses by necessary method RM_2 in Sec. 3-2.

The conditional response distribution for Y given $D(Y) = D(Y^o) = D^o$ is available from (8-149) together with (8-103), (8-181), and (8-183). For the exponent of the normal density we have

$$\begin{aligned} \operatorname{tr} Z'Z &= \operatorname{tr} \left[(\bar{y} - \mu)\mathbf{1}' + C(Y)D \right]' \Sigma^{-1} \left[(\bar{y} - \mu)\mathbf{1}' + C(Y)D \right] \\ &= \operatorname{tr} n(\bar{y} - \mu)' \Sigma^{-1} (\bar{y} - \mu) + \operatorname{tr} C(Y)C'(Y)\Sigma^{-1} \\ &= \operatorname{tr} n(\bar{y} - \mu)' \Sigma^{-1} (\bar{y} - \mu) + \operatorname{tr} S(Y)\Sigma^{-1} \end{aligned} \tag{8-189}$$

where

$$\begin{aligned} S(Y) &= C(Y)C'(Y) = C(Y)D(Y)D'(Y)C'(Y) \\ &= T(Y)T'(Y) \\ &= (Y - \bar{y}\mathbf{1}')(Y - \bar{y}\mathbf{1}')' \end{aligned} \tag{8-190}$$

is the inner product matrix for the row residuals of Y; recall (8-102). The response density function for $[Y] = [\bar{y}, C(Y)] = [\bar{y}, T(Y)O(Y)]$ given $D(Y) = D(Y^0) = D^0$ is

$$\frac{A_{n-1}^{(p)}}{A_p^{(p-1)}(2\pi)^{np/2}} \exp\left[-\frac{n}{2}(\bar{y}-\mu)'\Sigma^{-1}(\bar{y}-\mu)\right] \operatorname{etr}\left[-\tfrac{1}{2}S(Y)\Sigma^{-1}\right]$$

$$\times \frac{|C(Y)|^n}{|\Gamma|^n} \frac{d\sqrt{n}\,\bar{y}\,dC(Y)}{|C(Y)|^{p+1}}$$

$$= \frac{A_{n-1}^{(p)}}{A_p^{(p-1)}(2\pi)^{np/2}} \exp\left[-\frac{n}{2}(\bar{y}-\mu)'\Sigma^{-1}(\bar{y}-\mu)\right] \operatorname{etr}\left[-\tfrac{1}{2}S(Y)\Sigma^{-1}\right]$$

$$\times \frac{|S(Y)|^{n/2}}{|\Sigma|^{n/2}} \frac{d\sqrt{n}\,\bar{y}}{|T(Y)|} \frac{dT(Y)}{|T(Y)|_\Delta} dO(Y)$$

$$= \frac{|\Sigma|^{-1/2}}{(2\pi)^{p/2}} \exp\left[-\frac{n}{2}(\bar{y}-\mu)'\Sigma^{-1}(\bar{y}-\mu)\right] d\sqrt{n}\,\bar{y}$$

$$\times \frac{A_{n-1}^{(p)}}{(2\pi)^{(n-1)p/2}} \operatorname{etr}\left[-\tfrac{1}{2}S(Y)\Sigma^{-1}\right] \frac{|S(Y)|^{(n-1)/2}}{|\Sigma|^{(n-1)/2}} \frac{dT(Y)}{|T(Y)|_\Delta} \frac{dO(Y)}{A_p^{(p-1)}}$$

$$= \frac{|\Sigma|^{-1/2}}{(2\pi)^{p/2}} \exp\left[-\frac{n}{2}(\bar{y}-\mu)'\Sigma^{-1}(\bar{y}-\mu)\right] d\sqrt{n}\,\bar{y}$$

$$\times \frac{A_{n-1}^{(p)}}{A_p^{(p-1)}(2\pi)^{(n-1)p/2}} \operatorname{etr}\left[-\tfrac{1}{2}S(Y)\Sigma^{-1}\right] \frac{|S(Y)|^{(n-1)/2}}{|\Sigma|^{(n-1)/2}} \frac{dC(Y)}{|C(Y)|^p} \quad (8\text{-}191)$$

where we follow the pattern of calculation in formula (8-183). This distribution agrees with (8-103) for the progression model, but has the additional uniform distribution for the rotation matrix $O(Y)$.

8-5-3 The Scale Parameter

Consider the scale parameter Γ. As noted in Sec. 8-4-3 this parameter indexes left cosets on the parameter space G and we have

$$[\mu, \Gamma] = [0, \Gamma][\delta, I] \quad (8\text{-}192)$$

where δ is the coefficient of variation. Correspondingly the equation $[Y] = \theta[Z]$ has the scale component

$$\Gamma^{-1}C(Y) = C \quad (8\text{-}193)$$

The marginal distribution for C is available from (8-163) and trivially from the independence in (8-183). The probability differential is

$$\frac{A_{n-1}^{(p)}}{A_p^{(p-1)}(2\pi)^{(n-1)p/2}} \operatorname{etr}(-\tfrac{1}{2}CC') |C|^{n-1} \frac{dC}{|C|^p} \quad (8\text{-}194)$$

202 INFERENCE AND LINEAR MODELS

This distribution is called the matrix chi with $(n-1)$ degrees of freedom and designated $\chi_p(n-1)$.

As noted in Sec. 8-3-3 the analysis of the multivariate normal scale is usually in terms of the covariance matrix $\Sigma = \Gamma\Gamma'$ and the inner product matrix $S(Y) = C(Y)C'(Y)$; the corresponding inner product matrix for the variation is $S = CC'$. The factorizations

$$C = TO \qquad C(Y) = T(Y)O(Y) \qquad (8\text{-}195)$$

in (8-176) and preceding (8-191) show that the distributions of T in (8-183) and $T(Y)$ in (8-191) are exactly as for T in (8-98) and $T(Y)$ in (8-103). Accordingly, the distributions for $S = CC' = TOO'T' = TT'$ and $S(Y) = C(Y)C'(Y) = T(Y)O(Y)O'(Y)T'(Y) = T(Y)T'(Y)$ are the same as given for the normal progression model in (8-112) and (8-113) in Sec. 8-3.

From a superficial viewpoint we could think of using the distribution (8-194) with equation (8-193) for tests and confidence regions for Γ. Some complications are present, however.

The multivariate response distribution for Y depends just on μ and Σ. The scale parameter Γ is a square root of Σ but there are many square roots. Suppose we factor $\Gamma = \Upsilon\Omega$ into positive lower triangular Υ and positive orthogonal Ω. Then

$$\Sigma = \Gamma\Gamma' = \Upsilon\Omega\Omega'\Upsilon' = \Upsilon\Upsilon' \qquad (8\text{-}196)$$

We noted in formula (8-108) the one-one equivalence between Σ and Υ. It then follows that we can view Υ in $\Gamma = \Upsilon\Omega$ as an essential parameter equivalent to Σ.

Let us now examine the group structure of Υ within the present linear group. We have

$$[\mu, \Gamma] = [0, \Gamma][\delta, I]$$
$$= [0, \Upsilon][0, \Omega][\delta, I] \qquad (8\text{-}197)$$
$$= [0, \Upsilon][\eta, \Omega]$$

where $\eta = \Omega\delta = \Omega\Omega^{-1}\Upsilon^{-1}\mu = \Upsilon^{-1}\mu$.

For the variation-based inference methods we need the left coset property for the component parameters. Accordingly, consider

$$G_L^* = \{[\eta, \Omega] : \eta \in \mathbb{R}^p, \Omega \in G_R\} \qquad (8\text{-}198)$$

where the rotation group G_R was defined in (8-178); this is closed under products and inverses and is accordingly a group. Thus we have

$$[\mu, \Gamma]G_L^* = [0, \Upsilon]G_L^* \qquad (8\text{-}199)$$

The group coordinates for the sample space can be separated in the reverse order:

$$[\bar{z}, C] = [\bar{z}, I][0, O][0, T]$$
$$= [\bar{z}, O][0, T] \qquad (8\text{-}200)$$

Thus T indexes the orbits

$$G_L^*[\bar{z}, C] = G_L^*[0, T] \tag{8-201}$$

For this we have used the factorization $C = OT$ into positive orthogonal times positive lower triangular; note carefully that the present definitions for T and O are *different* from those used earlier and defined by (8-176).

We now determine the effect of the change of variable $C = OT$ for the differentials. From (7-127) we have

$$\frac{dC}{|C|^p} = dO\, H\, \frac{1}{|T|_v / |T|_\Delta\, |T|_\Delta} \frac{dT}{}$$

$$= dO\, \frac{dT}{|T|_v} \tag{8-202}$$

This uses formula (8-83) for the positive lower triangular group and takes $H = 1$ on the basis of the special interpretation of dO given after formula (8-181).

We now use the change of variable $C = OT$ with the final expression in formula (8-183):

$$\frac{1}{(2\pi)^{p/2}} \text{etr}\left(-\frac{n}{2}\bar{z}\bar{z}'\right) d\sqrt{n}\,\bar{z}\, \frac{A_{n-1}^{(p)}}{A_p^{(p-1)}(2\pi)^{(n-1)p/2}} \text{etr}\left(-\tfrac{1}{2}OTT'O'\right)|T|^{n-1}\, dO\, \frac{dT}{|T|_v}$$

$$= \frac{1}{(2\pi)^{p/2}} \text{etr}\left(-\frac{n}{2}\bar{z}\bar{z}'\right) d\sqrt{n}\,\bar{z}\, \frac{dO}{A_p^{(p-1)}}$$

$$\times \frac{A_{n-1}^{(p)}}{(2\pi)^{(n-1)p/2}} \text{etr}\left(-\tfrac{1}{2}TT'\right) s_1^{n-1-p} \cdots s_p^{n-2}\, dT \tag{8-203}$$

Thus the present T has the distribution described by

$$T = \begin{bmatrix} s_1 & & & 0 \\ t_{21} & & & \\ \vdots & \ddots & & \\ t_{p1} & \cdots & t_{p,p-1} & s_p \end{bmatrix} = \begin{bmatrix} \chi_{n-p} & & & 0 \\ z_{21} & & & \\ \vdots & \ddots & & \\ z_{p1} & \cdots & z_{p,p-1} & \chi_{n-1} \end{bmatrix}$$

where the z values here designate independent standard normals and the χ values designate independent chi values with degrees of freedom as subscribed. The marginal distribution for the present T is

$$\frac{A_{n-1}^{(p)}}{(2\pi)^{(n-1)p/2}} \text{etr}\left(-\tfrac{1}{2}TT'\right) \frac{|T|^{n-1}}{|T|_v}\, dT \tag{8-204}$$

The preceding distribution (8-204) together with equations (8-193), (8-197), and (8-200) provides tests and confidence regions for Υ and thus for Σ; this follows the pattern in Sec. 7-3-5. Equation (8-193) can be rewritten as

$$G_R \Omega^{-1} \Upsilon^{-1} C(Y) = G_R OT$$
$$G_R \Upsilon^{-1} C(Y) = G_R T \tag{8-205}$$

Thus the tests and confidence intervals are based on the right PLT component of $\Upsilon^{-1}C(Y)$ being equal to T [see formula (7-103)].

8-5-4 The Location Component

Now consider the location parameter μ. Again as in Sec. 8-3-4 we can use integration results from the scale analysis. The parameter μ indexes left cosets and we have

$$[\mu, \Gamma] = [\mu, I][0, \Gamma] \tag{8-206}$$

The presentation $[Y] = \theta[Z]$ has the location component

$$C^{-1}(Y)(\bar{y} - \mu) = C^{-1}\bar{z} = \mathbf{t} \tag{8-207}$$

We now derive the marginal distribution for $\mathbf{t} = C^{-1}\bar{z}$ using (8-156) together with the normal case distribution in (8-183). We record this in terms of the density for $\sqrt{n}\,\mathbf{t}$:

$$\int_{G_s} \frac{A_{n-1}^{(p)}}{A_p^{(p-1)}(2\pi)^{np/2}} \, \mathrm{etr}\left[-\tfrac{1}{2}C(n\mathbf{tt'} + I)C'\right] |C|^{n-p}\, dC$$

$$= \int_{G_s} \frac{A_n^{(p)}}{A_p^{(p-1)}(2\pi)^{np/2}} \, \mathrm{etr}\left[-\tfrac{1}{2} C(n\mathbf{tt'} + I)C'\right] |C|^{n-p} \left|n\mathbf{tt'} + I\right|^{n/2} dC$$

$$\times \frac{A_{n-1}^{(p)}}{A_n^{(p)}} \left|n\mathbf{tt'} + I\right|^{-n/2}$$

$$= \frac{A_{n-1}^{(p)}}{A_n^{(p)}} \frac{1}{\left|I + n\mathbf{tt'}\right|^{n/2}}$$

$$= \frac{A_{n-p}}{A_n} \frac{1}{\left|I + n\mathbf{tt'}\right|^{n/2}}$$

$$= \frac{A_{n-p}}{A_n} (1 + \Sigma n t_j^2)^{-n/2} \tag{8-208}$$

For this we have used $|I + AB| = |I + BA|$ where A is $r \times s$ and B is $s \times r$. Note that this distribution corresponds closely to (8-117) but without a final "disguising" factor; it is called a *matrix Student* $t_{p \times 1}(n-1)$ *distribution on* $(n-1)$ *degrees of freedom*.

This distribution (8-208) with equation (8-207) gives tests and confidence regions for the location parameter μ.

For some background material on the models and analysis in this chapter see Fraser (1968, 1973).

REFERENCES AND BIBLIOGRAPHY

Bishop, L.: "The Jacobian of a Matrix Transformation," Department of Statistics, University of Wisconsin, 1977.

Deemer, W. L., and L. Olkin: The Jacobians of Certain Matrix Transformations Useful in Multivariate Analysis, *Biometrika*, vol. 38, pp. 345–367, 1951.

Fisher, R. A.: Frequency Distribution of the Values of the Correlation Coefficient in Samples from an Indefinitely Large Population, *Biometrika*, vol. 10, pp. 507–521, 1915.

Fraser, D. A. S.: Data Transformations and the Linear Model, *Ann. Math. Stat.*, vol. 38, pp. 1456–1465, 1967.

————: "The Structure of Inference," Krieger Publishing Company, Huntington, N.Y., 1968.

————: Inference and Redundant Parameters in Multivariate Analysis-III, in P. R. Krishnaiah (ed.), "Proc. of the Third International Symposium on Multivariate Analysis," Academic Press, Dayton, Ohio, 1973.

————: "Probability and Statistics, Theory and Applications," Duxbury Press, North Scituate, Mass., 1976.

———— and M. S. Haq: Structural Probability and Prediction for the Multivariate Models, *Jour. Roy. Stat. Soc.*, ser. B, vol. 31, pp. 317–331, 1969.

———— and ————: Inference and Prediction for the Multilinear Model, *Jour. Stat. Res.*, vol. 4, pp. 93–109, 1970.

Hotelling, H.: New Light on the Correlation Coefficient and Its Transformations, *Jour. Roy. Stat. Soc.*, ser. B, vol. 15, pp. 193–218, 1953.

James, A. T.: Normal Multivariate Analysis and the Orthogonal Group, *Ann. Math. Stat.*, vol. 25, pp. 40–75, 1954.

Ng, K.-W.: "Inference for Component Parameters in Multivariate Models," Unpublished Thesis, University of Toronto, 1975.

Tan, W.-Y., and I. Guttman: A Disguised Wishart Variable and a Related Theorem, *Jour. Roy. Stat. Soc.*, ser. B, vol. 33, pp. 147–152, 1971.

Watson, G. S.: Analysis of Dispersion on a Sphere, *Monthly Notices, Roy. Astron. Soc., Geophys. Suppl.*, vol. 7, pp. 153–159, 1956.

———— and E. J. Williams: On the Construction of Significance Tests on the Circle and the Sphere, *Biometrika*, vol. 43, pp. 344–352, 1956.

CHAPTER
NINE

DISTRIBUTIONS ON THE CIRCLE AND SPHERE

Response measurements can sometimes be directions on the plane, or in three or more dimensions. For example, a surveyor measures the direction of a distant object; a geophysicist measures the direction of the horizontal component of a magnetic field; a zoologist records the direction of travel for a type of large turtle after laying its eggs. These examples involve directions on the plane. As another example, a mineralogist measures the direction of magnetization in a rock stratum. This example involves directions in three dimensions.

A direction on the plane or in three dimensions can be recorded as a point on the unit circle or on the unit sphere. In this chapter we investigate distributions on the circle and on the sphere. A survey of background material may be found in Stephens (1962) and Mardia (1972). The approach here is derived from that in Fraser (1968).

9-1 THE CIRCLE

Consider a random system with a response that is a direction on the plane.

9-1-1 The Model

We can record a direction on the plane by means of a unit vector:

$$\mathbf{y} = \begin{pmatrix} y_1 \\ y_2 \end{pmatrix} \qquad \mathbf{y}'\mathbf{y} = 1 \tag{9-1}$$

lying on the unit circle in $\mathbb{R}^2$. Alternatively, we can record the direction by giving the angle y measured, say, positively from a reference direction, for example, $(1, 0)'$; thus

$$\mathbf{y} = \begin{pmatrix} y_1 \\ y_2 \end{pmatrix} = \begin{pmatrix} \cos y \\ \sin y \end{pmatrix} \quad y \text{ in } [0, 2\pi) \quad (9\text{-}2)$$

Suppose that the location for the response direction is unknown but that otherwise the distribution form is known or known up to a shape parameter λ.

For this let $f_\lambda(z_1, z_2) = f_\lambda(z)$ designate the distribution form as appropriately standardized and let

$$\mathbf{z} = \begin{pmatrix} z_1 \\ z_2 \end{pmatrix} = \begin{pmatrix} \cos z \\ \sin z \end{pmatrix} \quad z \text{ in } [0, 2\pi) \quad (9\text{-}3)$$

designate the corresponding objective variation. We treat f_λ as a density function on the unit circle or on the line segment $[0, 2\pi)$ using dz for length tangentially on the circle or, equivalently, dz as length on the line segment. A reasonable standardization is to locate the mode of the distribution in a standard direction $(1, 0)'$.

An example of a distribution form is the *normal* distribution on the circle proposed by von Mises (1918):

$$f_\lambda(z_1, z_2) = \frac{1}{2\pi I_0(\lambda)} \exp(\lambda z_1)$$

$$= \frac{1}{2\pi I_0(\lambda)} \exp(\lambda \cos z) \quad (9\text{-}4)$$

where

$$I_0(\lambda) = \frac{1}{2\pi} \int_0^{2\pi} \exp(\lambda \cos z) \, dz \quad (9\text{-}5)$$

is the imaginary Bessel function of zero order. For $\lambda = 0$ the distribution is uniform; for λ large it is tightly concentrated about $(1, 0)'$. Note that this distribution can be obtained from the standard bivariate normal located at $(\lambda, 0)'$ by conditioning to the circle of unit radius, for we can write the sum of squares in the exponent as

$$(z_1 - \lambda)^2 + z_2^2 = 1^2 + \lambda^2 - 2\lambda \cos z$$

for a point $(z_1, z_2) = (\cos z, \sin z)$ on the unit circle.

In applications the parameter of primary interest is the location or underlying direction of the response measurements. With the standardization just mentioned this location becomes the direction or angle from the reference direction $(1, 0)'$. For this let

$$\theta = \begin{pmatrix} \theta_{11} & \theta_{12} \\ \theta_{21} & \theta_{22} \end{pmatrix} = \begin{pmatrix} \theta_1 & -\theta_2 \\ \theta_2 & \theta_1 \end{pmatrix} = \begin{pmatrix} \cos\theta & -\sin\theta \\ \sin\theta & \cos\theta \end{pmatrix} \quad (9\text{-}6)$$

be the rotation through the angle θ from the reference direction. The primary parameter is then represented as the rotation θ from the reference direction or, equivalently, as the angle θ from the reference direction; this double use for the same symbol θ will cause no confusion in actual context.

The response presentations θ can be assembled as

$$G = \left\{ g = \begin{pmatrix} \cos a & -\sin a \\ \sin a & \cos a \end{pmatrix} : a \in [0, 2\pi) \right\} \tag{9-7}$$

These transformations are closed under products and inverses; they form a group, the rotation group on $\mathbb{R}^2$. This is the two-dimensional version of the group G_R of $p \times p$ matrices mentioned briefly in formula (8-178).

For n performances of the system we obtain $\mathbf{y}_1, \ldots, \mathbf{y}_n$ for the response and $\mathbf{z}_1, \ldots, \mathbf{z}_n$ for the variation. We assemble these as $2 \times n$ matrices:

$$Y = (\mathbf{y}_1, \ldots, \mathbf{y}_n) \qquad Z = (\mathbf{z}_1, \ldots, \mathbf{z}_n)$$
$$= \begin{pmatrix} y_{11} & \cdots & y_{1n} \\ y_{21} & \cdots & y_{2n} \end{pmatrix} \qquad = \begin{pmatrix} z_{11} & \cdots & z_{1n} \\ z_{21} & \cdots & z_{2n} \end{pmatrix} \tag{9-8}$$

The response presentation then has the form

$$Y = \theta Z \tag{9-9}$$

and the distribution for variation has the form

$$f_\lambda(Z) = \Pi f_\lambda(\mathbf{z}_i) \tag{9-10}$$

We thus obtain the structural model

$$\mathcal{M}_V = (\Omega; \mathbb{R}^{2n}, \mathcal{B}^{2n}, \mathcal{V}, G) \tag{9-11}$$

where $\Omega = G \times \Lambda$, $\mathcal{B}^{2n}$ is the Borel class in $\mathbb{R}^{2n}$, $\mathcal{V}$ is the class of densities $f_\lambda(\cdot)$ in (9-10) on the Cartesian product of the unit circles, and G is the transformation group (9-7). We will see from the construction procedure in Sec. 9-1-2 that the exactness Assumption 7-2 holds.

9-1-2 The Analysis

A transformation in G takes n points $\mathbf{z}_1, \ldots, \mathbf{z}_n$ on the unit sphere and rotates them. Accordingly, for position, we consider a rotation to obtain the n given points from the n points as oriented toward the standard direction $(1, 0)'$. For this we examine the sum vector

$$\Sigma \mathbf{z}_i = Z\mathbf{1} = \begin{pmatrix} \Sigma z_{1i} \\ \Sigma z_{2i} \end{pmatrix} = l(Z)\mathbf{a}(Z) \tag{9-12}$$

where

$$l^2(Z) = (\Sigma z_{1i})^2 + (\Sigma z_{2i})^2$$
$$\mathbf{a}(Z) = \begin{bmatrix} \cos a(Z) \\ \sin a(Z) \end{bmatrix} = l^{-1}(Z)\Sigma \mathbf{z}_i \tag{9-13}$$

are the squared length and unit vector for the sum vector. The sum vector gives an average direction for the n points and a large value for its length indicates that the points are tightly clustered around the average direction. Alternatively, we could use the first vector $\mathbf{z}_1$ as giving the position of the n points; this would avoid the problem (set of measure zero) in which the length $l(Z) = 0$.

We now take $[Z]$ to be the rotation from $(1, 0)'$ to $\mathbf{a}(Z)$:

$$[Z] = \begin{bmatrix} a_1(Z) & -a_2(Z) \\ a_2(Z) & a_1(Z) \end{bmatrix} = \begin{bmatrix} \cos a(Z) & -\sin a(Z) \\ \sin a(Z) & \cos a(Z) \end{bmatrix} \tag{9-14}$$

and for $D(Z)$ we then obtain

$$D(Z) = [Z]^{-1} Z$$
$$= [\mathbf{d}_1(Z), \ldots, \mathbf{d}_n(Z)] \tag{9-15}$$

Note, of course, that

$$\Sigma \mathbf{d}_i(Z) = l(Z) \begin{pmatrix} 1 \\ 0 \end{pmatrix} \tag{9-16}$$

is a vector in the standard direction $(1, 0)'$.

The Jacobians and measures from Secs. 7-2-2 and 7-2-3 are trivially available. A rotation applied separately to each column of Z is a rotation on $\mathbb{R}^{2n}$. Rotations do not change length, relative angle, or volume. Accordingly, we have

$$dM(Z) = dZ = \Pi \, d\mathbf{z}_i \tag{9-17}$$

where each $d\mathbf{z}_i$ is length on the appropriate unit circle and

$$d\mu(g) = dv(g) = dg = da \tag{9-18}$$

where dg is just length on the line segment $[0, 2\pi)$. The special Jacobian $J(D)$ is the ratio of length on the orbit to length on the group. Consider a change da in the angle a. The effect on

$$\begin{pmatrix} \cos a & -\sin a \\ \sin a & \cos a \end{pmatrix} [\mathbf{d}_1(Z), \ldots, \mathbf{d}_n(Z)] \tag{9-19}$$

is a change da for each of the vectors on the right; the linear change is then $\sqrt{n}\, da$. Accordingly,

$$J(D) = \sqrt{n} \tag{9-20}$$

We are, of course, taking dD to be volume orthogonal to the orbit GD at D (see Sec. 7-2-6).

Now consider the inference base $(\mathcal{M}_V, Y^0)$. We follow the analysis in Secs. 7-1 and 7-2. The observed orbit gives the observed value for the variation

$$D(Z) = D(Y^0) = D^0 \tag{9-21}$$

or

$$\mathbf{d}_i(Z) = \mathbf{d}_i(Y^0) = \mathbf{d}_i^0 \qquad i = 1, \ldots, n \tag{9-22}$$

This gives the inference base (a) from Sec. 7-1-4:

$$(\mathcal{M}_D, D^0) \tag{9-23}$$

The distributions in the model $\mathcal{M}_D$ are available from (7-46):

$$h_\lambda(D) = \int_0^{2\pi} \Pi f_\lambda(d_{1i} \cos a - d_{2i} \sin a, d_{1i} \sin a + d_{2i} \cos a) \sqrt{n} \, da \tag{9-24}$$

This, of course, leads to the observed likelihood function for λ from (7-69):

$$L(D^0; \lambda) = ch_\lambda(D^0) \tag{9-25}$$

The integration (9-24) is trivial for a computer.

The unobservable characteristics of the variation lead to the inference base (b) from Sec. 7-1-4:

$$(\mathcal{M}_V^{D^0}, [Y^0]) \tag{9-26}$$

The distributions for the structural model $\mathcal{M}_V^{D^0}$ are available from (7-47):

$$h_\lambda^{-1}(D^0) \Pi f_\lambda(d_{1i}^0 \cos a - d_{2i}^0 \sin a, d_{1i}^0 \sin a + d_{2i}^0 \cos a) \sqrt{n} \, da \tag{9-27}$$

in $[0, 2\pi)$. This distribution describing the variation a is used with the transformation

$$a(Y) = \theta + a \quad \text{modulo } 2\pi \tag{9-28}$$

The corresponding observed response value is $a(Y^0)$.

The response distribution corresponding to the identified $D(Y) = D(Y^0) = D^0$ is available from (7-49):

$$h_\lambda^{-1}(D_0) \Pi f_\lambda[(\theta_1 a_1 + \theta_2 a_2) d_{1i}^0 + (-\theta_1 a_2 + \theta_2 a_1) d_{2i}^0, \\ (-\theta_2 a_1 + \theta_1 a_2) d_{1i}^0 + (\theta_2 a_2 + \theta_1 a_1) d_{2i}^0] \sqrt{n} \, da \tag{9-29}$$

where $a = a(Y)$ designates the location for Y.

The inference base (9-23) is used for inferences concerning the shape parameter λ. The inference base (9-26) provides the inferences, tests, and confidence intervals for the parameter θ. For this we have a real-valued distribution (9-27) for the angle a and we have the equation (9-28), $a(Y) = \theta + a$, with observed value $a(Y) = a(Y^0)$. This is perhaps the simplest and most direct inference situation in statistics.

9-1-3 The Normal Example

The normal distribution on the circle is recorded in formula (9-4). It provides a simple bell-shaped curve for the variation on the circle. The parameter λ acts as a scale parameter. The circle does not have a continuous two parameter group without a fixed point, and thus does not have a location-scale group. A three-parameter location-scale skewness group is examined in Sec. 9-3.

The marginal density function for the orbit as given by $D(Y) = D(Z) = D$ is available by direct integration:

$$h_\lambda(D) = \int [2\pi I_0(\lambda)]^{-n} \exp[\lambda \Sigma(a_1 d_{1i} - a_2 d_{2i})] \sqrt{n}\, da$$

$$= [2\pi I_0(\lambda)]^{-n} \int_0^{2\pi} \exp[\lambda l(D) \cos a]\, da\, \sqrt{n}$$

$$= \frac{2\pi I_0[\lambda l(D)]}{[2\pi I_0(\lambda)]^n} \sqrt{n} \qquad (9\text{-}30)$$

The observed likelihood function $ch_\lambda(D^0)$ is easily plotted from (9-30).
The conditional distribution for the position a is then

$$\frac{1}{2\pi I_0[\lambda l(D^0)]} \exp[\lambda l(D^0) \cos a] \qquad (9\text{-}31)$$

on the line segment $[0, 2\pi)$. This is the normal distribution on the circle but with the scale λ replaced by $\lambda l(D^0)$. The preceding distribution together with

$$a(Y) = \theta + a$$

gives tests and confidence regions for θ. Note that the distribution depends on the length $l(Y^0) = l(D^0)$ of the sum vector $\Sigma \mathbf{y}_i^0$ and not otherwise on D^0; larger values for $l(Y^0)$ mean higher precision.

This is one of the few examples where the distribution for the essential characteristics on the orbit is available. As just noted, only the length $l(Y^0) = l(D^0)$ is needed concerning the orbit D^0.

The null distribution ($\lambda = 0$) for the length l is the distribution for the sum of n unit vectors independent and uniformly distributed on the unit circle. This is readily available from probability theory:

$$h_0^*(l) = l \int_0^\infty J_0(lt) J_0^n(t) t\, dt$$

where $J_0(\cdot)$ is the Bessel function of zero order.

The function $l(D)$ is the likelihood statistic for the distribution (9-30). Accordingly (9-30), as a function of λ, provides the likelihood function for the likelihood statistic distribution itself. Thus we obtain

$$h_\lambda^*(l) = h_0^*(l) \frac{h_\lambda(D)}{h_0(D)}$$

$$= l \int_0^\infty J_0(lt) J_0^n(t) t\, dt \, \frac{2\pi I_0(\lambda) \sqrt{n}/[2\pi I_0(\lambda)]^n}{2\pi I_0(0) \sqrt{n}/[2\pi I_0(0)]^n}$$

$$= \frac{I_0(\lambda l)}{I_0^n(\lambda)} l \int_0^\infty J_0(lt) J_0^n(t) t\, dt$$

Tests and confidence intervals are readily available for λ, for we are in the relatively simple position of having a real observed value $l^0 = l(D^0)$ and a distribution $h_\lambda^*(l)$ involving a single real parameter λ.

9-2 THE SPHERE

Now consider a random system with a response that is a direction in three dimensions. For example, the response could be the direction of magnetization in a rock stratum, the direction of incoming electromagnetic radiation, or the direction of a temperature gradient.

9-2-1 The Model

We record direction by means of a unit vector

$$\mathbf{y} = \begin{bmatrix} y_1 \\ y_2 \\ y_3 \end{bmatrix} \qquad \mathbf{y'y} = 1 \tag{9-32}$$

a vector on the unit sphere in $\mathbb{R}^3$.

Suppose that the location for the response distribution on the sphere is unknown, that the orientation of the distribution about that direction is unknown, but that the distribution form is otherwise known or known up to a shape parameter λ.

For this let $f_\lambda(\mathbf{z})$ designate the distribution form as appropriately standardized and let

$$\mathbf{z} = \begin{bmatrix} z_1 \\ z_2 \\ z_3 \end{bmatrix} \qquad \mathbf{z'z} = 1 \tag{9-33}$$

designate the corresponding objective variation. We treat f_λ as a density function on the unit sphere using Euclidean area tangential to the sphere. As a possible standardization, suppose we locate the mode of the distribution in a standard direction, say $(1, 0, 0)'$. Then orient some rotational characteristic with a second standard direction, say $(0, 1, 0)'$.

A *normal* distribution on the sphere has been proposed by Fisher (1953):

$$f_\lambda(\mathbf{z}) = \frac{\lambda}{4\pi \sinh \lambda} \exp(\lambda z_1) \tag{9-34}$$

For $\lambda = 0$ the distribution is uniform; for large λ it becomes tightly concentrated about $(1, 0, 0)'$. This distribution can be obtained from the standard three-variate normal located at $(\lambda, 0, 0)'$ by conditioning to the unit sphere; see the parallel discussion for the circle in Sec. 9-1-1.

The primary parameter for applications is the "location" or direction of the response distribution. A secondary parameter is the orientation of this distribution about the location direction. These parameters can be combined in the following rotation matrix:

$$\boldsymbol{\theta} = \begin{bmatrix} \theta_{11} & \theta_{12} & \theta_{13} \\ \theta_{21} & \theta_{22} & \theta_{23} \\ \theta_{31} & \theta_{32} & \theta_{33} \end{bmatrix} = (\boldsymbol{\theta}_1, \boldsymbol{\theta}_2, \boldsymbol{\theta}_3) \tag{9-35}$$

with $\theta\theta' = I, |\theta| = 1$. This matrix gives direction and orientation by rotation from the standard direction $(1, 0, 0)'$ with secondary direction $(0, 1, 0)'$. Note that θ carries the standard $(1, 0, 0)'$ into θ_1 and the secondary $(0, 1, 0)'$ into θ_2.

The presentations θ can be assembled as

$$G = \left\{ g = \begin{bmatrix} a_{11} & a_{12} & a_{13} \\ a_{21} & a_{22} & a_{23} \\ a_{31} & a_{32} & a_{33} \end{bmatrix} = (\mathbf{a}_1, \mathbf{a}_2, \mathbf{a}_3) : \begin{array}{l} gg' = I \\ |g| = 1 \end{array} \right\} \qquad (9\text{-}36)$$

These transformations are closed under products and inverses; they form a group, the rotation or positive orthogonal group on $\mathbb{R}^3$. This group G is three dimensional but the notation (9-36) presents the group as a manifold embedded in $\mathbb{R}^9$; note that this is the three-dimensional version of the group G_R mentioned in formula (8-178).

For n performances of the system we have the responses $\mathbf{y}_1, \ldots, \mathbf{y}_n$ and the corresponding variations $\mathbf{z}_1, \ldots, \mathbf{z}_n$. We assemble these as $3 \times n$ matrices:

$$Y = (\mathbf{y}_1, \ldots, \mathbf{y}_n) \qquad Z = (\mathbf{z}_1, \ldots, \mathbf{z}_n)$$

$$= \begin{bmatrix} y_{11} & \cdots & y_{1n} \\ y_{21} & \cdots & y_{2n} \\ y_{31} & \cdots & y_{3n} \end{bmatrix} \qquad = \begin{bmatrix} z_{11} & \cdots & z_{1n} \\ z_{21} & \cdots & z_{2n} \\ z_{31} & \cdots & z_{3n} \end{bmatrix} \qquad (9\text{-}37)$$

The response presentation then has the form

$$Y = \theta Z \qquad (9\text{-}38)$$

and the distribution for variation is

$$f_\lambda(Z) = \Pi f_\lambda(\mathbf{z}_i) \qquad (9\text{-}39)$$

We thus obtain the structural model

$$\mathcal{M}_V = (\Omega; \mathbb{R}^{3n}, \mathcal{B}^{3n}, \mathcal{V}, G) \qquad (9\text{-}40)$$

where $\Omega = G \times \Lambda$, $\mathcal{B}^{3n}$ is the Borel class on $\mathbb{R}^{3n}$, $\mathcal{V}$ is the class of densities $f_\lambda(\cdot)$ in (9-39) on the Cartesian product of the spheres, and G is the transformation group (9-36). We will see from the construction procedure in Sec. 9-2-2 that the exactness assumption holds provided $n \geq 2$ and at least two of the vectors are linearly independent.

9-2-2 The Analysis

A transformation in G takes n points $\mathbf{z}_1, \ldots, \mathbf{z}_n$ and rotates them. Accordingly, we define a direction and orientation for such a sequence of points and then relate these to the standard directions $(1, 0, 0)'$ and $(0, 1, 0)'$.

For a direction for the n points we follow the pattern for the circle and choose the sum vector

$$\Sigma \mathbf{z}_i = Z\mathbf{1} = \begin{bmatrix} \Sigma z_{1i} \\ \Sigma z_{2i} \\ \Sigma z_{3i} \end{bmatrix} = l(Z)\mathbf{a}_1(Z) \qquad (9\text{-}41)$$

where

$$l^2(Z) = (\Sigma z_{1i})^2 + (\Sigma z_{2i})^2 + (\Sigma z_{3i})^2$$
$$\mathbf{a}_1(Z) = l^{-1}(Z)\Sigma \mathbf{z}_i \tag{9-42}$$

are the squared length and unit vector for the sum vector. For a second direction as an orientation relative to $\mathbf{a}_1(Z)$ we do not have an immediate and natural choice. An easily described choice, however, is $\mathbf{a}_2(Z)$, the unit residual for, say, $\mathbf{z}_1$ after regression on $\mathbf{a}_1(Z)$. Then let $\mathbf{a}_3(Z)$ be the third unit vector forming the following positive rotation matrix:

$$[Z] = O(Z) = [\mathbf{a}_1(Z), \mathbf{a}_2(Z), \mathbf{a}_3(Z)] \tag{9-43}$$

The sum vector gives an average direction for the n points and a large value for its length indicates that the points are tightly clustered around the average direction. Alternatively, we could use the first vector and any other linearly independent vector to record direction and orientation; this would avoid the problem (set of measure zero) in which the preceding construction procedure fails.

We now verify that the transformation $[Z]$ does relate to the reference directions. For $D(Z)$ we have

$$D(Z) = [Z]^{-1}Z = [\mathbf{d}_1(Z), \ldots, \mathbf{d}_n(Z)]$$
$$= \begin{bmatrix} d_{11}(Z) & d_{12}(Z) & \cdots & d_{1n}(Z) \\ d_{21}(Z) & d_{22}(Z) & \cdots & d_{2n}(Z) \\ 0 & d_{32}(Z) & \cdots & d_{3n}(Z) \end{bmatrix} \tag{9-44}$$

Note from the definition of $[Z]$ that we have

$$\Sigma \mathbf{d}(Z) = l(Z) \begin{bmatrix} 1 \\ 0 \\ 0 \end{bmatrix} \tag{9-45}$$

and that $d_{31}(Z) = 0$ since $\mathbf{z}_1$ lies in the plane of $\mathbf{a}_1(Z)$, $\mathbf{a}_2(Z)$ and correspondingly, $\mathbf{d}_1(Z)$ lies in the plane of $(1, 0, 0)'$, $(0, 1, 0)'$.

The Jacobians and measures from Secs. 7-2-2 and 7-2-3 are trivially available. A rotation applied separately to each column of Z is a rotation of $\mathbb{R}^{3n}$. Rotations do not change length, relative angle, or volume. Accordingly, we have

$$dM(Z) = dZ = \Pi\, d\mathbf{z}_i \tag{9-46}$$

where each $d\mathbf{z}_i$ is area on the appropriate unit sphere and

$$d\mu(g) = dv(g) = dg \tag{9-47}$$

where dg is surface volume on the manifold G embedded in $\mathbb{R}^9$. The special Jacobian $J(D)$ is not readily available, but as we have noted it is not needed for most of the formulas used for the inference analysis.

Now consider the inference base $(\mathcal{M}_V, Y^0)$. We follow the analysis for the general case in Secs. 7-1 and 7-2 and for the circle in Sec. 9-1.

The observed orbit gives the observed value for the variation

$$D(Z) = D(Y^0) = D^0 \tag{9-48}$$

$$\mathbf{d}_i(Z) = \mathbf{d}_i(Y^0) = \mathbf{d}_i^0 \qquad i = 1, \ldots, n \tag{9-49}$$

This gives the inference base (a) from Sec. 7-1-4:

$$(\mathcal{M}_D, D^0) \tag{9-50}$$

The distributions in the model $\mathcal{M}_D$ are available from (7-46):

$$h_\lambda(D) = \int_G f_\lambda(OD) J(D)\, dO$$

$$= \int_G \Pi f_\lambda(Od_i) J(D)\, dO$$

$$= \int_G f_\lambda(OD)\, dO\; J(D)$$

$$= k_\lambda(D) J(D) \tag{9-51}$$

This, of course, leads to the observed likelihood function for λ:

$$L(D^0; \lambda) = ch_\lambda(D^0)$$

$$= c \int_G f_\lambda(OD^0)\, dO\; J(D^0)$$

$$= c \int_G f_\lambda(OD^0)\, dO$$

$$= ck_\lambda(D^0) \tag{9-52}$$

The integration for (9-51) or (9-52) is three-dimensional.

The unobservable characteristics of the variation lead to the inference base (b) from Sec. 7-1-4:

$$(\mathcal{M}_V^{D^0}, [Y^0]) \tag{9-53}$$

The distributions for the model $\mathcal{M}_V^{D^0}$ are available from (7-47):

$$h_\lambda^{-1}(D^0) f_\lambda(OD^0) J(D^0)\, dO = k_\lambda^{-1}(D^0) f_\lambda(OD^0)\, dO$$

$$= k_\lambda^{-1}(D^0) \Pi f_\lambda(Od_i^0)\, dO \tag{9-54}$$

This three-dimensional distribution describing the variation O is used with the transformation

$$O(Y) = \theta O$$

$$\begin{bmatrix} a_{11}(Y) & a_{12}(Y) & a_{13}(Y) \\ a_{21}(Y) & a_{22}(Y) & a_{23}(Y) \\ a_{31}(Y) & a_{32}(Y) & a_{33}(Y) \end{bmatrix} = \begin{bmatrix} \theta_{11} & \theta_{12} & \theta_{13} \\ \theta_{21} & \theta_{22} & \theta_{23} \\ \theta_{31} & \theta_{32} & \theta_{33} \end{bmatrix} \begin{bmatrix} a_{11} & a_{12} & a_{13} \\ a_{21} & a_{22} & a_{23} \\ a_{31} & a_{32} & a_{33} \end{bmatrix} \tag{9-55}$$

The corresponding observed value is $O(Y^0)$. Note that the transformation (9-55) relates the three-dimensional response $O(Y)$, the three-dimensional parameter θ, and the three-dimensional variation O; of course each is three-dimensional but with coordinates in nine dimensions.

The response distribution corresponding to the identified $D(Y) = D(Y^0) = D^0$ is available from (7-49):

$$h_\lambda^{-1}(D^0)f_\lambda(\theta^{-1}O(Y)D^0)J(D^0)\,dO(Y) = k_\lambda^{-1}(D^0)\Pi f_\lambda(\theta^{-1}O(Y)\mathbf{d}_i^0)\,dO(Y) \quad (9\text{-}56)$$

where $Y = O(Y)D^0$ for responses consistent with the observed Y^0.

The inference base (9-50) is used for inferences concerning λ. In applications, only the likelihood function (9-52) may be available computationally.

The inference base (9-53) provides tests and confidence regions for the direction and orientation recorded in the matrix θ. In contrast with the circle in Sec. 9-1, the rotation group is now three dimensional rather than one dimensional and our notation embeds it in $\mathbb{R}^9$. The spherical geometry of astronomy can, however, facilitate calculations.

9-2-3 The Normal Example

The normal distribution (9-34) proposed by Fisher (1953),

$$f_\lambda(\mathbf{z}) = \frac{\lambda}{4\pi \sinh \lambda} \exp(\lambda z_1) \quad (9\text{-}57)$$

is a direct parallel of the von Mises (1918) distribution in Sec. 9-1. The normalizing constant is easily verified by integration over the unit sphere:

$$\int \exp(\lambda z_1)\,d\mathbf{z} = \int_0^\pi \exp(\lambda \cos t)2\pi \sin t \, dt$$

$$= \int_{-1}^{+1} \exp(\lambda u)2\pi\,du$$

$$= \frac{2\pi}{\lambda}[\exp(\lambda) - \exp(-\lambda)]$$

$$= \frac{4\pi \sinh \lambda}{\lambda} \quad (9\text{-}58)$$

where t is the angle between $\mathbf{z}$ and $(1, 0, 0)'$ and $u = \cos t$.

The marginal density for the orbit as given by $D(Y) = D(Z) = D$ is available, except for $J(D)$, by direct integration:

$$\frac{h_\lambda(D)}{J(D)} = \int_G f_\lambda(OD)\,dO$$

$$= \int \frac{\lambda^n}{(4\pi \sinh \lambda)^n} \exp[\Sigma\lambda(a_{11}, a_{12}, a_{13})\mathbf{d}_i]\,dO$$

$$= \frac{\lambda^n}{(4\pi)^n \sinh^n \lambda} \int \exp\left[\lambda l(D)a_{11}\right] dO \qquad (9\text{-}59)$$

From Sec. 8-3-5 and the discussion near (8-182) we find that the surface volume of G is

$$A_3^{(2)} 2^{3 \cdot 2/4} = A_3 A_2 2^{3/2}$$

$$= \frac{2\pi^{3/2}}{\Gamma(\tfrac{3}{2})} \frac{2\pi}{\Gamma(1)} 2^{3/2}$$

$$= 4\pi \cdot 2\pi \cdot 2^{3/2} \qquad (9\text{-}60)$$

On the other hand, $\mathbf{a}_1$ takes values in the unit sphere which has surface area 4π. Accordingly, the integration of O for given $\mathbf{a}_1$ produces

$$\frac{4\pi \cdot 2\pi \cdot 2^{3/2}}{4\pi} = 2\pi \cdot 2^{3/2} \qquad (9\text{-}61)$$

Thus

$$k_\lambda(D) = \frac{h_\lambda(D)}{J(D)}$$

$$= \frac{\lambda^n 2\pi \cdot 2^{3/2}}{(4\pi)^n \sinh^n \lambda} \int \exp\left[\lambda l(D)a_{11}\right] d\mathbf{a}_1$$

$$= \frac{\lambda^n 2\pi \cdot 2^{3/2}}{(4\pi)^n \sinh^n \lambda} \cdot \frac{4\pi \sinh\left[\lambda l(D)\right]}{\lambda l(D)} \qquad (9\text{-}62)$$

It follows that the likelihood function for λ is

$$L(D; \lambda) = c \begin{cases} \dfrac{\lambda^{n-1} \sinh[\lambda l(D)]}{\sinh^n \lambda} & \text{if } \lambda \neq 0 \\ l(D) & \text{if } \lambda = 0 \end{cases} \qquad (9\text{-}63)$$

which can easily be plotted for the observed D^0. Note that the likelihood function depends only on the function $l(D)$ from D.

The conditional distribution for the position O is then available from (9-54) with (9-59) and (9-62):

$$k_\lambda^{-1}(D^0) f_\lambda(OD^0) \, dO = \frac{\lambda l(D^0)}{2\pi \cdot 2^{3/2} \cdot 4\pi \sinh\left[\lambda l(D^0)\right]} \exp\left[\lambda l(D^0)a_{11}\right] dO$$

$$= \frac{\lambda l(D^0)}{16\sqrt{2} \cdot \pi^2 \sinh\left[\lambda l(D^0)\right]} \exp\left[\lambda l(D^0)a_{11}\right] dO \qquad (9\text{-}64)$$

The density depends only on a_{11}. Accordingly, we can work with, say, the first column of O and integrate out the remaining coordinates; for this we use the result (9-61) as in the integration for (9-62). The marginal distribution for $\mathbf{a}_1$ is

$$\frac{\lambda l(D^o)}{4\pi \sinh [\lambda l(D^o)]} \exp [\lambda l(D^o) a_{11}] \, d\mathbf{a}_1 \qquad (9\text{-}65)$$

The conditional distribution of $\mathbf{a}_2$ given $\mathbf{a}_1$ is, of course, uniform on the unit circle orthogonal to $\mathbf{a}_1$. We can then write

$$\frac{\lambda l(D^o)}{4\pi \sinh [\lambda l(D^o)]} \exp [\lambda l(D^o) a_{11}] \, d\mathbf{a}_1 \cdot \frac{d\mathbf{a}_2}{2\pi} \qquad (9\text{-}66)$$

The distribution for $\mathbf{a}_1$ is the normal distribution on the sphere with the original parameter λ replaced by $\lambda l(D^o)$, and the distribution of $\mathbf{a}_2$ given $\mathbf{a}_1$ is uniform. The preceding distribution together with the equation

$$O(Y) = \theta O \qquad (9\text{-}67)$$

give tests or confidence regions for θ. Note that the distribution depends on the length $l(Y^o) = l(D^o)$ of the sum vectors $\Sigma \mathbf{y}_i^o$ and not otherwise on D^o.

Consider further the distribution (9-66) in relation to the presentation equation (9-67):

$$\theta^{-1} O(Y) = O$$

$$\begin{bmatrix} \theta'_1 \\ \theta'_2 \\ \theta'_3 \end{bmatrix} [\mathbf{a}_1(Y), \mathbf{a}_2(Y), \mathbf{a}_3(Y)] = \begin{bmatrix} O_1 \\ O_2 \\ O_3 \end{bmatrix} \qquad (9\text{-}68)$$

where we now use O_1, O_2, O_3 for the first, second, and third rows of O. The parameter θ_1 is the primary parameter that gives the direction of the response distribution. For it, we have

$$\theta'_1 \mathbf{a}_1(Y) = a_{11} \qquad \theta'_1 \mathbf{a}_2(Y) = a_{12} \qquad \theta'_1 \mathbf{a}_3(Y) = a_{13} \qquad (9\text{-}69)$$

Then, from the remarks preceding (9-65) and (9-66), we note for formula (9-66) that we could equally have used O_1 in place of $\mathbf{a}_1$ and O_2 in place of $\mathbf{a}_2$. Thus the Fisher distribution with parameter $\lambda l(D^o)$ for O_1 provides the tests and confidence regions for the primary direction parameter θ_1.

In particular, if $(v, 1)$ is, say, a 95 percent confidence interval of values for a_{11}:

$$\int_v^1 \frac{\lambda l(D^o)}{4\pi \sinh [\lambda l(D^o)]} \exp [\lambda l(D_o) a_{11}] 2\pi \, da_{11} = 0.95 \qquad (9\text{-}70)$$

based on (9-58), then

$$S(Y) = \{\theta_1 : \theta'_1 \mathbf{a}_1(Y) \geq v\} \qquad (9\text{-}71)$$

is the 95 percent confidence region for the direction θ_1. The observed region is

$$S(Y^o) = \{\theta_1 : \theta'_1 \mathbf{a}_1(Y^o) \geq v\} \qquad (9\text{-}72)$$

Note that the region consists of all the directions that are within an angle having cosine v with $a_1(Y)$.

Just as in Sec. 9-1-3, this is one of the few examples where the distribution for the essential characteristic of the orbit is readily available.

The null distribution ($\lambda = 0$) for the length l is the distribution for the sum of n unit vectors independent and random on the unit sphere. This is readily available from probability theory:

$$h_0^*(l) = \frac{2l}{\pi} \int_0^\infty \frac{\sin^n t \sin(lt)}{t^{n-1}} dt$$

$$= \frac{l}{2^{n-1}} \phi_n(l) \qquad (9\text{-}73)$$

where

$$\phi_n(l) = \frac{1}{(n-2)!} \sum_{s=0}^{n} \binom{n}{s} (-1)^s (n-l-2s)_+^{n-2} \qquad (9\text{-}74)$$

$$t_+ = \begin{cases} t & \text{if } t \geq 0 \\ 0 & \text{otherwise} \end{cases}$$

The function $l(D)$ is the likelihood statistic for the distribution (9-59). Accordingly (9-63), as a function of λ, provides the likelihood function for the likelihood statistic distribution itself. Thus we obtain the density:

$$h_\lambda^*(l) = h_0^*(l) \frac{L(D; \lambda)}{L(D; 0)}$$

$$= \frac{l}{2^{n-1}} \phi_n(l) \frac{\lambda^{n-1} \sinh(\lambda l)/\sinh^n \lambda}{l}$$

$$= \frac{\sinh(\lambda l)}{\sinh^n \lambda} \left(\frac{\lambda}{2}\right)^{n-1} \phi_n(l)$$

for l on $(0, \infty)$. Tests and confidence intervals are then available for λ, for we are again in the same relatively simple position as for the case of the circle: a real observed value $l^0 = l(D^0)$ and a distribution $h_\lambda^*(l)$ involving a single real parameter.

9-3 THE CIRCLE: GENERALIZED DISTRIBUTION FORM†

Consider a random system with a response that is a direction on the plane. As before, we record a response value as a unit vector or as a point in the unit circle.

In Sec. 9-1 we developed a model for such a system. The model allows for an unknown rotation of a basic distribution form that is known or known up to a

† With Malcolm Cairns.

shape parameter λ. The shape parameter λ could be a scale parameter as in Sec. 9-1-3 or it could be more general. The specification, however, may allow for scaling and even skewness of the response distribution. Can such scaling and skewness parameters be included in a transformation presentation of a more basic distribution form?

Cairns (1975) has developed a model for directional data that provides a transformation presentation for rotation, scaling, and skewness. We examine this more general model in this section. An analogous model exists for the sphere in $\mathbb{R}^3$ and for the sphere more generally.

9-3-1 The Model

As before we record a direction on the plane by means of a unit vector

$$\mathbf{y} = \begin{pmatrix} y_1 \\ y_2 \end{pmatrix} = \begin{pmatrix} \cos y \\ \sin y \end{pmatrix} \qquad \begin{matrix} \mathbf{y}'\mathbf{y} = 1 \\ y \text{ in } [0, 2\pi). \end{matrix} \tag{9-75}$$

lying on the unit circle in $\mathbb{R}^2$. We suppose that the response distribution is unknown in location, scaling, and skewness, but that otherwise the distribution form is known or known up to a shape parameter λ.

We let $f_\lambda(z_1, z_2) = f_\lambda(\mathbf{z})$ designate the basic distribution form as appropriately standardized and let

$$\mathbf{z} = \begin{pmatrix} z_1 \\ z_2 \end{pmatrix} = \begin{pmatrix} \cos z \\ \sin z \end{pmatrix} \qquad \begin{matrix} \mathbf{z}'\mathbf{z} = 1 \\ z \text{ in } [0, 2\pi) \end{matrix} \tag{9-76}$$

designate the objective variation. We treat f_λ as a density function on the unit circle or on the line segment $[0, 2\pi)$ using $d\mathbf{z} = dz$ for length tangentially on the unit circle. A reasonable approach to standardization is to locate the mode of the distribution in a standard direction $(1, 0)'$ and then symmetrize and in some way standardize the scale.

A useful example of a distribution form is the *projected normal distribution* $PN((\lambda, 0)', I)$:

$$\begin{aligned}
f_\lambda(\mathbf{z}) &= \int_0^\infty \phi_2[rz_1, rz_2 | (\lambda, 0)', I] r \, dr \\
&= \int_0^\infty \frac{1}{2\pi} \exp\left[-\frac{1}{2}(rz_1 - \lambda)^2 - \frac{1}{2}(rz_2)^2\right] r \, dr \\
&= \frac{1}{2\pi} \exp\left[-\frac{1}{2}\lambda^2 + \frac{1}{2}(\lambda z_1)^2\right] \int_0^\infty \exp\left[-\frac{1}{2}(r - \lambda z_1)^2\right] r \, dr \qquad (9\text{-}77) \\
&= \frac{1}{2\pi} \exp(-\lambda^2 z_2^2/2) \left(\int_{-\lambda z_1}^\infty e^{-t^2/2} t \, dt + \lambda z_1 \int_0^\infty e^{-(r - \lambda z_1)^2/2} \, dr \right) \\
&= \frac{1}{2\pi} \exp(-\lambda^2 z_2^2/2) \left[\exp(-\lambda^2 z_1^2/2) + \lambda z_1 \sqrt{2\pi} \Phi(\lambda z_1)\right] \\
&= \frac{1}{\sqrt{2\pi}} \phi(\lambda) + \lambda z_1 \Phi(\lambda z_1) \phi(\lambda z_2),
\end{aligned}$$

where $\phi_2(\mathbf{z}|\boldsymbol{\mu}, \Sigma)$ is the bivariate normal $(\boldsymbol{\mu}; \Sigma)$ density, and ϕ and Φ are the standard normal density and distribution functions. In a similar way the bivariate normal $(\boldsymbol{\mu}; \Sigma)$ gives the projected normal $PN(\boldsymbol{\mu}; \Sigma)$. The von Mises normal distribution (9-4) can be obtained by taking a rotationally symmetrical normal on the plane and conditioning to the circle with radius 1; the present normal is obtained by projecting radially to the circle with radius 1. The transformations we examine produce the general projected normal that has a spectrum of possibilities including bimodal distributions—with different sized modes at various angles of separation.

For applications, the primary parameter typically records the location of the response distribution; the remaining parameters record the scaling and skewness. For this consider a 2×2 matrix

$$\theta = \begin{pmatrix} \cos \alpha & -\sin \alpha \\ \sin \alpha & \cos \alpha \end{pmatrix} \begin{pmatrix} 1 & \tau \\ 0 & 1 \end{pmatrix} \begin{pmatrix} \sigma & 0 \\ 0 & \sigma^{-1} \end{pmatrix}$$

$$= \theta_3 \theta_2 \theta_1 = \begin{pmatrix} \theta_{11} & \theta_{12} \\ \theta_{21} & \theta_{22} \end{pmatrix}. \tag{9-78}$$

The parameter θ_3 is a 2×2 rotation matrix; θ_2 is a matrix that skews the plane parallel to the first axis; and θ_1 scales in the direction of the first axis by σ and in the direction of the second axis by σ^{-1}. For a transformation on the unit circle we apply θ to a point $\mathbf{z}$ by matrix multiplication and then project the resulting point radially onto the unit sphere:

$$\theta \circ \mathbf{z} = \theta \mathbf{z}/|\theta \mathbf{z}|. \tag{9-79}$$

The transformation θ applied to the projected normal $PN[(\lambda, 0)', I]$ gives the projected normal $PN(\lambda \theta(1, 0)', \theta \theta')$.

The response presentations θ can be assembled as

$$G = \{\theta = \theta_3 \theta_2 \theta_1 : \alpha \in [0, 2\pi), \tau \in \mathbb{R}, \sigma \in \mathbb{R}^+\}. \tag{9-80}$$

The product $\theta_2 \theta_1$ is an arbitrary positive upper triangular matrix with determinant $+1$; θ_3 is an arbitrary rotation matrix. The product $\theta_3 \theta_2 \theta_1$ is thus an arbitrary 2×2 matrix with determinant $+1$. Such matrices are closed under products and inverses; they form a group, the special linear group $SL_2(\mathbb{R})$ on the plane.

The group G on $\mathbb{R}^2$ clearly carries rays from the origin into rays from the origin. Accordingly, the action (9-79) is a transformation group on the rays or equivalently on the (representative) points on the unit sphere; in fact it is a transformation on the "double-ended rays" or one-dimensional linear subspaces.

For n performances of the system we obtain $\mathbf{y}_1, \ldots, \mathbf{y}_n$ for the response and $\mathbf{z}_1, \ldots, \mathbf{z}_n$ for the variation:

$$Y = (\mathbf{y}_1, \ldots, \mathbf{y}_n) \qquad Z = (\mathbf{z}_1, \ldots, \mathbf{z}_n)$$

$$= \begin{pmatrix} y_{11} & \cdots & y_{1n} \\ y_{21} & \cdots & y_{2n} \end{pmatrix} \qquad = \begin{pmatrix} z_{11} & \cdots & z_{1n} \\ z_{21} & \cdots & z_{2n} \end{pmatrix} \tag{9-81}$$

The response presentation then has the form

$$Y = \theta \circ Z \qquad (9\text{-}82)$$

where θ operates column by column in accord with (9-79). The distribution for variation has the form

$$f_\lambda(Z) = \Pi f_\lambda(\mathbf{z}_i) \qquad (9\text{-}83)$$

We thus obtain the structural model

$$\mathcal{M}_V = (\Omega; \mathbb{R}^{2n}, \mathcal{B}^{2n}, \mathcal{V}, G) \qquad (9\text{-}84)$$

where $\Omega = G \times \Lambda$, $\mathcal{B}^{2n}$ is the Borel class on $\mathbb{R}^{2n}$, $\mathcal{V}$ is the class of densities (9-83) on the product of the unit spheres, and G is the transformation group (9-80) with action (9-82) using (9-79). We will see the action is exact, provided $n \geq 3$ and a trivial set of measure zero is excluded.

9-3-2 The Transformations

A transformation in G takes n points $\mathbf{z}_1, \ldots, \mathbf{z}_n$ on the unit sphere and distorts their relative position—without any change in their relative ordering.

Now consider the Jacobians and measures needed for the analysis. For the transformation on $\mathbb{R}^n$ we first examine a single coordinate, in effect, the case $n = 1$:

$$g \circ \mathbf{z} = \frac{g\mathbf{z}}{|g\mathbf{z}|} = \frac{g\mathbf{z}}{c(g, \mathbf{z})}. \qquad (9\text{-}85)$$

Consider a two-dimensional neighborhood of a point

$$\mathbf{z} = \begin{pmatrix} r \cos z \\ r \sin z \end{pmatrix} \qquad (9\text{-}86)$$

in $\mathbb{R}^2$ and the corresponding neighborhood at $\tilde{\mathbf{z}} = g\mathbf{z}$; the volume change is unity

$$\left| \frac{\partial g\mathbf{z}}{\partial \mathbf{z}} \right| = 1, \qquad (9\text{-}87)$$

as g is a matrix transformation belonging to the special linear group of matrices with determinant 1. The effect of the transformation on the radial coordinate is simple,

$$\tilde{r} = \left| g \begin{bmatrix} \cos z \\ \sin z \end{bmatrix} \right| r \qquad (9\text{-}88)$$

as is seen by noting that

$$\tilde{\mathbf{z}} = gr \begin{bmatrix} \cos z \\ \sin z \end{bmatrix} = rg \begin{bmatrix} \cos z \\ \sin z \end{bmatrix};$$

thus

$$\left| \frac{\partial \tilde{r}}{\partial r} \right| = \left| g \begin{bmatrix} \cos z \\ \sin z \end{bmatrix} \right|. \qquad (9\text{-}89)$$

Also note that the coordinates r and $\mathbf{z}$ provide orthogonal contours on $\mathbb{R}^2$.

Now consider the succession of transformations from a two-dimensional neighborhood of z on the unit circle, to the corresponding neighborhood of gz, to the resulting neighborhood of $g \circ z$. The first transformation preserves volume; by (9-89) radial distance is changed by the factor $|gz|$; accordingly, distance on the circle through gz is changed by the factor $|gz|^{-1}$. The second transformation is a radial projection using the factor $|gz|^{-1}$ to obtain a point on the unit circle; accordingly, distance on the circles is changed by the same factor $|gz|^{-1}$. The two transformations together change distance on the unit circle by the factor

$$\left|\frac{dg \circ z}{dz}\right| = |gz|^{-2} \tag{9-90}$$

where we are now viewing z as a point on the unit circle.

We thus obtain the invariant measures

$$dm(\mathbf{z}) = |gz|^2\, d\mathbf{z},\, dM(Z) = \Pi |gz_i|^2\, dZ \tag{9-91}$$

using

$$J_n(Z) = \Pi |gz_i|^{-2} = \Pi c^{-2}(g, z_i). \tag{9-92}$$

For the transformations on the group we depart from our usual pattern of embedding the matrices in the Euclidean space for all the elements of the matrix and instead use coordinates based on the reverse of the factorization (9-78):

$$G = \begin{bmatrix} s & 0 \\ 0 & s^{-1} \end{bmatrix} \begin{bmatrix} 1 & t \\ 0 & 1 \end{bmatrix} \begin{bmatrix} \cos a & -\sin a \\ \sin a & \cos a \end{bmatrix} \tag{9-93}$$

where $s \in \mathbb{R}^+$, $t \in \mathbb{R}$, and $a \in [0, 2\pi)$; specifically we examine G as

$$G = \mathbb{R}^+ \times \mathbb{R} \times [0, 2\pi).$$

The left group transformation

$$\tilde{g} = gg^*$$

from g^* to $\tilde{g}$ can be examined in the pattern used for J_n and we obtain

$$J_3(g) = s; \tag{9-94}$$

for the details see Cairns (1975, p. 64). Thus we have

$$d\mu(g) = \frac{ds\, dt\, da}{s}. \tag{9-95}$$

9-3-3 The Analysis

Now consider group and reference point notation for a point Z in the Cartesian product of the n unit circles. For this we need to have three column vectors in Z that are essentially different: we call two vectors *essentially different* if they are not identical nor the negative of each other, in other words if they are not linearly dependent. For notation, suppose these are the first three vectors z_1, z_2, z_3 in Z.

224 INFERENCE AND LINEAR MODELS

We then define $[Z]$ so that

$$(\mathbf{z}_1, \mathbf{z}_2, \mathbf{z}_3) = [Z] \circ \left[\begin{pmatrix} 1 \\ 0 \end{pmatrix}, \begin{pmatrix} 0 \\ \pm 1 \end{pmatrix}, \begin{pmatrix} \pm 1/\sqrt{2} \\ \pm 1/\sqrt{2} \end{pmatrix} \right] \tag{9-96}$$

where the signs are chosen so that a positive linear distortion can carry the three vectors directions on the left to the three on the right. It is easily verified that

$$[Z] = [k_1(Z)\mathbf{z}_1, k_2(Z)\mathbf{z}_2] \tag{9-97}$$

where

$$k_1(z) > 0, \qquad |k_1(Z)\mathbf{z}_1, k_2(Z)\mathbf{z}_2| = 1, \tag{9-98}$$

and where

$$[k_1(Z)\mathbf{z}_1, k_2(Z)\mathbf{z}_2] \begin{pmatrix} \pm 1/\sqrt{2} \\ \pm 1/\sqrt{2} \end{pmatrix} \tag{9-99}$$

is a positive multiple of $\mathbf{z}_3$. Computationally, we would find a linear combination of $\mathbf{z}_1$ and $\mathbf{z}_2$ that generates $\mathbf{z}_3$ (see (9-99)) and then scale (positively or negatively) the columns of a trial (9-97) to satisfy (9-98).

We thus obtain

$$Z = [Z] \circ D(Z) \tag{9-100}$$

with

$$D(Z) = [Z]^{-1} \circ Z$$

$$= \left[\begin{pmatrix} 1 \\ 0 \end{pmatrix}, \begin{pmatrix} 0 \\ \pm 1 \end{pmatrix}, \begin{pmatrix} \pm 1/\sqrt{2} \\ \pm 1/\sqrt{2} \end{pmatrix}, \ldots \right]. \tag{9-101}$$

We do not calculate the Jacobian $J(D)$ as it is not needed for our present analyses; recall the discussion following (7-43).

Now consider the inference base $(\mathcal{M}_Y, Y^0)$ with model $\mathcal{M}_Y$ in (9-84) and observed response matrix Y^0.

The observed orbit gives the observed value for the variation

$$D(Z) = D(Y^0) = D^0 \tag{9-102}$$

or $\mathbf{d}_i(Z) = \mathbf{d}_i(Y^0) = \mathbf{d}_i^0$ for $i = 1, \ldots, n$. This gives us the inference base (a):

$$(\mathcal{M}_D, D^0) \tag{9-103}$$

from Sec. 7-1-4. The distributions in the model $\mathcal{M}_D$ are available from (7-46):

$$h_\lambda(D) = \int_{-\infty}^{\infty} \int_{0}^{\infty} \int_{0}^{2\pi} \Pi f_\lambda(\mathbf{z}_i) \Pi c_i^{-2} s^{-1} \, da \, ds \, dt \tag{9-104}$$

where

$$\mathbf{z}_i = \begin{pmatrix} s & 0 \\ 0 & s^{-1} \end{pmatrix} \begin{pmatrix} 1 & t \\ 0 & 1 \end{pmatrix} \begin{pmatrix} \cos a & -\sin a \\ \sin a & \cos a \end{pmatrix} \mathbf{d}_i / c_i,$$

$$c_i = \left| \begin{pmatrix} s & 0 \\ 0 & s^{-1} \end{pmatrix} \begin{pmatrix} 1 & t \\ 0 & 1 \end{pmatrix} \begin{pmatrix} \cos a & -\sin a \\ \sin a & \cos a \end{pmatrix} \mathbf{d}_i \right|, \tag{9-105}$$

and we are in effect using the modified measure $J(D)\, dD$ on the reference point manifold Q; recall that we did not calculate $J(D)$. This, of course, leads to the observed likelihood for λ:

$$L(D^0; \lambda) = ch_\lambda(D^0). \tag{9-106}$$

The unobserved characteristics of the variation lead to the inference base (b) in Sec. 7-1-4:

$$(\mathcal{M}_V^{D^0}, [Y^0]) \tag{9-107}$$

The distribution for g in the model $\mathcal{M}_V^{D^0}$ is available from (7-47) and (9-105):

$$h_\lambda^{-1}(D^0)\Pi f_\lambda(z_i)\Pi c_i^{-2} s^{-1}\, da\, ds\, dt, \tag{9-108}$$

a distribution on the group G. This distribution for the unidentified variation is used with the transformation

$$\begin{pmatrix} s(Y) & 0 \\ 0 & s^{-1}(Y) \end{pmatrix} \begin{pmatrix} 1 & t(Y) \\ 0 & 1 \end{pmatrix} \begin{pmatrix} \cos a(Y) & -\sin a(Y) \\ \sin a(Y) & \cos a(Y) \end{pmatrix}$$

$$= \begin{pmatrix} \cos \alpha & -\sin \alpha \\ \sin \alpha & \cos \alpha \end{pmatrix} \begin{pmatrix} 1 & \tau \\ 0 & 1 \end{pmatrix} \begin{pmatrix} \sigma & 0 \\ 0 & \sigma^{-1} \end{pmatrix} \begin{pmatrix} s & 0 \\ 0 & s^{-1} \end{pmatrix} \begin{pmatrix} 1 & t \\ 0 & 1 \end{pmatrix} \begin{pmatrix} \cos a & -\sin a \\ \sin a & \cos a \end{pmatrix}.$$

$$\tag{9-109}$$

The observed values for the response coordinates are obtained from $Y = Y^0$.

The response distribution corresponding to the identified $D(Y) = D(Y^0) = D^0$ is available from (7-49) and can be calculated routinely.

In fact inference for θ would naturally be in terms of the components of θ. In the pattern of Sec. 7-4 we could make tests or form confidence intervals for the principal direction α. And then for a given value of α—perhaps, reasonably, the maximum likelihood value—make tests or form confidence intervals for the skewness τ. And then for a given value of α and τ—perhaps, reasonably, the maximum likelihood values—make tests or form confidence intervals for the scaling σ.

The distributions for the preceding can be calculated in a very direct manner by taking the reference point at the observed Y^0. Suppose we use Y^0 in place of D^0 in the notation (9-96) and (9-101); then Y and Z can be expressed in terms of transformations

$$Y = [Y]Y^0 \qquad Z = [Z]Y^0$$

relative to Y^0. The equation (9-109) for the observed response Y^0 then becomes

$$\begin{pmatrix} 1 & 0 \\ 0 & 1 \end{pmatrix} = \begin{pmatrix} \cos \alpha & -\sin \alpha \\ \sin \alpha & \cos \alpha \end{pmatrix} \begin{pmatrix} 1 & \tau \\ 0 & 1 \end{pmatrix} \begin{pmatrix} \sigma & 0 \\ 0 & \sigma^{-1} \end{pmatrix}$$

$$\times \begin{pmatrix} s & 0 \\ 0 & s^{-1} \end{pmatrix} \begin{pmatrix} 1 & t \\ 0 & 1 \end{pmatrix} \begin{pmatrix} \cos a & -\sin a \\ \sin a & \cos a \end{pmatrix}.$$

The marginal distribution for a can be obtained from the revised (9-108) by integrating out s and t:

$$h_\lambda^3(a)\, da = \int_0^\infty \int_{-\infty}^\infty h_\lambda^{-1}(Y^0) \Pi f_\lambda(\mathbf{z}_i) \Pi c_i^{-2} s^{-1}\, ds\, dt \cdot da;$$

this is a distribution on $[0, 2\pi)$. The preceding distribution can be compared with the equation

$$0 = \alpha + a \quad \text{modulo } 2\pi$$

involving the "observed value."

For a chosen value for the parameter $\alpha = -a$, say the maximum density value $\hat{\alpha} = -\hat{a}$, the conditional distribution for t can be obtained from (9-108) by integrating out s:

$$h_\lambda^2(t)\, dt = \frac{1}{h_\lambda^3(a)} \int_0^\infty h_\lambda^{-1}(Y^0) \Pi f_\lambda(\mathbf{z}_i) \Pi c_i^{-2} s^{-1}\, ds\, dt$$

with the particular value for a substituted. This is a distribution on $\mathbb{R}$ which can be compared with the equation

$$0 = \tau + t$$

involving the "observed value."

Then for chosen values for the parameters $\alpha = -a$, $\tau = -t$, say the maximum density values $\hat{\alpha}$, $\hat{\beta}$, the conditional distribution for s can be obtained directly from (9-108):

$$h_\lambda^1(s)\, ds = \frac{1}{h_\lambda^3(a) h_\lambda^2(t)} h_\lambda^{-1}(Y^0) \Pi f_\lambda(\mathbf{z}_i) \Pi c_i^{-2} s^{-1} \cdot ds$$

with the particular values for a, t substituted. This is a distribution on $\mathbb{R}^+$ which can be compared with the equation

$$1 = \sigma s$$

involving the "observed value."

9-3-4 The Turtle Data

Dr. E. Gould of John Hopkins University School of Hygiene collected data recording the direction taken by sea turtles after laying their eggs. The data are recorded in Table 9-1 and plotted in Fig. 9-1.

The usual pattern has a predominant direction which involves returning to the sea and a secondary direction tending to be opposite to the preceding. We record the results of the computer analysis obtained in Cairns (1975); the analysis uses the projected normal and the three-parameter transformation group.

The marginal likelihood for the basic-shape parameter was calculated by three-dimensional computer integration and is plotted in Fig. 9-2. The likelihood

Table 9-1 The directions taken by turtles

Angle	y_1	y_2	Angle	y_1	y_2
8.0	0.139	0.990	83.0	0.993	0.122
9.0	0.156	0.988	88.0	0.999	0.035
13.0	0.225	0.974	88.0	0.999	0.035
13.0	0.225	0.974	88.0	0.999	0.035
14.0	0.242	0.970	90.0	1.000	0.000
18.0	0.309	0.951	92.0	0.999	−0.035
22.0	0.375	0.927	92.0	0.999	−0.035
27.0	0.454	0.891	93.0	0.999	−0.052
30.0	0.500	0.866	95.0	0.996	−0.087
34.0	0.559	0.829	96.0	0.995	−0.105
38.0	0.616	0.788	98.0	0.990	−0.139
38.0	0.616	0.788	100.0	0.985	−0.174
40.0	0.643	0.766	103.0	0.974	−0.225
44.0	0.695	0.719	106.0	0.961	−0.276
45.0	0.707	0.707	113.0	0.921	−0.391
47.0	0.731	0.682	118.0	0.883	−0.469
48.0	0.743	0.669	138.0	0.669	−0.743
48.0	0.743	0.669	153.0	0.454	−0.891
48.0	0.743	0.669	153.0	0.454	−0.891
48.0	0.743	0.669	155.0	0.423	−0.906
50.0	0.766	0.643	204.0	−0.407	−0.914
53.0	0.799	0.602	215.0	−0.574	−0.819
56.0	0.829	0.559	223.0	−0.682	−0.731
57.0	0.839	0.545	226.0	−0.719	−0.695
58.0	0.848	0.530	237.0	−0.839	−0.545
58.0	0.848	0.530	238.0	−0.848	−0.530
61.0	0.875	0.485	243.0	−0.891	−0.454
63.0	0.891	0.454	244.0	−0.899	−0.438
64.0	0.899	0.438	250.0	−0.940	−0.342
64.0	0.899	0.438	251.0	−0.946	−0.326
64.0	0.899	0.438	257.0	−0.974	−0.225
65.0	0.906	0.423	268.0	−0.999	−0.035
65.0	0.906	0.423	285.0	−0.966	0.259
68.0	0.927	0.375	319.0	−0.656	0.755
70.0	0.940	0.342	343.0	−0.292	0.956
73.0	0.956	0.292	350.0	−0.174	0.985
78.0	0.978	0.208			
78.0	0.978	0.208			
78.0	0.978	0.208			
83.0	0.993	0.122			

is rather sharply discriminating and has a maximum at $\hat{\lambda} = 0.7$. This value is used in the following analysis of the presentation parameters.

The distribution for a is plotted in Fig. 9-3. The maximum density value is $\hat{a} = -0.46$ yielding

$$\hat{\alpha} = -\hat{a} = 0.46.$$

The 95 percent confidence interval for α is (0.26, 0.66).

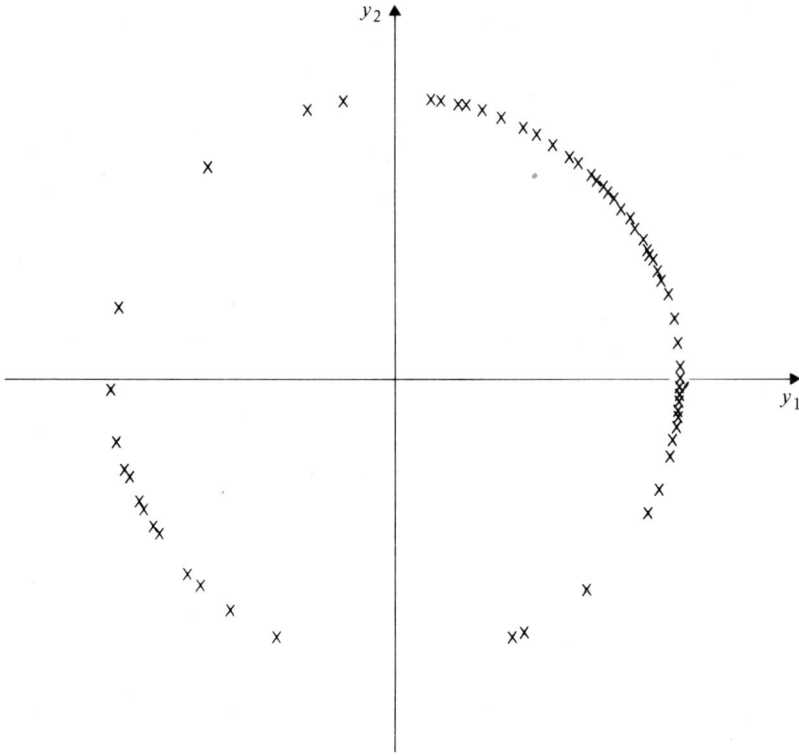

Figure 9-1 The $n = 76$ observations on turtle direction.

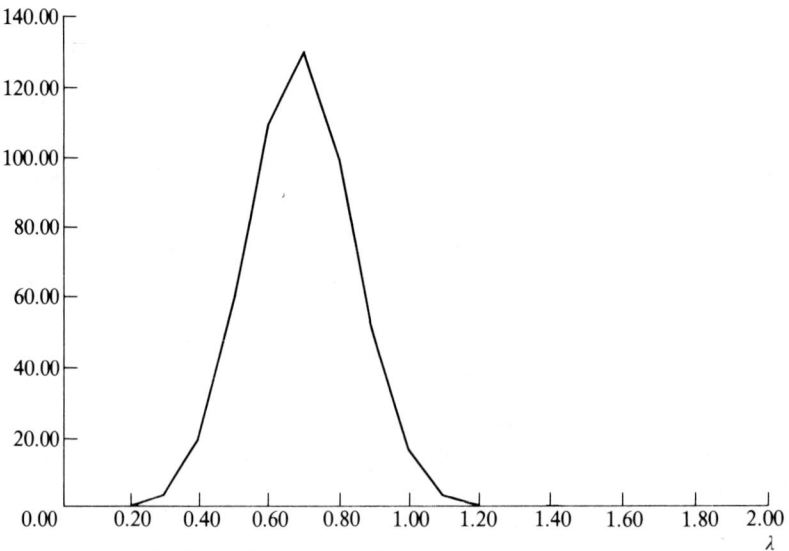

Figure 9-2 The likelihood for the shape λ.

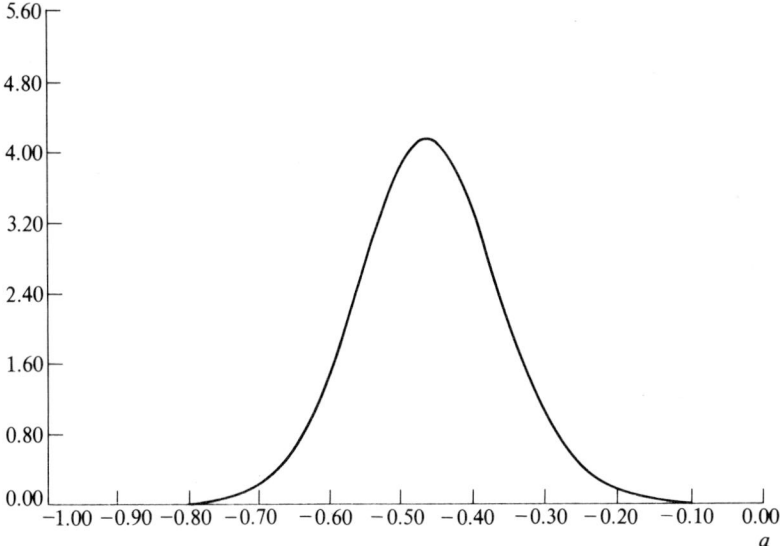

Figure 9-3 The density function for a.

The distribution for t given a is plotted in Fig. 9-4. The maximum density value is $\hat{t} = -0.08$ yielding

$$\hat{\tau} = -\hat{t} = 0.08.$$

The 95 percent confidence interval for τ is $(-0.53, 0.72)$.

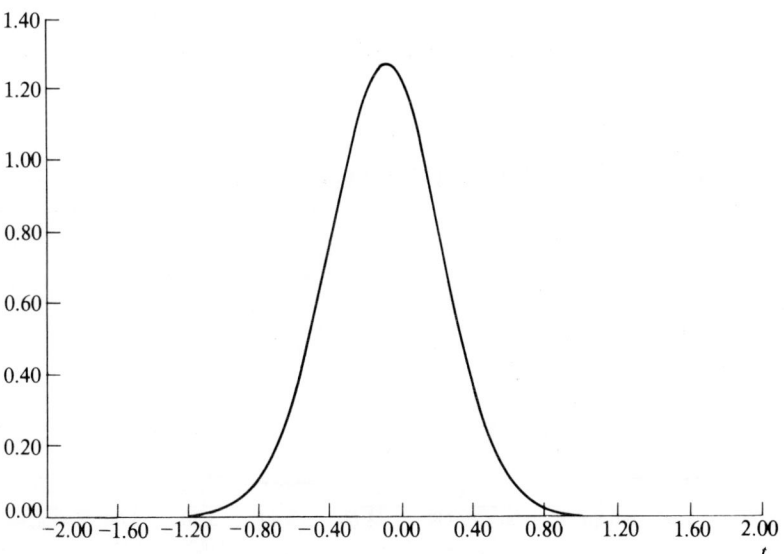

Figure 9-4 The density function for t.

230 INFERENCE AND LINEAR MODELS

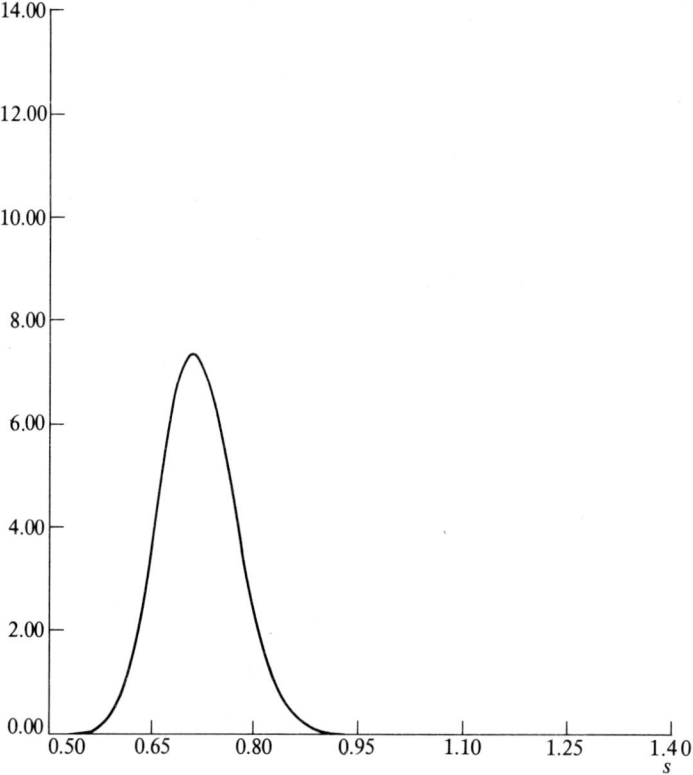

Figure 9-5 The density function for s.

The distribution for s given a, t is plotted in Fig. 9-5. The maximum density value is $\hat{s} = 0.71$ yielding

$$\hat{\sigma} = \hat{s}^{-1} = 1.38.$$

The 95 percent confidence interval for σ is (1.20, 1.62).

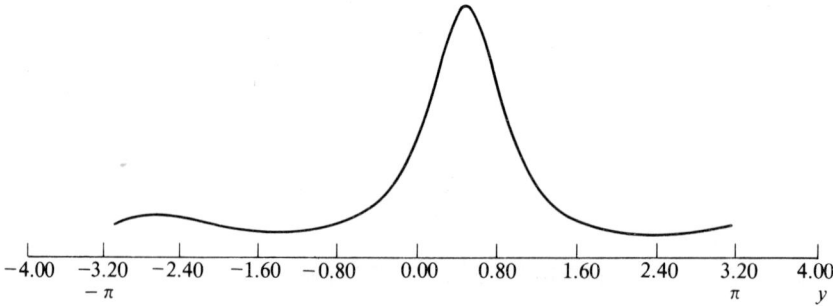

Figure 9-6 The fitted projected-normal response distribution.

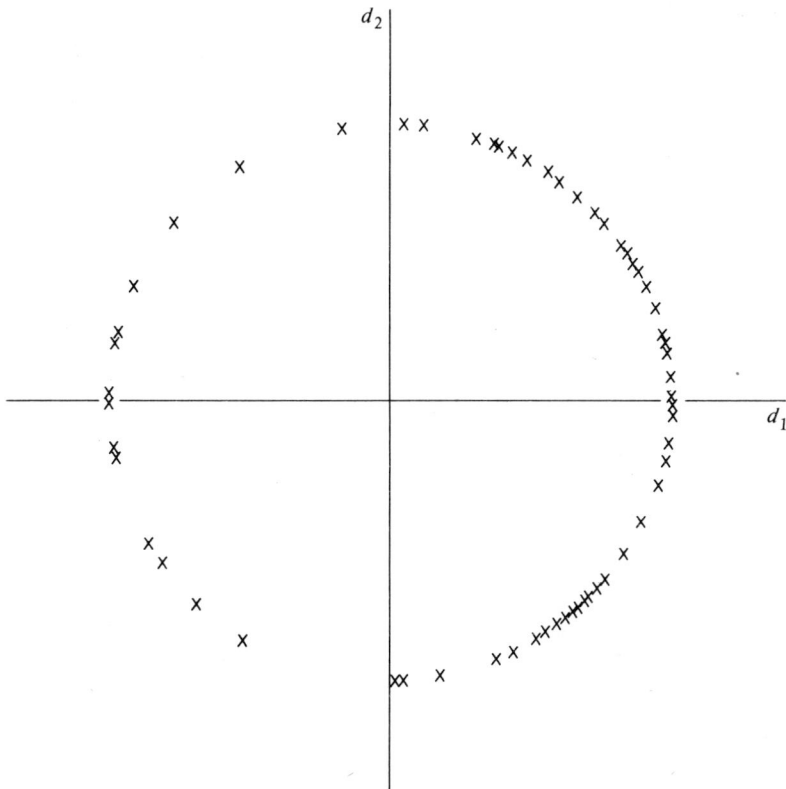

Figure 9-7 The residuals, the vectors in D_0.

The maximum likelihood values are
$$(\hat{\alpha}; \hat{\tau}, \hat{\sigma}) = (0.46, 0.08, 1.38).$$
The corresponding projected normal $PN(\hat{\lambda}\hat{\theta}(1,0)'; \hat{\theta}\hat{\theta}')$ is plotted in Fig. 9-6. It is bimodal with modes at 61° and 241°. The modes are diametrically opposite supporting the view "that the turtles have a preferred direction but some are confusing forwards with backwards" (see Stephens, 1969).

The residuals, recorded as columns in D^0, are plotted in Fig. 9-7. They should appear approximately as a sample from the projected normal $PN(0.7(1,0)', I)$ using the maximum likelihood value $\hat{\lambda} = 0.7$.

The variation-based model provides likelihood analysis for the distribution form parameter λ. It also provides tests, confidence intervals, and, of course, estimates for the presentation parameters α, τ, σ; the figures record some of the available distributions.

A response-model analysis using routine maximum likelihood analysis with a probability mixture of von Mises' normals gives modes at 63.5° and 241.2° (see Mardia, 1975).

REFERENCES AND BIBLIOGRAPHY

Cairns, M. B.: "A Structural Model for the Analysis of Directional Data," Ph.D. thesis, University of Toronto, 1975.

Fisher, R. A.: Dispersion on a Sphere, *Proc. Roy. Soc. London*, ser. A, vol. 217, pp. 295–305, 1953.

Fraser, D. A. S.: "The Structure of Inference," Huntington Krieger Publishing Company, New York, 1968.

Gumbel, E. J., J. A. Greenwood, and D. Durand: The Circular Normal Distribution: Theory and Tables, *Jour. Amer. Stat. Assoc.*, vol. 48, pp. 131–152, 1953.

Hammersley, J. M., and D. C. Hanscomb: "Monte Carlo Methods," Methuen, London, 1964.

Hartman, P., and G. S. Watson: "Normal" Distribution Functions on Spheres and the Modified Bessel Functions, *Ann. Prob.*, vol. 2, pp. 593–607, 1974.

Kendall, D. G.: Pole Seeking Brownian Motion and Bird Navigation, *Jour. Roy. Stat. Soc.*, ser. B, vol. 36, pp. 365–417, 1974.

Mardia, K. V.: "Statistics of Directional Data," Academic Press, London, 1972.

———: Statistics of Directional Data, *Jour. Roy. Stat. Soc.*, ser. B, 1975.

Mises, R. von: Uber die Ganzzahligkeit der Atomgewichte und Verwandte Fragen, *Physik. Z.*, vol. 19, pp. 490–500, 1918.

Stephens, M. A.: "The Statistics of Directions, the Fisher and von Mises Distributions," Ph.D. Thesis, University of Toronto, 1962.

———: Tests for the von Mises Distribution, *Biometrika*, vol. 56, 149–160, 1969.

Watson, G. S.: Analysis of Dispersion on a Sphere, *Monthly Notices, Roy. Astron. Soc., Geophys. Suppl.*, vol. 7, pp. 153–159, 1956.

———: Goodness-of-Fit Tests on a Circle, *Biometrika*, vol. 48, pp. 109–114, 1961.

———: Goodness-of-Fit on a Circle II, *Biometrika*, vol. 49, pp. 57–63, 1962.

——— and E. J. Williams: On the Construction of Significance Tests on the Circle and the Sphere, *Biometrika*, vol. 43, pp. 344–352, 1956.

Wheeler, S., and G. S. Watson: A Distribution-Free Two-Sample Test on a Circle, *Biometrika*, vol. 51, pp. 256–257, 1964.

CHAPTER
TEN
BIOASSAY AND DILUTION SERIES

A familiar problem in biological and pharmacological investigations is to determine the effective strength of a drug, chemical, or other stimulus administered to a living animal. A similar and somewhat related problem is to determine the concentration of organisms, say bacteria, in a suspension or solution. Sometimes these can be investigated directly—by increasing the strength until a reaction occurs in an animal or by counting the organisms in a sample of the solution. In other cases such direct methods may not be feasible, or perhaps even possible—and indeed not advisable if measurements or counts cannot be made accurately. An indirect approach for the case of a drug is to administer various doses to different animals and to record for each whether or not a reaction occurs; this is called *bioassay*. A parallel approach for the case of organisms in a solution is to sample various dilutions and, for each sample, test with a nutrient and record whether or not there is sterility; this is called a *dilution series* assessment.

In this chapter we investigate statistical inference for bioassay and dilution series. For these problems the traditional methods are less than satisfactory and involve calculating essentially just the maximum likelihood estimate for the parameters. Here we examine more incisive methods (Fraser and Prentice, 1971) that provide tests and confidence intervals for the primary parameter and the appropriate likelihood function for the remaining parameter. In effect this is an accurate split of the traditional likelihood function into two appropriately specific parts—of which one is amplified to give the tests and confidence intervals.

In Sec. 10-1 we discuss the background and the model. We find that the distribution form apart from its location can be modeled directly. This gives us an objective distribution for the variation and a location presentation for the response. In Sec. 10-2 we consider the analysis together with its applications to data from bioassay and dilution series.

10-1 THE MODEL

In this section we discuss the background and model for bioassay and dilution series.

10-1-1 Bioassay

Consider a drug that can be administered to a certain type of animal. Typically the response of the animal may be quite complex. For the cases considered here, however, the observed response is taken to be just "reaction" or "no reaction." In various contexts the reaction of interest may be death, or recovery from a disease, or remission, or something simpler or more specific.

The amount of drug administered to an animal by weight or volume is usually called the *dose* and designated X. Often, however, the effect of a drug can relate more naturally to the dose in multiplicative units—to the logarithm of the dose. The more natural measure is called the *dosage* and designated x; then we have $x = \log X$, where typically $\log X = \log_{10} X$.

Optimistically we might think of giving a progressively increasing dosage until reaction occurs and then recording the corresponding dosage x. In practice the preceding would rarely work; there is usually a time lag after administration until possible effect. Also, a progressive dosage is typically different from a dosage at one time. For example, if water temperature is gradually increased during a shower the effect goes almost unnoticed, but the same final temperature would be intolerable initially. Ideally, however, we can think of the threshold dosage x that gives a reaction for an individual, and then let $P(x)$ be the distribution function for x in the population of animals under investigation. This is called the *tolerance distribution*.

An investigator would like to know the distribution function P or at least its salient features. In particular, he or she often focuses on the dosage that gives a 50 percent reaction rate in the population, the *effective dosage* 50 percent; we call this the ED50 and designate it θ. Note that the ED50 is the median value for x and

$$P(\text{ED50}) = P(\theta) = 0.50 \tag{10-1}$$

The scaling or variability of x may be of interest.

The ED50 is a *standard* administration of drug that produces a 50 percent reaction rate. We define the *strength* s of an administration of a drug to be the amount that the dosage x exceeds the ED50; thus

$$s = x - \theta \tag{10-2}$$

With a reasonable choice of dosage scale the distribution form for x is often approximately normal or logistic. Thus

$$P(x) = G\left(\frac{x-\theta}{\sigma}\right) = G\left(\frac{s}{\sigma}\right) \qquad (10\text{-}3)$$

where θ is the ED50, σ is the scaling, and for the normal

$$G(z) = \int_{-\infty}^{z} \frac{1}{\sqrt{2\pi}} e^{-t^2/2} \, dt \qquad (10\text{-}4)$$

and for the logistic

$$G(z) = \frac{1}{1+e^{-z}} \qquad (10\text{-}5)$$

More generally we can allow a parametric family $\{G_\lambda : \lambda \in \Lambda\}$ for the distribution function G.

As we have noted, the progressive dosage is typically unrealistic. Rather, an animal is given a specific dosage and the investigator records whether or not reaction occurs. For an experimental design consider k dosages $x_1, \ldots, x_k$ and suppose that each is administered to n different animals. The total of kn animals would be randomly sampled from the population and the dosage levels randomly assigned to the sampled animals. If the available animals are not relatively homogeneous with respect to uncontrollable factors, a better arrangement would be a randomized block design using, say, litters as the blocking factor. Let y_j be the number of reactions at dosage x_j. With independence we then have the following probability function for $(y_1, \ldots, y_k)$:

$$\prod_{j=1}^{k} \binom{n}{y_j} P^{y_j}(x_j) Q^{n-y_j}(x_j) \qquad (10\text{-}6)$$

where we let $Q(x) = 1 - P(x)$.

In a typical investigation the dosages $x_1, \ldots, x_k$ are taken to be equally spaced, ranging from a "no reaction" dosage up to a "sure reaction" dosage; these limits are usually available from preliminary tests and the investigation is the formal determination of the precise location of the tolerance distribution. Thus with a spacing interval h we would have the dosages, say, $x_0, x_0 + h, \ldots, x_0 + (k-1)h$.

With equal spacing an important type of experimental randomization is available. Rather than start from some reference dosage x_0 the experimenter randomly chooses a number v from the uniform $[0, h)$ distribution and uses the dosages

$$\ldots, v-h, v, v+h, \ldots \qquad (10\text{-}7)$$

In practice there will be just k of these dosages. However, with the range from "no reaction" to "sure reaction" we can of course think of the open-ended series of dosages, but with the appropriate certain results for each end.

Table 10-1

Dose	Dosage x_j	Number of reactions y_j
10^3	3	0
10^4	4	1
10^5	5	5
10^6	6	6
10^7	7	7
10^8	8	10
10^9	9	9

Consider the data in Table 10-1 investigating the effect of a pneumonia organism on mice. The data with $k = 7$, $n = 10$ were made available by D. B. W. Reid of the University of Toronto; the initial randomization on the interval $[0, 1)$ was not formally included in the original design.

On the basis of ascribed sure reactions at the extremes we can present this in an open-ended form as Table 10-2. In Table 10-3 we record the y values beside the corresponding strengths $s = x - \theta$ of the administrations. These data are analyzed in Sec. 10-2.

The probability differential covering the randomization v and the binomial responses $y_j = y(v + jh)$ at dosages $x_j = v + jh$ has the following form:

$$\prod_{j=-\infty}^{\infty} \binom{n}{y_j} P^{y_j}(v+jh) Q^{n-y_j}(v+jh) \, dv = \prod \binom{n}{y_j} G^{y_j}\left(\frac{v+jh-\theta}{\sigma}\right)$$

$$\times \left[1 - G\left(\frac{v+jh-\theta}{\sigma}\right)\right]^{n-y_j} dv$$

$$= f_\sigma(x - \theta, y) \, dv$$

$$= f_\sigma(s, y) \, dv \qquad (10\text{-}8)$$

where, for example, (s, y) is an abbreviation for the extended vector $[(s_j, y_j): j = -\infty, \ldots, +\infty]$. The randomization v takes values in $[0, h)$ and the y_j take values in $\{0, 1, \ldots, n\}$. For small dosages we have $P(x)$ near zero and $1 - P(x)$ near 1, giving $y = 0$ with near certainty; with large dosages we have $P(x)$ near 1 and $1 - P(x)$ near zero giving $y_j = n$ with near certainty. The theoretical aspects of convergence of the double-ended product are examined in Fraser and Prentice (1971).

We can display observed data in several ways. An obvious first display is to plot the reaction proportions y_j/n against the dosages x_j. This is called a *stimulus response curve*. Note that it is an empirical version of the tolerance distribution function $P(x) = G[(x - \theta)/\sigma]$.

A second method of display is oriented toward the location-scale nature of the dosage; the function G^{-1} is applied to the proportions y_j/n. Note that the function G^{-1} applied to the probabilities gives the standardized dosages $(x - \theta)/\sigma$; correspondingly, the function G^{-1} applied to the proportions gives an "estimate"

Table 10-2	
x_j	y_j
.	.
.	.
.	.
2	0
3	0
4	1
5	5
6	6
7	7
8	10
9	9
10	10
11	10
.	.
.	.
.	.

Table 10-3	
s_j	y_j
.	.
.	.
.	.
$2-\theta$	0
$3-\theta$	0
$4-\theta$	1
$5-\theta$	5
$6-\theta$	6
$7-\theta$	7
$8-\theta$	10
$9-\theta$	9
$10-\theta$	10
$11-\theta$	10
.	.
.	.
.	.

of the standardized dosages. The second display involves plotting $G^{-1}(y_j/n)$ against the dosage x_j; a fitted line then gives estimates for θ and σ. For the normal, the inverse G^{-1} is called the *probit* function and is available from tables; to avoid negative numbers the probit is usually taken to be $5 + G^{-1}$. For the logistic, the inverse G^{-1} is called the *logit* function and is available explicitly:

$$G^{-1}(u) = \ln \frac{u}{1-u} \tag{10-9}$$

For applications in other than the biological areas consider the following: testing of material under various levels of explosive force; testing of electronic equipment under various levels of voltage surge; reconviction of a released convict under various related conditions.

10-1-2 Dilution Series

Consider a solution or suspension containing bacteria or some other organism of interest. This could be a water supply containing coliform bacteria; or it could be a vaccine containing some live viruses; or it could be particular bacilli in flour. The problem is to determine the concentration of the organism in the solution. We assume that the organism is uniformly and randomly distributed throughout the solution.

An obvious first approach is to count the organisms directly, as, for example, white cells are counted on a blood slide. However, with bacteria in water or viruses in a vaccine such a visual approach is typically unavailable or, if available, prone to gross counting errors.

An alternative approach having the same dichotomous 0, 1 nature as the bioassay design is to take samples at various dilutions and, for each sample, provide a nutrient under growth conditions and record whether the sample is sterile as evidenced by no growth or fertile as evidenced by growth. By taking low-dilution samples with almost sure fertility through to high-dilution samples with almost sure sterility it is possible to bracket the critical region and to estimate the density of the organism in the solution.

We now discuss this dilution series method following very closely the pattern for the bioassay method. However, it is convenient to retain one major distinction concerning direction. For bioassay we examine a sequence with increasing dosages; for dilution series assessment we examine a sequence with *increasing dilution*, decreasing concentration, but with increasing dilution and associated increasing probability of sterility.

Consider a given solution containing a living organism distributed uniformly and at random throughout the solution. For a derived solution obtained by dilution we define the *dilution factor* X to be the final volume divided by the volume before dilution. We then define the *dilution dosage* $x = \log X$ to be the dilution factor expressed in logarithmic units, typically base 10.

Ideally we can think of taking an initial sample unit; subjecting it to a progressive dilution dosage x, retaining just a sample unit; and stopping when the retained unit becomes sterile. Let $P(x)$ be the distribution function for this threshold dilution dosage. We derive the form of $P(x)$.

Let λ be the average number of organisms per unit volume for the original solution and let $\theta = \log \lambda$ be the *log-concentration*; the average number of organisms per unit volume is then 10^θ. For a derived solution obtained by dilution dosage x the average number per unit volume is

$$10^{\theta-x} = 10^{-(x-\theta)} = 10^{-s} \qquad (10\text{-}10)$$

We now define the *dilution strength* of a solution. For a solution with an average of one organism per unit volume we say the *dilution strength* is 0 and for a solution obtained from this by a dilution dosage s we say the *dilution strength* is s. Accordingly we see that the dilution strength is the negative of the log-concentration.

We now see that the original solution has dilution strength $-\theta$ and we then note that a dilution dosage θ applied to the original solution produces a derived solution with dilution strength 0. For a solution of dilution strength s the average number of particles per unit volume is

$$10^{-s} \qquad (10\text{-}11)$$

Accordingly, from the Poisson distribution we have the probability

$$\exp(-10^{-s}) \qquad (10\text{-}12)$$

that a unit volume is sterile.

Now consider a dilution dosage x applied to the original solution. From formulas (10-10) and (10-12) the probability that a unit volume is sterile is

$$P(x) = \exp(-10^{-(x-\theta)})$$
$$= G(x - \theta) \tag{10-13}$$

where

$$G(s) = \exp(-10^{-s}) \tag{10-14}$$

Note that this is the distribution function for the extreme-value distribution.

The present definitions thus give us a direct correspondence to bioassay. The formula (10-3) for the tolerance distribution with $\sigma = 1$ becomes the present formula (10-13), and the normal and logistic distributions, (10-4) and (10-5), are replaced by the extreme-value distribution (10-14).

The remaining pattern for dilution series assessment corresponds directly with that for bioassay. For an experimental design we consider k dilutions $x_1, \ldots, x_k$ and from each obtain n samples of unit volume; we then have kn samples. Note that the dilutions should be large with respect to the samples in order that the samples be "independent" samples from the diluted version of the original solution. Each sample is then tested with a nutrient to determine if the "reaction" sterility occurs. Let y_j be the number sterile at dosage x_j. We then have the probability function

$$\prod_{j=1}^{k} \binom{n}{y_j} P^{y_j}(x_j) Q^{n-y_j}(x_j) \tag{10-15}$$

where, of course, $Q(x) = 1 - P(x)$.

In the standard investigation the dilution dosages $x_1, \ldots, x_k$ are taken to be equally spaced, ranging from a "sure fertile" dosage up to a "sure sterile" dosage. Thus with spacing interval h we have the dosages

$$x_0, x_0 + h, \ldots, x_0 + (k-1)h$$

With equal spacing we have available the randomization discussed for bioassay. In fact, the randomization was proposed originally by Fisher (1935a) for the dilution series application. The experimenter randomly chooses a number v from the uniform $[0, h)$ distribution and uses the dilution dosages

$$\ldots, v - h, v, v + h, \ldots$$

As before we consider the open-ended series of dosages.

Consider the data shown in Table 10-4 investigating the density of rope spores in potato flour. The data were reported by Fisher and Yates (1963); we have $k = 10$, $n = 5$, and $h = \log 2 = 0.301$. The initial randomization on the interval $[0, h)$ was not formally included in the original design. The dilutions are based on a basic unit of 1 gram although examined in a mixture having volume 100 cm^3. Thus we can have negative dilution dosages and the density is recorded as a count per gram. We view this in the open-ended form used for the bioassay example in Sec. 10-1-1. These data are analyzed in Sec. 10-2.

The probability differential covering the randomization v and the binomial

240 INFERENCE AND LINEAR MODELS

Table 10-4

Dilution factor	Dilution dosage x_j	Number sterile y_j
$\frac{1}{4}$	$-2h$	0
$\frac{1}{2}$	$-h$	0
1	0	0
2	h	0
4	$2h$	1
8	$3h$	2
16	$4h$	3
32	$5h$	3
64	$6h$	5
128	$7h$	5

responses $y_j = y(v + jh)$ at dosages $x_j = v + jh$ has the following form corresponding to (10-8):

$$\prod_{j=-\infty}^{\infty} \binom{n}{y_j} P^{y_j}(v+jh) Q^{n-y_j}(v+jh)\, dv = \prod \binom{n}{y_j} G^{y_j}(v+jh-\theta)$$

$$\times [1 - G(v+jh-\theta)]^{n-y_j}\, dv$$

$$= f(x-\theta, y)\, dv$$

$$= f(s, y)\, dv \qquad (10\text{-}16)$$

where v takes values in $[0, h)$, the y_j take values in $\{0, 1, \ldots, n\}$, and (s, y) designates the extended vector as in formula (10-8). For a discussion of convergence, again see Fraser and Prentice (1971). The methods of display for bioassay are, of course, available here, but with some simplification resulting from having $\sigma = 1$.

10-2 THE ANALYSIS: THEORY AND EXAMPLES

We now consider the analysis of an inference base involving the bioassay and dilution model from Sec. 10-1.

10-2-1 The Formal Model

Consider the bioassay model in formulas (10-8) and (10-16) together with the illustrations provided by Tables 10-2 and 10-4 in the preceding section.

For this model, it is useful to envisage a long laboratory table marked out with strength scale s. Points along the scale at intervals of length h each have n independent tests at the corresponding strengths. The randomization v on the interval $[0, h]$ has the effect of randomly locating the lattice of points on the strength scale. The statistical model (10-8) has the form

$$f_\sigma(s, y)\, dv = \prod_{j=-\infty}^{\infty} \binom{n}{y_j} G^{y_j}\!\left(\frac{s_j}{\sigma}\right)\left[1 - G\!\left(\frac{s_j}{\sigma}\right)\right]^{n-y_j} dv \qquad (10\text{-}17)$$

The investigator, however, does not see the strength scale itself. Rather, he or she sees the corresponding dosage scale

$$x = \theta + s \qquad (10\text{-}18)$$

which is a displacement $+\theta$ on the strength scale.

We have used the term variation for the objective description of distribution form. Here the basic variation can be pictured directly on the laboratory table. For the investigator the only thing not observable is the *strength scale* itself; he or she sees the dosage scale, a simple translation on the strength scale.

The randomization was formally applied to randomly locate the lattice of points on the dosage scale. The purpose of the randomization, however, is to randomly locate the lattice of strength values on the strength scale. Recall the discussion in Sec. 3-2-4. Accordingly we do not condition on the "observed" randomization because the *essential* randomization on the strength scale is unobservable.

For notation let $\mathscr{V}$ designate the distributions represented by formula (10-17). The sample space is

$$\mathscr{S} = [0, h) \times \{0, 1, \ldots, n\}^k \qquad (10\text{-}19)$$

where $[0, h)$ is the interval for the randomization and $\{0, 1, \ldots, n\}$ is the space for y_j with j running through the integers $1, \ldots, k$. However, to be formally correct for the theory developed below we should record a sample space $\{0, 1, \ldots, n\}$ for each y_j with j running through integers from $-\infty$ to $+\infty$. Let $\mathscr{G}$ be the location transformations

$$\mathscr{G} = \{[\theta, 1]: \theta \in \mathbb{R}\} \qquad (10\text{-}20)$$

which carry the strength scale into the dosage scale (10-18). We then have the variation-based model

$$\mathscr{M}_V = \{\Omega; \mathscr{S}, \mathscr{A}, \mathscr{V}, \mathscr{G}\} \qquad (10\text{-}21)$$

where $\Omega = \{(\theta, \sigma)\} = \mathscr{G} \times \mathbb{R}^+$, $\mathscr{A}$ is the appropriate σ-algebra for the sample space $\mathscr{S}$ in (10-19), $\mathscr{V}$ is the class of distributions (10-17) with parameter σ, and $\mathscr{G}$ is the location group; or more generally we would have σ replaced by λ covering both scale and form. This is a very simple structural model based on the location group.

10-2-2 The Analysis

Consider the analysis of an inference base

$$(\mathscr{M}_V, \{x^0, y^0\}) \qquad (10\text{-}22)$$

using the model $\mathscr{M}_V$ in (10-21) and data $\{x^0, y^0\}$ such as in Tables 10-2 or 10-4.

242 INFERENCE AND LINEAR MODELS

This is a location analysis, a much simplified version of the location-scale analysis in Chap. 2. For the analysis we need a location statistic and a reference point.

First consider the location statistic and envisage the long laboratory table. The values of y_j for strengths negatively on the scale are surely zero; let $r(s, y)$ be the first strength coming up the scale at which the corresponding count is different from zero; and, of course, let $r(x, y)$ be the first dosage coming up the scale at which the count is different from zero. For example, from Tables 10-2 and 10-3 we have $r(s, y)$ recorded as $4 - \theta$ and we have $r(x, y) = 4$. Clearly, in general, we have

$$r(x, y) = \theta + r(s, y) \tag{10-23}$$

Now consider the corresponding reference value. This is obtained by a location transformation $[-r, 1]$. For the variation this adjusts the scale so that the zero point is opposite the first count y_j that is different from zero. The corresponding transformation for the responses gives the same result: the zero point is opposite the first count y_j that is different from zero. For this adjusted scale, let

$$\begin{aligned} d &= s - r(s, y) \\ &- x - r(x, y) \end{aligned} \tag{10-24}$$

Consider the data in Table 10-2; in Table 10-5 we record the counts opposite this new d scale.

Table 10-5

d_j	y_j
.	.
.	.
.	.
−2	0
−1	0
0	1
1	5
2	6
3	7
4	10
5	9
6	10
7	10
.	.
.	.
.	.

The observed orbit D^0 is represented by Table 10-5. The distributions for the orbit are obtained by integrating over the location statistic $r = r(s, y)$:

$$h_\sigma(D) = \int_{-\infty}^{\infty} f_\sigma(r + d, y) \, dr$$

$$= \int_{-\infty}^{\infty} \Pi \binom{n}{y_j} G^{y_j}\left(\frac{r+d_j}{\sigma}\right)\left[1 - G\left(\frac{r+d_j}{\sigma}\right)\right]^{n-y_j} dr \quad (10\text{-}25)$$

In the pattern of preceding chapters we then have the inference base

$$(\mathscr{M}_D, D^0) \quad (10\text{-}26)$$

If we insert the observed values in (10-25) we obtain the observed likelihood function for σ:

$$ch_\sigma(D^0) = c \int_{-\infty}^{\infty} \Pi G^{y_j^0}\left(\frac{r+d_j^0}{\sigma}\right)\left[1 - G\left(\frac{r+d_j^0}{\sigma}\right)\right]^{n-y_j^0} dr \quad (10\text{-}27)$$

The unobservable characteristic of the variation is the location of the first nonzero count on the strength scale. The distribution for this unobservable characteristic is

$$g_\sigma(r: D^0) = h_\sigma^{-1}(D^0) \Pi \binom{n}{y_j} G^{y_j}\left(\frac{r+d_j^0}{\sigma}\right)\left[1 - G\left(\frac{r+d_j^0}{\sigma}\right)\right]^{n-y_j} \quad (10\text{-}28)$$

This distribution for the variation is used with the transformation

$$r(x, y) = \theta + r \quad (10\text{-}29)$$

The corresponding observed value for the location of the response is $r(x, y^0)$. This gives the inference base

$$(\mathscr{M}_V^{D^0}, r(x, y^0)) \quad (10\text{-}30)$$

in the pattern of preceding chapters.

The response distribution for $r^* = r(x, y)$ corresponding to the identified variation is

$$h_\sigma^{-1}(D^0) \Pi \binom{n}{y_j} G^{y_j}\left(\frac{r^* - \theta + d_j^0}{\sigma}\right)\left[1 - G\left(\frac{r^* - \theta + d_j^0}{\sigma}\right)\right]^{n-y_j} \quad (10\text{-}31)$$

on the real line.

The usual bioassay analysis is based on the overall likelihood function

$$c \Pi \binom{n}{y_j} G^{y_j}\left(\frac{x_j - \theta}{\sigma}\right)\left[1 - G\left(\frac{x_j - \theta}{\sigma}\right)\right]^{n-y_j} \quad (10\text{-}32)$$

Our analysis here has factored this likelihood function into two very specific components. The first component is the proper likelihood function (10-27)

$$ch_\sigma(D^0)$$

for the analysis of the parameter σ. The second component is the likelihood version of (10-31). Our analysis goes far beyond a simple likelihood analysis of (10-31); it admits tests and confidence intervals using (10-31) or using the distribution (10-28) with the equation (10-29).

Thus, rather than just a single likelihood function for (θ, σ), we have a separation giving the specific likelihood for σ and something *more* than just the specific likelihood for the parameter of interest θ.

10-2-3 Bioassay Example[†]

Consider an investigation concerning the effect of a certain strain of pneumonia organism on mice. The main purpose of the investigation is one of standardization—to find the ED50 dosage of pneumonia organism that produces a 50 percent mortality rate in mice; with mortality, the ED50 is usually called lethal dosage fifty and written LD50.

The investigation produced the data in Table 10-6; these data were made available by D. B. W. Reid of the University of Toronto. Note that these are the data used for Tables 10-1, 10-2, and 10-3 in Sec. 10-1-1. We have $k = 7, n = 10$.

Table 10-6

Number of organisms	Mortality rate
10^3	0/10
10^4	1/10
10^5	5/10
10^6	6/10
10^7	7/10
10^8	10/10
10^9	9/10

The usual method of bioassay analysis is based on the likelihood function and its large sample approximation. The method involves an iterative solution for the maximum likelihood estimate and was proposed by Fisher (1935b). For a general survey, see Finney (1971).

The method involves an initial display of the data as a plot of probit or logit against dosage (see the end of Sec. 10-1-1). A straight line is fitted. This provides an estimate of standardized dosage

$$\frac{x - \hat{\theta}}{\hat{\sigma}} = -\frac{\hat{\theta}}{\hat{\sigma}} + \frac{1}{\hat{\sigma}}x = \hat{\alpha} + \hat{\tau}x$$

as a function of dosage x, and it gives a preliminary estimate for (θ, σ). The preliminary estimate is then used with an iterative procedure for obtaining the exact maximum likelihood estimate $(\hat{\mu}, \hat{\sigma})$ based on the probability function (10-32). The iterative procedure involves the usual local linearization method for solving nonlinear equations; see, for example, Draper and Smith (1966). This maximum likelihood procedure is available as a bioassay package at most computer centres.

For the present data the logistic model was used; in practice there seems to be very little difference between the use of the logistic and the use of the normal.

[†] With Daryl Pregibon and Allen McIntosh.

The weighted straight line regression of logit on dosage followed by several iterations yielded the following line:

$$\hat{l} = -5.788 + 1.019x = \hat{\alpha} + \hat{\tau}x$$

From this we obtain the maximum likelihood estimates

$$\hat{\theta} = -\frac{\hat{\alpha}}{\hat{\tau}} = 5.683 \qquad \hat{\tau} = \frac{1}{\hat{\sigma}} = 1.019$$

To a first-order approximation the estimated standard deviation for $\hat{\theta}$ is 0.3224. An approximate 95 percent confidence interval for θ is obtained by taking ± 1.96 standard deviations:

$$(5.05, 6.31)$$

This is based on the large-sample theory for likelihood functions.

Now consider the analysis as based on the theory in Chap. 7. As indicated in Sec. 10-2-2 we have a separation of the likelihood function giving the specific likelihood component for σ or $\tau = 1/\sigma$, and something more—in fact a full variation-based model for the primary parameter θ.

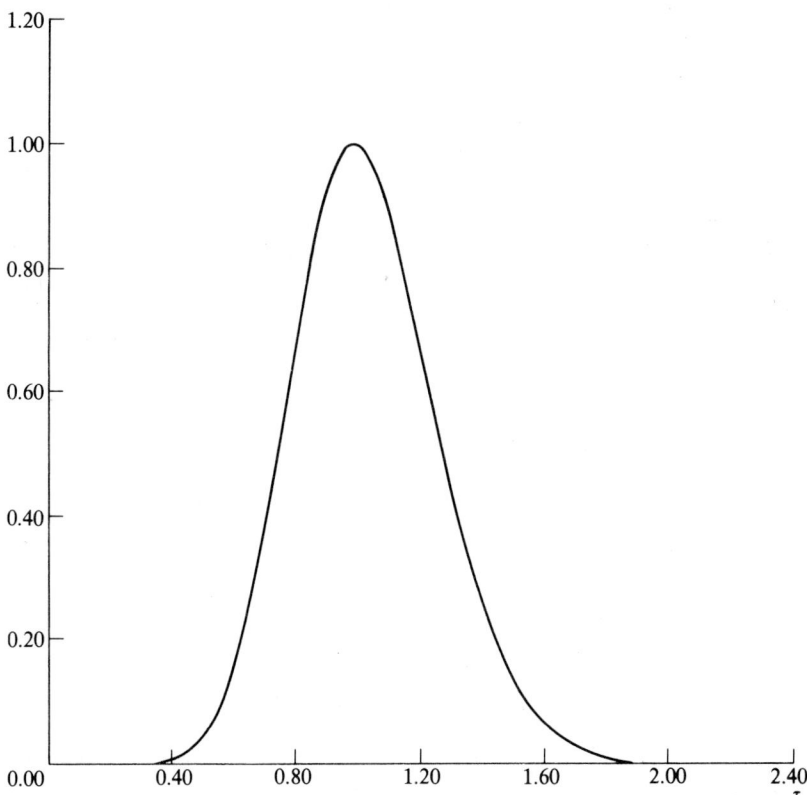

Figure 10-1 Likelihood function for τ in the bioassay example.

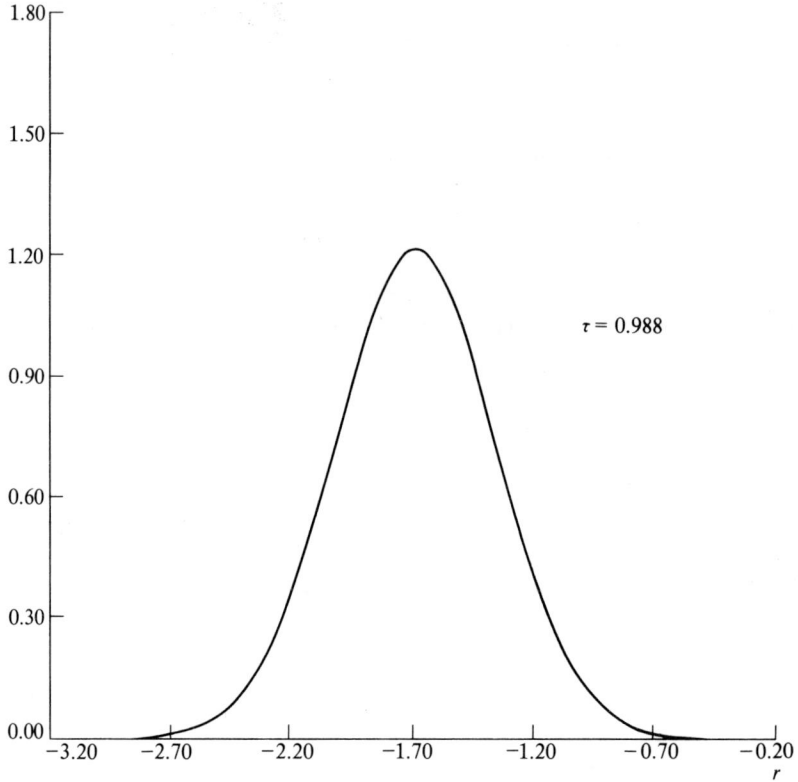

Figure 10-2 Density function for the location statistic r in the bioassay example $\tau = 0.988$.

The likelihood function for $\tau = 1/\sigma$ is plotted in Fig. 10-1. The parameter τ indicates precision and its reciprocal σ is the standard deviation for the tolerance distribution. The maximum likelihood estimate of τ is 0.988; this value is smaller than the value 1.019 obtained from the analysis of the joint likelihood function.

Table 10-7

τ	Level in percentage	Lower limit	Upper limit
0.6	90	4.83	6.36
	95	4.67	6.49
	99	4.20	6.71
1.0	90	5.13	6.20
	95	5.02	6.30
	99	4.75	6.46
1.4	90	5.24	6.11
	95	5.14	6.18
	99	4.41	6.28

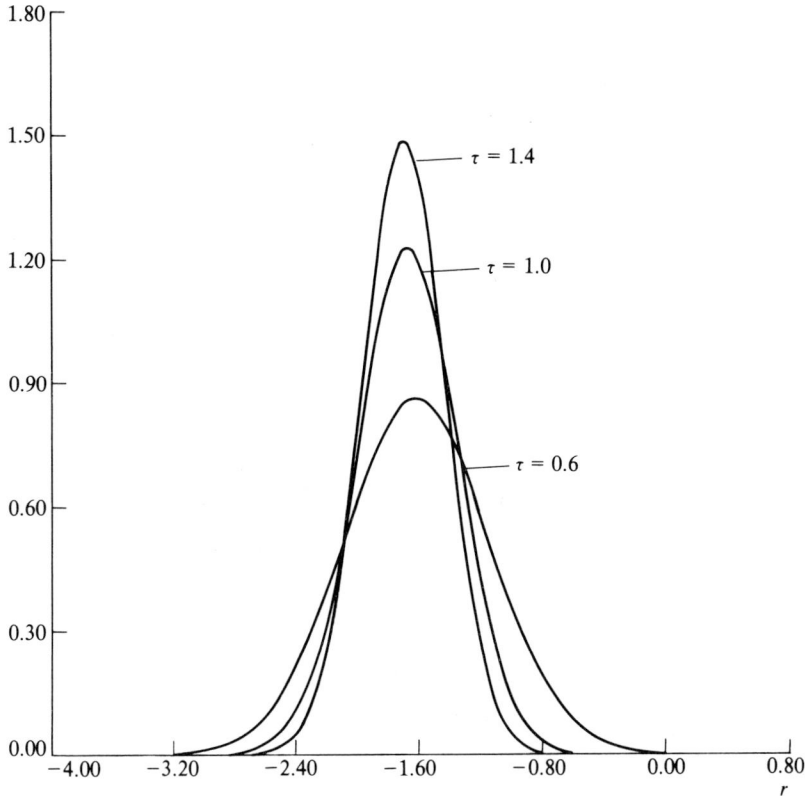

Figure 10-3 Density function for the location statistic r in the bioassay example $\tau = 0.6, 1.0, 1.4$.

The conditional distribution for r is plotted for $\tau = 0.988$ in Fig. 10-2. The distributions for other values of τ have a similar shape, but are characterized by the fact that as τ increases the distribution becomes more concentrated; see Fig. 10-3. Again, interpreting τ as precision we see that as precision increases the inferences re θ get sharper.

The 90, 95, and 99 percent confidence intervals for the LD50 θ are recorded for several values of τ in Table 10-7. Note that the 95 percent confidence interval obtained here with $\tau = 1.0$ is similar to that obtained with the large-sample likelihood method. In general, this is not the case, however, as the density function for r is usually asymmetric.

10-2-4 Dilution Series Example†

Consider an investigation concerning the density of rope spores (*Bacillus mesentericus*) in a particular batch of potato flour. The purpose is to estimate the number

† With Daryl Pregibon and Allen McIntosh.

248 INFERENCE AND LINEAR MODELS

Table 10-8

Dilution factor (re 1 gram/100 cm³)	Number of sterile samples out of $n = 5$
$\frac{1}{4}$	0
$\frac{1}{2}$	0
1	0
2	0
4	1
8	2
16	3
32	3
64	5
128	5

λ of spores per gram of flour. The investigation produced the data in Table 10-8; the data were reported by Fisher and Yates (1963). Note that these are the data recorded in Table 10-4 of Sec. 10-1-2. We have $k = 10$, $n = 5$. Also recall that the

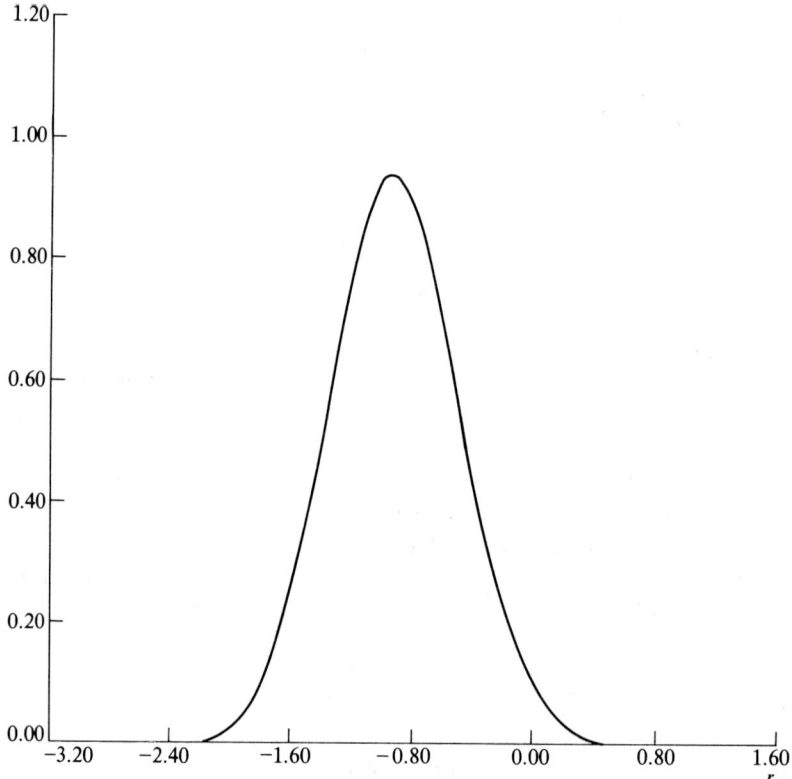

Figure 10-4 Density for location statistic in the dilution series example.

Table 10-9

Confidence level, in percentage	Lower limit for λ	Upper limit for λ
90	459	1205
95	415	1312
98	370	1445
99	341	1542

dilutions are based on a standard unit of one gram although examined in a mixture having volume 100 cm³.

The usual method is based on the likelihood function and its large-sample approximation. The likelihood maximization with a single real parameter is straightforward; see, for example, Finney (1971, chap. 21). The maximum likelihood estimate of λ is 766 spores per gram and the approximate 95 percent confidence interval based on large-sample theory is

$$(431, 1363)$$

Now consider the exact analysis as based on the theory for the variation-based model in Sec. 10-2-2. The distribution for the location statistic r is recorded in Fig. 10-4; note the slight asymmetry of the distribution. The connection with the parameter λ is given by $r(x, y) - \theta = r$ where $\theta = \log \lambda$ is the log-concentration.

Exact confidence intervals for θ and then for λ can be obtained from central probability intervals for the distribution of r. The median estimate corresponding to the center of the r distribution is 757 spores per gram—in fact, 756.58.

Note that the approximate 95 percent confidence interval based on asymptotic normality is (431, 1363) whereas the exact 95 percent confidence interval taking into account the asymmetry and nonnormality is (415, 1312).

10-2-5 Conclusions

The preceding examination of the dilution and bioassay problems is by no means intended to be complete. Many books have been written on the subject and no doubt many more will be written. Our central objective is to describe how drug strength and bacterial density can be assessed by *exact* methods using a variation-based model.

Some natural extensions from our development here are: (1) estimation for percentage points of the tolerance distribution other than the median; (2) assessment of the relative potency of two competing drugs; (3) examination of a more exhaustive family of tolerance distributions (robustness); (4) indifference to maverick observations (resistance); and (5) methods for model assessment (for example, are bacteria uniformly distributed throughout the suspension?).

REFERENCES AND BIBLIOGRAPHY

Draper, N. R., and H. Smith: "Applied Regression Analysis," John Wiley and Sons, New York, 1966.
Finney, D. J.: "Statistical Method in Biological Assay," 2nd imp., Hafner Press, New York, 1971.
Fisher, R. A.: On the Mathematical Foundations of Theoretical Statistics, *Phil. Trans. Roy. Soc. London*, ser. A, vol. 222, pp. 309–368, 1922. Also as paper 10 in Fisher (1950).
———: Theory of Statistical Estimation. *Proc. Camb. Phil. Soc.*, vol. XXII, pt. 5, pp. 700–725, 1925. Also as paper 11 in Fisher (1950).
———: The Logic of Inductive Inference, *Jour. Roy. Stat. Soc.*, vol. 98, pp. 39–54, 1935a. Also as paper 26 in Fisher (1950).
———: "Contributions to Mathematical Statistics," John Wiley and Sons, New York, 1950.
——— and F. Yates: "Statistical Tables for Biological Agricultural and Medical Research," 6th ed., Oliver and Boyd, Edinburgh, 1963.
Fraser, D. A. S., and R. L. Prentice: Randomized Models and the Dilution and Bioassay Problems, *Ann. Math. Stat.*, vol. 42, no. 1, pp. 141–146, 1971.

CHAPTER
ELEVEN
EXTENDED LIKELIHOOD METHODS

The more parameters there are in a statistical model, the more complex the problem is of finding reasonable inference procedures. Indeed, the question often is less one of finding reasonable, let alone optimum, procedures and more one of finding some seemingly relevant procedure.

In preceding chapters we have seen how the presence of a variation-based model provides incisive access to reasonable, indeed the absolute, inference procedures for many parameters. Specifically, we obtain a separation of an inference base by means of necessary method RM_3 in Sec. 3-3 (see Sec. 7-1). For the shape parameter we obtain an ordinary statistical model with an available likelihood function and for the remaining location and structuring parameter we obtain an unequivocal determination of tests and confidence regions.

With the ordinary response-based model the observed likelihood function is the particular entity commonly available for inference. The corresponding variation-based model in effect splits this likelihood into two precise components. As we have noted, one component is the proper likelihood for the shape parameter and the other component is the likelihood for the location and structuring parameters — but the variation-based model makes much more available for these location parameters, specifically the distributions and relations that give unequivocal tests and confidence regions.

With the ordinary response-based model the observed likelihood function is, as just noted, the available entity for inference. The applied context, however, may support a generalized version of the variation-based model. Some separation of the likelihood function may then be possible. For this we maintain our interpretation of likelihood as the probability for what is observed. In this chapter we examine some extended likelihood methods; the methods are *extended* by using a somewhat more general interpretation of what is observable.

11-1 SOME LIKELIHOOD COMPONENTS

As a preamble to investigating extended likelihood we first examine two common methods for obtaining components of the likelihood function. Typically these methods do not give a component that can be viewed as a probability for what is observed. Indeed, some examples suggest that these components can be quite misleading.

An observed likelihood function $L(y;\theta)$ provides an assessment for the full parameter θ. In the multiparameter case an assessment for a component parameter is not immediately available; we briefly examine two common methods for obtaining component "likelihoods."

11-1-1 Profile Likelihood

The first method is a commonly used method and involves *profiling out* unwanted parameters. Let $\theta_1 = h(\theta)$ be a parameter component of interest. Then the *profile* "likelihood" is

$$L_1(\theta_1) = \sup_{h(\theta) = \theta_1} L(y;\theta) \tag{11-1}$$

This involves taking the supremum over all θ values corresponding to the value θ_1 of the parameter component—that is, over the preimage set $h^{-1}(\theta_1)$. If the full likelihood function is on a product space with one coordinate θ_1, then $L(\theta_1)$ is the *profile* of the full likelihood as viewed from the θ_1 axis.

Note that the maximum likelihood value for the profile likelihood is the θ_1 value associated with the maximum likelihood value for the full likelihood.

Even a simple normal example can illustrate how misleading the profile likelihood can be. Consider a sample $y_1, \ldots, y_n$ from the normal (μ, σ). The profile likelihood for σ^2 is

$$\begin{aligned}L_1(\sigma^2) &= \sup_\mu c(\sigma^2)^{-n/2} \exp\left[-\frac{(n-1)s_y^2}{2\sigma^2} - \frac{n(\bar{y}-\mu)^2}{2\sigma^2}\right] \\ &= c(\sigma^2)^{-n/2} \exp\left[-\frac{(n-1)s_y^2}{2\sigma^2}\right]\end{aligned} \tag{11-2}$$

This "likelihood" function for σ^2 depends only on the sample variance s_y^2. The likelihood function for this component variable, however, is

$$L(s_y^2;\sigma^2) = c(\sigma^2)^{-(n-1)/2} \exp\left[-\frac{(n-1)s_y^2}{2\sigma^2}\right] \tag{11-3}$$

Note that the maximizing value for profile likelihood is $(n-1)s_y^2/n$, whereas the maximizing value for the proper likelihood (11-3) from s_y^2 is s_y^2 itself. With normal regression models the discrepancy $(n-1)/n$ becomes $(n-r)/n$ where r is the number of regression vectors. For larger r the discrepancy can be very serious and misleading.

As a second example consider the bioassay example in Sec. 10-2-3. The maximum profile likelihood value for the secondary parameter τ is 1.019. By contrast, the maximum likelihood value for the proper likelihood from the component variable is 0.988—showing that the precision for the primary parameter is in fact lower than that indicated by the profile likelihood.

11-1-2 Integrated Likelihood

The second method is also a commonly used method and involves *integrating out* unwanted parameters. Let $\theta_1 = h(\theta)$ be a parameter component of interest and $\psi_1 = k(\theta)$ be a complementary parameter. Then the *integrated "likelihood"* is

$$L_2(\theta_1) = \int L(y;\theta_1,\psi_1)\,d\mu(\psi_1|\theta_1) \tag{11-4}$$

where $d\mu(\psi_1|\theta_1)$ designates a measure for ψ_1 given θ_1 chosen in some convenient or insightful way. In certain ways this is even farther from true likelihood than the profile likelihood.

If $\mu(\psi_1|\theta_1)$ happens to be a probability measure on ψ_1 values then certain difficulties with L_2 are avoided. However, for most problems there does not exist a probability measure that, say, eliminates the effect of a parameter—as, for example, by eliminating some corresponding variable. On the other hand, if $\mu(\psi_1|\theta_1)$ is allowed to be a measure uniform in a certain sense then it may be possible to eliminate the effect of a parameter by eliminating a corresponding variable; but then those certain difficulties are not avoided.

An integrated likelihood is not in general a likelihood. This raises the question: what is it? With a full likelihood function that turns out to be hard to analyze, it is natural to try any plausible techniques. But trying something does not make that something good. Justifications are needed from somewhere, and the examples for integrated likelihood are mostly negative.

The integrated likelihood has a place in Bayesian theory. The development in this book, however, finds no place for Bayesian theory. For some general comments on these integration methods and on Bayesian methods in general, see Fraser (1972a, 1974).

11-2 EXTENDED LIKELIHOOD

In Sec. 7-1 we examined structural models and obtained the separation of an inference base by using necessary method RM_3 from Sec. 3-3. At the beginning of this chapter we noted how this corresponded to a separation of the full likelihood function.

We now examine some generalized structural models, models that are structural if a particular component parameter is specified in value. We then investigate what is observable and derive a separation of the likelihood function. For further details on the development of the method see Fraser (1972b).

11-2-1 Generalized Structural Model

We use most of the basic notation for a structural model. Let $\mathscr{S}$ be the sample space with σ-algebra $\mathscr{A}$ and let $\mathscr{V}$ be a class of distributions for the variation with parameter λ in Λ. As before, we suppose that there is a group G.

For a first generalization we suppose that the *action* of the group G on the space $\mathscr{S}$ depends on a parameter κ. Thus for θ in the group G we have the transformation θ_κ on the space $\mathscr{S}$; we let G_κ designate the class of transformations θ_κ. In our abbreviated notation we then have:

$$Y = \theta_\kappa Z \text{ with } \theta \text{ in } G$$
$$Z \text{ has distribution in the class } \mathscr{V} \qquad (11\text{-}5)$$

The full parameter is $(\theta, \lambda, \kappa)$ and $\Omega = G \times \Lambda \times K$. The model has a parameter λ for distribution form and a parameter κ for the type of response expression; the primary parameter θ gives the particular response presentation.

As a simple example consider the regression model

$$\mathbf{y} = X_\kappa \boldsymbol{\beta} + \sigma \mathbf{z}$$
$$\mathbf{z} \text{ has distribution } f_\lambda(\mathbf{z})\, d\mathbf{z} \qquad (11\text{-}6)$$

The group here is the regression scale group examined in Chap. 6. The application of the group to the response space depends on the vectors of the design matrix; we suppose that the design matrix X_κ depends on the parameter κ. This model can be appropriate if the proper mode of expression for the input variables is in doubt; the parameter κ would allow some reasonable range of possibilities.

For a second generalization we suppose that there is a structural model

$$Y = \theta Z \text{ with } \theta \text{ in } G$$
$$Z \text{ has distribution in the class } \mathscr{V} \qquad (11\text{-}7)$$

for some natural response variable for the system being investigated, a variable giving, say, the additivity and near normality expected with some regression models. However, the proper mode of expression for the response is not clear and the given response variable is some reexpression ρ of the natural response variable; we suppose the transformations ρ are bijections on $\mathscr{S}$. This gives the model

$$Y = \rho \theta Z \text{ with } \theta \text{ in } G, \rho \text{ in } \Pi$$
$$Z \text{ has a distribution in the class } \mathscr{V} \qquad (11\text{-}8)$$

The full parameter is (θ, λ, ρ) and $\Omega = G \times \Lambda \times \Pi$ where Π is the class of possible response expressions in terms of the natural variable. The model has a parameter

λ for distribution form and a parameter ρ for the reexpression of the natural response; the primary parameter θ gives the particular natural response.

As a simple example consider the regression model but with uncertainty as to the proper mode of expression for the response. Consider

$$\rho^{-1}\mathbf{y} = \begin{bmatrix} \rho^{-1}y_1 \\ \vdots \\ \rho^{-1}y_n \end{bmatrix} = X\boldsymbol{\beta} + \sigma\mathbf{z} \qquad (11\text{-}9)$$

$\mathbf{z}$ has distribution $f_\lambda(\mathbf{z})\,d\mathbf{z}$

where ρ^{-1} applied to a coordinate for the given response variable produces the natural response variable that has the particular linear structure; the possible transformations ρ would be bijections on the real line. As an example for ρ consider the power transformations

$$\rho^{-1}y = \begin{cases} y^\rho & \rho \neq 0 \\ \ln y & \rho = 0 \end{cases} \qquad (11\text{-}10)$$

but note that these transformations operate primarily on the positive axis and thus the bijective property could be an approximation for some appropriate range.

The second generalization can be formally treated as a special case of the first generalization. For this we write $Y = \rho\theta Z = \rho\theta\rho^{-1}\rho Z = \theta_\rho \rho Z$, and use $\rho G \rho^{-1}$ as the transformation group and ρZ as a modified form for the variation.

The two generalizations, however, can be taken as they stand and combined to form the *generalized structural model*. For this it is notationally convenient to use λ for the full parameter other than θ:

$$Y = \lambda \theta_\lambda Z \text{ with } \theta \text{ in } G \qquad (11\text{-}11)$$
$$Z \text{ has distribution in the class } \mathscr{V}$$

The parameter λ covers the form of the distribution in $\mathscr{V}$, the action of the group G, and the reexpression of the natural response. For the natural response variable we find it convenient to write

$$Y_\lambda = \lambda^{-1}Y \qquad (11\text{-}12)$$

Note that for a specified value of λ the model is essentially a structural model (Sec. 7-1) and can be put in the usual form as

$$Y_\lambda = \theta_\lambda Z \text{ with } \theta \text{ in } G \qquad (11\text{-}13)$$
$$Z \text{ has a distribution in the class } \mathscr{V}$$

11-2-2 The Analysis

For the model (11-11) the parameter λ is more complex than with the ordinary model in preceding chapters. Besides distribution form, the parameter λ now involves group application and response reexpression. However, for given λ the model is essentially the structural model (7-8) and (7-9) in Sec. 7-1; for the continuous case we will assume the differential properties in Sec. 7-2.

Consider an inference base involving the generalized structural model (11-11) together with an observed response value Y^0. Our approach is to examine this for a chosen λ and see what is observable and what is unobservable. As in preceding chapters we obtain an observed orbit among possible orbits and a conditional model along the observed orbit. We do, however, reexpress this and examine it on the given response space.

For chosen λ the observed Y determines the natural response value $Y_\lambda^0 = \lambda^{-1} Y^0$. In the pattern of our earlier analysis, this value produces the observed orbit for the variation

$$G_\lambda Z = G_\lambda Y_\lambda^0 = G_\lambda \lambda^{-1} Y^0 \qquad (11\text{-}14)$$

This observed orbit is one orbit among the partition of $\mathscr{S}$ into orbits by the transformation group G_λ. We can reexpress the preceding on the given response space by using the transformation λ. We then obtain the observed *post-orbit*

$$\lambda G_\lambda Z = \lambda G_\lambda \lambda^{-1} Y^0 \qquad (11\text{-}15)$$

among the partition into such post-orbits.

For our analysis we will calculate the probability for the observed post-orbit (11-15) and express it suitably at the observed response point Y^0. In the continuous case this will give us an observed likelihood function. The model for possible likelihood functions we have found to be generally inaccessible.

Also for our analysis we will have the conditional model for $[Z]_\lambda$ given D_λ and the equation

$$Y = \lambda \theta_\lambda [Z]_\lambda D_\lambda$$

together with the observed value Y^0. For given λ this part of the analysis directly follows the pattern in the preceding chapters, with the obvious adjustments of notation.

This leaves us essentially with just the problem of calculating the probability for the observed post-orbit (11-15). We investigate this in Sec. 11-2-3.

11-2-3 Calculating Likelihood for λ

We now consider the probability for a post-orbit in terms of a volume measure at the observed response Y. For this we examine the continuous case and introduce the assumptions in Sec. 7-2-1. The resulting likelihood function is recorded as (11-27); the reader not particularly interested in the details of the calculations can proceed directly to (11-27) and Sec. 11-2-4.

The probability element for the variation Z is

$$f_\lambda(Z)\, dZ \qquad (11\text{-}16)$$

From Sec. 7-2 we obtain the following ingredients for the calculations. The *marginal distribution* in terms of the volume dD_λ orthogonal to the orbit at the reference point D_λ is

$$h_\lambda(D_\lambda)\, dD_\lambda \tag{11-17}$$

The *conditional distribution* for the position $g_\lambda = [Z]_\lambda$ is

$$h_\lambda^{-1}(D_\lambda) f_\lambda(g_\lambda D_\lambda) J_N(g_\lambda D_\lambda) J_L^{-1}(g_\lambda) J_\lambda(D_\lambda)\, dg_\lambda \tag{11-18}$$

where dg_λ is Euclidean volume on the group G; the corresponding equation is $Y = \lambda \theta_\lambda g_\lambda D_\lambda$ with the observed value Y^0.

We now pursue a reexpression of (11-17) in terms of volume at the observed response Y. For this consider dZ as a measure but restricted to a set formed as a compound of a neighborhood of values near D_λ with a neighborhood of values for g_λ. We then re-record (11-17) as a quotient of (11-16) by (11-18):

$$\frac{f_\lambda(Z)\, dZ}{h_\lambda^{-1}(D_\lambda) f_\lambda(g_\lambda D_\lambda) J_N(g_\lambda D_\lambda) J_L^{-1}(g_\lambda) J_\lambda(D_\lambda)\, dg_\lambda} = h_\lambda(D_\lambda) \cdot \frac{J_N^{-1}(g_\lambda D_\lambda)}{J_L^{-1}(g_\lambda)} \cdot \frac{dZ}{J_\lambda(D_\lambda)\, dg_\lambda} \tag{11-19}$$

The use of quotient differentials is straightforward and is based on standard methods of advanced calculus and linear forms. The reason for using quotients should now become clear. Under a change of variable both the numerator and denominator are relatively easily transformed. By contrast a measure orthogonal to orbits will typically transform to something *non*-orthogonal. The quotient provides the means to remain with an orthogonal measure.

We first reexpress the probability (11-19) for the orbit in terms of coordinates at Y_λ by using the properties of the invariant measures

$$h_\lambda(D_\lambda) \frac{J_N^{-1}([Y_\lambda]_\lambda D_\lambda)\, dY_\lambda}{J_L^{-1}([Y_\lambda]_\lambda) J_\lambda(D_\lambda)\, d[Y_\lambda]_\lambda} \tag{11-20}$$

We now commence the transformation at the actual response variable Y. For the full coordinates of Y we have the vector differential

$$dY_\lambda = \frac{\partial \lambda^{-1} Y}{\partial Y}\, dY = J(\lambda^{-1}: Y)\, dY \tag{11-21}$$

and volume differential

$$dY_\lambda = |J(\lambda^{-1}: Y)|\, dY \tag{11-22}$$

where $J(\lambda^{-1}: Y)$ is the Jacobian *matrix* for the transformation λ^{-1} applied at the point Y. For the coordinates $Y = \lambda[Y_\lambda]_\lambda D_\lambda$ in terms of the orbit coordinates $[Y_\lambda]_\lambda$ on the group we have the vector differential

$$dY = J^{-1}(\lambda^{-1}: Y)\, dY_\lambda = J^{-1}(\lambda^{-1}: Y)\, W_\lambda(Y_\lambda)\, d[Y_\lambda]_\lambda \tag{11-23}$$

where

$$W_\lambda(g_\lambda D_\lambda) = \frac{\partial g_\lambda D_\lambda}{\partial g_\lambda} \tag{11-24}$$

is an $N \times L$ matrix of partial derivatives. Note that the calculation of $W_\lambda(g_\lambda D_\lambda)$ at D_λ is an obvious first step for the direct calculation of $J_\lambda(D_\lambda)$, a calculation we have avoided except in simple cases. Then from the footnote to the discussion preceding (6-25) we have the following relation for the volume differential *along* the orbit:

$$d\lambda[Y_\lambda]_\lambda D_\lambda = |(\ldots)'[J^{-1}(\lambda^{-1}:Y)W_\lambda(Y_\lambda)]|^{1/2} \, d[Y_\lambda]_\lambda \qquad (11\text{-}25)$$

where the first parentheses give the transpose of the brackets following. We then substitute in (11-20) and use (11-21) for the numerator and (11-25) for the denominator; the quotient $dv = dY/d\lambda[Y_\lambda]_\lambda D_\lambda$ gives volume orthogonal to the post-orbit at Y. We obtain

$$\frac{h_\lambda(D_\lambda)}{J_\lambda(D_\lambda)} \cdot \frac{J_N^{-1}([Y_\lambda]_\lambda D_\lambda)}{J_L^{-1}([Y_\lambda]_\lambda)} \cdot \frac{|J(\lambda^{-1}:Y)|}{|(\ldots)'[J^{-1}(\lambda^{-1}:Y)W_\lambda(Y_\lambda)]|^{-1/2}} \, dv \qquad (11\text{-}26)$$

We have calculated the probability for the observed post-orbit through the given response. Using volume orthogonal to the observed post-orbit we obtain the *likelihood for* λ:

$$L(\lambda) = c \frac{h_\lambda(D_\lambda)}{J_\lambda(D_\lambda)} \cdot \frac{J_N^{-1}([Y_\lambda]_\lambda D_\lambda)}{J_L^{-1}([Y_\lambda]_\lambda)} \cdot \frac{|J(\lambda^{-1}:Y)|}{|(\ldots)'[J^{-1}(\lambda^{-1}:Y)W_\lambda(Y_\lambda)]|^{-1/2}} \qquad (11\text{-}27)$$

11-2-4 Some Comparisons

Consider an inference base involving the generalized structural model (11-11) together with an observed response value Y^0. In Sec. 11-2-2 we saw that for given λ we have an observed post-orbit and in Sec. 11-2-3 we obtained the probability for this in terms of volume orthogonal to the post-orbit. From this we obtained the likelihood (11-27) for the parameter λ.

We now examine this likelihood in comparison with the profile likelihood and use a simple example that incorporates most of the complications of the generalized model.

For the example we use the distribution

$$f_\lambda(z_1) f_\lambda(z_2) \, dz_1 \, dz_2 \qquad (11\text{-}28)$$

for the variation on $\mathbb{R}^2$ and we let (y_1, y_2) be the given response variable and $(y_1^\lambda, y_2^\lambda) = \lambda^{-1}(y_1, y_2)$ be the natural response variable with the simple location model

$$\begin{aligned} y_1^\lambda &= z_1 \\ y_2^\lambda &= \theta + z_2 \end{aligned} \qquad \theta \in \mathbb{R} \qquad (11\text{-}29)$$

The transformations λ from natural response space to given response space are assumed to be continuously differentiable.

The likelihood (11-27) for λ can be calculated in a straightforward manner:

$$L(\lambda) = c f_\lambda(y_1^\lambda) \left| \frac{dy_1^\lambda}{dy_1} \right| \qquad (11\text{-}30)$$

The likelihood depends only on y_1. Note that the density function for y_1 is

$$f_\lambda(y_1^\lambda)\left|\frac{dy_1^\lambda}{dy_1}\right| \tag{11-31}$$

Thus (11-30) *is* the likelihood from the variable y_1 that identifies the orbit.

Now consider the profile likelihood L_1 given by (11-1). This is obtained by profiling out the parameter θ; we obtain

$$L_1(\lambda) = f_\lambda(y_1^\lambda)\left|\frac{dy_1^\lambda}{dy_1}\right| f_\lambda(a_\lambda)\left|\frac{dy_2^\lambda}{dy_2}\right| \tag{11-32}$$

where a_λ maximizes $f_\lambda(z_2)$. If there is no distortion along the orbit, $|dy_2^\lambda/dy_2| =$ constant, then the preceding depends only on y_1 and is the actual likelihood from y_1 combined with an extraneous factor $f_\lambda(a_\lambda)$ recording the modal density. If there is distortion along the orbit then there is an additional extraneous factor. For this example the likelihood (11-27) avoids some of the obvious difficulties that arise with profile likelihood.

On the basis of our discussion in Sec. 11-1 it hardly seems appropriate to bother with the integrated likelihood L_2 given by (11-4); for completeness, however, we record the comparison. The full likelihood function for (λ, θ) has the form

$$cf_\lambda(y_1^\lambda)\left|\frac{dy_1^\lambda}{dy_1}\right| \cdot f_\lambda(y_2^\lambda - \theta)\left|\frac{dy_2^\lambda}{dy_2}\right|$$

How can the extraneous factors be removed by integration? A uniform distribution is an obvious way of eliminating the θ factor; in some cases it may be the only way. To eliminate the differential factor $|dy_2^\lambda/dy_2|$ as well would then require the adjusted uniform measure

$$\left|\frac{dy_2}{dy_2^\lambda}\right| d\theta$$

for the parameter θ. But this measure depends on the observed response! This measure has been proposed within the Bayesian framework but runs counter to the Bayesian philosophy by having a *prior* measure that "depends" on a subsequent outcome! The integrated likelihood thus seems even less attractive in the context of the present example.

The example illustrates most of the complications that arise with the generalized model. It does not, however, illustrate the case where the post-orbit through Y changes its direction with λ. We consider now how this affects the likelihood (11-27) for Y.

For this we consider the generalized model and suppose that the orientation of the post-orbits depends on λ. We investigate the effect of this by examining a new response $X = h(Y)$ obtained by a diffeomorphism h of $\mathbb{R}^N$ on $\mathbb{R}^N$. Let

$$K(Y) = \frac{\partial Y}{\partial X} \tag{11-33}$$

be the Jacobian matrix of the old with respect to the new variable. The likelihood function (11-27) can be recalculated using the Euclidean volume measure for the new variable X:

$$L^*(\lambda) = \frac{ch_\lambda(D_\lambda)}{J_\lambda(D_\lambda)} \cdot \frac{J_N^{-1}([Y_\lambda]_\lambda D_\lambda)}{J_L^{-1}([Y_\lambda]_\lambda)} \cdot \frac{|K(Y)| \, |J(\lambda^{-1}:Y)|}{|(\ldots)'[K^{-1}(Y)J^{-1}(\lambda^{-1}:Y)W_\lambda(Y_\lambda)]|^{-1/2}} \quad (11\text{-}34)$$

This differs from the likelihood (11-27) by the factor

$$\frac{|K(Y)| \, |(\ldots)'[J^{-1}(\lambda^{-1}:Y) W_\lambda(Y_\lambda)]|^{-1/2}}{|(\ldots)'[K^{-1}(Y)J^{-1}(\lambda^{-1}:Y) W_\lambda(Y_\lambda)]|^{-1/2}}$$

This factor arises because orthogonality for X can be different from that for Y.

In conclusion, we note that the likelihood (11-27) handles most of the complications that arise with the generalized model. However, if the post-orbits vary with λ then the likelihood (11-27) can depend on the particular choice of variable to represent the response. One realistic approach to this phenomenon is to examine the likelihood (11-34) using a sequence of choices for the response X and to iterate using at each stage the variable suggested by the preceding stage. Another approach is available if the reexpressions λ form a group; we investigate this in Sec. 11-3 and also its application to the power transformed regression model.

For further discussions on the methods in this section see Fraser (1967, 1972b).

11-3 GROUP-BASED LIKELIHOOD AND THE TRANSFORMED REGRESSION MODEL

In the preceding section we examined the generalized structural model and determined the likelihood function for the secondary parameter λ. A key element in the determination involved the choice of the supporting volume element at the observed response; we used volume orthogonal to the post-orbit of the underlying structural model.

The generalized structural model (11-11) is a composite of two different generalizations of the ordinary structural model (7-9). One of these (11-5) involved a parameter for the application of the group G to the basic sample space. The other (11-8) involved a parameter for a class of reexpressions of the natural response variable. Arguments can be given that a reasonable class of reexpressions should form a group (Fraser, 1972b). If they do form a group then there can be a natural choice of supporting volume element at the observed response. This gives the group-based likelihood in Secs. 11-3-1 and 11-3-2. Then in Sec. 11-3-3 we illustrate the likelihood methods by examining the transformed regression model.

11-3-1 Group-Based Reexpressions

Consider the second generalization (11-8) of the structural model:

$$Y = \lambda\theta Z \text{ with } \theta \text{ in } G, \lambda \text{ in } \Lambda$$
$$Z \text{ has a distribution in the class } \mathscr{V} \quad (11\text{-}35)$$

and suppose that the class Λ of reexpression transformations λ is a transformation group. Note that we are using the combined notation of (11-11) but the parameter λ now covers just the form of the distribution in $\mathscr{V}$ and the reexpression of the response. In this section we will work with λ as a reexpression transformation, but allow that it may have "spare coordinates" that index the distribution forms in $\mathscr{V}$.

In the generalized model (11-35) we have $Y = \lambda \theta Z$ where θ is in a group G and now λ is in a group Λ. If these combine so that the elements $\lambda \theta$ form a group then we have just the ordinary structural model. Our concern here, however, is for the more general case where the elements $\lambda \theta$ do not form a combined group.

For the generalized model (11-35) we will view θ in G as giving the basic or natural response expression and view λ in Λ as allowing for uncertainty as to the appropriate or natural way of expressing the response.

Our main example later in this section is the transformed regression model. Let **y** be the response vector as recorded and $\lambda^{-1}\mathbf{y}$ be the natural response vector with a regression-type model:

$$\lambda^{-1}\mathbf{y} = \begin{bmatrix} \lambda^{-1}y_1 \\ \vdots \\ \lambda^{-1}y_n \end{bmatrix} = X\beta + \sigma \mathbf{z} \qquad (11\text{-}36)$$

z has distribution $f_\lambda(\mathbf{z})$

As a class of response reexpressions consider the power transformations:

$$(\lambda)^{-1}y = \begin{cases} y^\lambda & \lambda \neq 0 \\ \ln y & \lambda = 0 \end{cases} \qquad (11\text{-}37)$$

For this we are using λ as a real number in the exponent; thus to avoid confusion we write (λ) for the response reexpression transformation. The transformation $(\lambda)^{-1}$ maps $(0, \infty)$ onto $(0, \infty)$ for $\lambda \neq 0$ and onto $(-\infty, \infty)$ for $\lambda = 0$. A modified form

$$(\lambda)^{-1}y = \begin{cases} \lambda^{-1}(y^\lambda - 1) & \lambda \neq 0 \\ \ln y & \lambda = 0 \end{cases} \qquad (11\text{-}38)$$

has continuity at $\lambda = 0$, but the range now depends strongly on λ; with the usual regression model the location-scale adjustment can be absorbed into the parameters of the model. The power transformations form a group if $\lambda > 0$; they also form a group if $\lambda \neq 0$. Both cases omit the logarithmic transformation.

11-3-2 Group-Based Likelihood

Consider the generalized structural model (11-35) with a group Λ of reexpression transformations. We first examine the effects of the groups G and Λ on the given response space.

Let Y be an observed response. For a given λ, the value on the natural response space is $\lambda^{-1}Y$, the orbit through $\lambda^{-1}Y$ is $G\lambda^{-1}Y$, and the post-orbit through the given response Y is

$$\lambda G\lambda^{-1}Y \tag{11-39}$$

As λ varies this post-orbit typically changes its direction of orientation at the observed Y. In Sec. 11-2-3 we calculated the probability for the observed post-orbit in terms of a volume measure orthogonal to the post-orbit.

Now with a group Λ on the given space we have an orbit ΛY through the observed Y. This Λ orbit through Y provides an intrinsic direction or orientation at the point Y. It is appropriate then to calculate the probability for the observed post-orbit in terms of volume measures aligned with this intrinsic direction.

To carry out the preceding effectively we need coordinates at Y that are stable with respect to the group Λ. Accordingly, we now suppose that Λ acts individually on the coordinates $Y_1, \ldots, Y_N$ of Y and that it acts exactly on each coordinate. In a neighborhood of Y we are then able to use invariant coordinates $X_1, \ldots, X_n$.

For the ith coordinate let d_i be a reference value and $\langle Y_i \rangle$ be the transformation in Λ that gives Y_i:

$$Y_i = \langle Y_i \rangle d_i \tag{11-40}$$

Now let $J_i(Y_i)$ be the "Jacobian" from d_i to Y_i:

$$J_i(Y_i) = c_i \left| \frac{d\langle Y_i \rangle X}{dX} \right|_{X=d_i} \tag{11-41}$$

where c_i is a scale constant to be chosen; thus

$$dX_i(Y_i) = \frac{dY_i}{J_i(Y_i)} \tag{11-42}$$

is an invariant measure. We then standardize one axis with respect to another by considering a transformation λ near the identity and choosing the constant c_i so that λ has the same effect on each axis.

As a first possibility we can calculate likelihood in terms of orthogonal volume in these new coordinates. For this we have

$$K(Y) = \frac{\partial Y}{\partial X} = \begin{bmatrix} J_1(Y_1) & & 0 \\ & \ddots & \\ 0 & & J_N(Y_N) \end{bmatrix}$$

$$= \text{dia } J_i(Y_i) \tag{11-43}$$

from (11-33) and we can then substitute in (11-34). This gives an invariant likelihood:

$$L^*(\lambda) = \frac{ch_\lambda(D_\lambda)}{J(D_\lambda)} \cdot \frac{J_N^{-1}([Y_\lambda]D_\lambda)}{J_L^{-1}([Y_\lambda])} \cdot \frac{\Pi J_i(Y_i) |J(\lambda^{-1} : Y)|}{|(\ldots)'[\text{dia } J_i^{-1}(Y_i) J^{-1}(\lambda^{-1} : Y) W(Y_\lambda)]|^{-1/2}} \tag{11-44}$$

Note in comparison with (11-34) that λ does not appear in certain places in the above formula; this follows from the fact that with the present generalization the application of G does not depend on λ.

We can, however, be more incisive in our calculation of likelihood and use a volume measure aligned with the Λ orbit—specifically, volume tangent to the Λ orbit combined with that orthogonal to the Λ orbit and G post-orbit.

We have standardized the new coordinates so that a change in λ produces the same change on each axis. Accordingly, in these new coordinates the tangent vector to the Λ orbit is the one-vector $\mathbf{1} = (1, \ldots, 1)'$. We thus want a volume measure along the $\mathbf{1}$ vector combined with one orthogonal to the $\mathbf{1}$ vector and the G post-orbit.

For this we follow the pattern of argument preceding (11-26). For the denominator we will need the volume measure along the post-orbit but reexpressed orthogonal to the $\mathbf{1}$ vector. Let

$$P = I - n^{-1}\mathbf{1}\mathbf{1}' \tag{11-45}$$

be the projection matrix into $\mathscr{L}^\perp(\mathbf{1})$; note that P replaces a column vector by a deviation vector relative to the average. The analogue for formula (11-25) in terms of projected volume for the X coordinates is

$$|(\ldots)'[P \operatorname{dia} J_i^{-1}(Y_i) J^{-1}(\lambda^{-1} : Y) W(Y_\lambda)]|^{1/2} \, d[Y_\lambda] \tag{11-46}$$

Then in the pattern for (11-26) we obtain the following probability for the observed post-orbit:

$$\frac{h_\lambda(D_\lambda)}{J(D_\lambda)} \cdot \frac{J_N^{-1}([Y_\lambda] D_\lambda)}{J_L^{-1}([Y_\lambda])} \cdot \frac{\Pi J_i(Y_i) \, |J(\lambda^{-1} : Y)|}{|(\ldots)'[P \operatorname{dia} J_i^{-1}(Y_i) J^{-1}(\lambda^{-1} : Y) W(Y_\lambda)]|^{-1/2}} \, dv_t \tag{11-47}$$

where dv_t is the volume measure tangent to the Λ orbit combined with that orthogonal to the Λ orbit and G post-orbit. This gives the transit likelihood

$$L^t(\lambda) = \frac{ch_\lambda(D_\lambda)}{J(D_\lambda)} \cdot \frac{J_N^{-1}([Y_\lambda] D_\lambda)}{J_L^{-1}([Y_\lambda])} \cdot \frac{\Pi J_i(Y_i) \, |J(\lambda^{-1} : Y)|}{|(\ldots)'[P \operatorname{dia} J_i^{-1}(Y_j) J^{-1}(\lambda^{-1} : Y) W(Y_\lambda)]|^{-1/2}} \tag{11-48}$$

In Sec. 11-2-4 we noted advantages of the extended likelihood (11-27). We also noted, however, that if the post-orbits varied with λ then the likelihood expression could depend on the choice of variable to record the response. The use here of the group Λ has eliminated this dependence in an incisive way.

The likelihood (11-44) and its refinement (11-48) provide assessments of the various reexpressions λ in the generalized structural model (11-35). For given λ the analysis for the parameter θ would follow the pattern in Chaps. 6 and 7.

11-3-3 Transformed Regression Model

Consider the transformed regression model (11-36) in Sec. 11-3-1. We examine this first for a general reexpression λ in a class Λ and then specialize to the group case with the power transformations. We use the notation from Chap. 6.

We first record some preliminary calculations:

$$J_\lambda(\mathbf{d}_\lambda) = |X'X|^{1/2} \tag{11-49}$$

$$\frac{J_n^{-1}\{[\mathbf{b}(\mathbf{y}_\lambda), s(\mathbf{y}_\lambda)]\mathbf{d}_\lambda\}}{J_{r+1}^{-1}\{[\mathbf{b}(\mathbf{y}_\lambda), s(\mathbf{y}_\lambda)]\}} = \frac{1}{s^{n-r-1}(\mathbf{y}_\lambda)} \tag{11-50}$$

$$J(\lambda^{-1}:\mathbf{y}) = \frac{\partial \lambda^{-1}\mathbf{y}}{\partial \mathbf{y}} = \begin{bmatrix} \dfrac{d\lambda^{-1}y_1}{dy_1} & & 0 \\ & \ddots & \\ 0 & & \dfrac{d\lambda^{-1}y_n}{dy_n} \end{bmatrix}$$

$$= \operatorname{dia}\left(\frac{d\lambda^{-1}y_i}{dy_i}\right) \tag{11-51}$$

$$W_\lambda(\mathbf{y}_\lambda) = \frac{\partial(X\mathbf{b}(\mathbf{y}_\lambda) + s(\mathbf{y}_\lambda)\mathbf{d}_\lambda)}{\partial(\mathbf{b}'(\mathbf{y}_\lambda), s(\mathbf{y}_\lambda))} = (X, \mathbf{d}_\lambda) \tag{11-52}$$

$$= D_\lambda$$

We then obtain the likelihood (11-27):

$$L(\lambda) = c \frac{h_\lambda(\mathbf{d}_\lambda)}{|X'X|^{1/2}} \cdot \frac{1}{s^{n-r-1}(\mathbf{y}_\lambda)} \cdot \frac{|J(\lambda^{-1}:\mathbf{y})|}{|D'_\lambda J^{-2}(\lambda^{-1}:\mathbf{y})D_\lambda|^{-1/2}} \tag{11-53}$$

which for the normal case becomes

$$L(\lambda) = c \frac{1}{A_{n-r}|X'X|^{1/2} s^{n-r-1}(\mathbf{y}_\lambda)} \cdot \frac{|J(\lambda^{-1}:\mathbf{y})|}{|D'_\lambda J^{-2}(\lambda^{-1}:\mathbf{y})D_\lambda|^{-1/2}} \tag{11-54}$$

For comparison purposes we record the profile likelihood

$$L_1(\lambda) = c \frac{\sup_{b,s} s^n f_\lambda(X\mathbf{b} + s\mathbf{d}_\lambda)|J(\lambda^{-1}:\mathbf{y})|}{s^n(\mathbf{y}_\lambda)} \tag{11-55}$$

which for the normal case becomes

$$L_1(\lambda) = c \left(\frac{n}{2\pi e}\right)^{n/2} \frac{|J(\lambda^{-1}:\mathbf{y})|}{s^n(\mathbf{y}_\lambda)} \tag{11-56}$$

Now consider the frequently used power transformation group with $\lambda > 0$, or with $\lambda \neq 0$. Recall that the group operates on $\mathbb{R}^+$.

We first record some additional calculations:

$$\frac{d(\lambda)^{-1}y}{dy} = \frac{dy^\lambda}{dy} = \lambda y^{\lambda-1} \tag{11-57}$$

$$J[(\lambda)^{-1}:\mathbf{y}] = \Pi \lambda y_i^{\lambda-1} = \lambda^n \Pi y_i^{\lambda-1}$$

Toward finding the invariant measure we consider the transformations (λ) with $\lambda > 0$ or $\lambda \neq 0$. The transformation (λ) acts as

$$(\lambda)y = y^{1/\lambda}$$

To simplify the calculations we reparameterize the group with $\eta = 1/\lambda$; thus

$$(\eta)y = y^\eta \tag{11-58}$$

and, in particular,

$$\langle y \rangle = (\ln y) \tag{11-59}$$

carries e into y:

$$\langle y \rangle e = e^{\ln y} = y$$

We then have

$$J[(\eta):y] = \frac{d(\eta)y}{dy} = \frac{dy^\eta}{dy}$$

$$= \eta y^{\eta-1}$$

$$J(\langle y \rangle : e) = \ln y \, e^{\ln y - 1}$$

$$= \ln y \, y e^{-1}$$

For simplicity we choose $c = e$ and obtain

$$J(y) = e |\ln y| y e^{-1} = |\ln y| y \tag{11-60}$$

Also note that

$$J(y_i)J[(\lambda)^{-1}:y_i] = |\ln y_i| y_i \lambda y_i^{\lambda-1}$$

$$= \lambda |\ln y_i| y_i^\lambda$$

$$= y_i^\lambda |\ln y_i^\lambda| \tag{11-61}$$

We can now calculate the likelihood (11-48):

$$L(\lambda) = c \frac{h_\lambda(\mathbf{d}_\lambda)}{|X'X|^{1/2}} \cdot \frac{1}{s^{n-r-1}(\mathbf{y}_\lambda)} \cdot \frac{\Pi y_i^\lambda |\ln y_i^\lambda|}{|(\ldots)'(P \operatorname{dia}^{-1} y_i^\lambda |\ln y_i^\lambda| D_\lambda)|^{-1/2}} \tag{11-62}$$

which for the normal case becomes

$$L(\lambda) = c \frac{1}{A_{n-r} |X'X|^{1/2} s^{n-r-1}(\mathbf{y}_n)} \cdot \frac{\Pi y_i^\lambda |\ln y_i^\lambda|}{|(\ldots)'(P \operatorname{dia}^{-1} y_i^\lambda |\ln y_i^\lambda| D_\lambda)|^{-1/2}} \tag{11-63}$$

This likelihood provides an assessment of the various reexpression powers λ for the transformed regression model. For given λ the analysis of the parameters β, σ proceeds as in Chap. 6.

The profile likelihood method has been applied to the transformed regression problem by Box and Cox (1964); an integrated likelihood method was also used. The extended likelihood in its standard form (11-53) and transit form (11-62) may be found in Klass (1970) and Fraser (1967, 1972b).

REFERENCES AND BIBLIOGRAPHY

Box, G. E. P., and D. R. Cox: An Analysis of Transformations, *Jour. Roy. Stat. Soc.*, ser. B, vol. 26, pp. 211–243, 1964.

Fraser, D. A. S.: Data Transformations and the Linear Models, *Ann. Math. Stat.*, vol. 38, pp. 1456–1465, 1967.

———: Bayes, Likelihood, or Structural, *Ann. Math. Stat.*, vol. 43, pp. 777–790, 1972a.

———: The Determination of Likelihood and the Transformed Regression Model, *Ann. Math. Stat.*, vol. 43, pp. 898–916, 1972b.

———: Comparison of Inference Philosophies, in G. Menges (ed.), "Information Inference and Decision," D. Reidel Publishing Company, Dordrecht, Holland, 1974.

Klass, W.: Extended Marginal Likelihood, Ph.D. thesis, University of Toronto, 1970.

CHAPTER
TWELVE
MULTIVARIATE REGRESSION MODELS

We commenced this book with a detailed investigation of the ordinary location-scale model. In Chap. 6 we replaced the simple location parameter by a regression dependence on a variety of input variables. Then in Chap. 8 we generalized in a different direction and replaced the simple scale component by a matrix dependence on a vector variation. In this concluding chapter we bring together these two generalizations; we examine a vector response—with regression dependence on input variables and with matrix dependence on the basic variation.

The *location-scale* multivariate model from Sec. 8-1 can be combined in a straightforward manner with the regression dependence on input variables; we do not bother to record this combination.

The multivariate model with *progressive* variation from Secs. 8-2 and 8-3 can be combined with the regression model from Chap. 6; we examine this in Secs. 12-1 and 12-2.

The more general multivariate model with *linear* variation from Secs. 8-4 and 8-5 can be combined with the regression model; we examine this in Secs. 12-3 and 12-4.

12-1 MULTIVARIATE REGRESSION MODEL WITH PROGRESSIVE VARIATION

In this section we examine the multivariate model with progressive variation from Secs. 8-2 and 8-3 as combined with the regression model from Chap. 6. In the next section we specialize this to the normal case.

Consider a random system with p response variables $y_1, \ldots, y_p$ and with input variables that can be changed by the investigator; the investigator wants information on how changes in the input variables affect the responses.

Suppose that the background information on the system specifies that the only effect from the input variable is on the location of the response and is linear over the appropriate range for the input variables. Also suppose that the distribution describing the variation has the rather special progressive form discussed at the beginning of Sec. 8-2. For the first variable the distribution has known form apart from location and scale; for the second variable, the distribution has known form apart from location, scale, and regression on the preceeding variable; and so on.

12-1-1 On Notation

We follow the pattern in Sec. 8-2, and let $\mathbf{y} = (y_1, \ldots, y_p)'$ designate the p-variate response and let

$$Y = (\mathbf{y}_1, \ldots, \mathbf{y}_n) = \begin{bmatrix} y_{11} & \cdots & y_{1n} \\ \vdots & & \vdots \\ y_{p1} & \cdots & y_{pn} \end{bmatrix} \quad (12\text{-}1)$$

designate the compound response for n performances of the system. As before we will use Y_1 for the first row vector in Y, Y_2 for the second row vector, and so on. Recall from Sec. 8-2 that we altered our notation from that with the sample index 1 to n down a column to that with the sample index 1 to n across a row. This was a shift from the usual convention of regression analysis to a convention of multivariate analysis most appropriate to our development here, and it provides a reasonable accommodation between the ideal of a fully unified notation and the usefulness of a notation for a particular area.

Now let $x_1, \ldots, x_r$ be input variables; some of these variables may actually be combinations of other input variables, thus allowing the usual polynomial and interactive regression models. For the response we let $\mathbf{x} = (x_1, \ldots, x_r)'$ designate the r-variate input and let

$$X = (\mathbf{x}_1, \ldots, \mathbf{x}_n) = \begin{bmatrix} x_{11} & \cdots & x_{1n} \\ \vdots & & \vdots \\ x_{r1} & \cdots & x_{rn} \end{bmatrix} \quad (12\text{-}2)$$

designate the compound input for n performances of the system. We will use X_1 for the first row vector in X, X_2 for the second row vector, and so on. Note that the matrix (12-2) is the transpose of the matrix introduced as (6-2) for the regression model. The matrix X is called the design matrix. Note that the inner product matrix here is XX', in contrast to $X'X$ in the Chap. 6 notation. This change in notation is simple and straightforward and has the advantage of keeping X as the "design matrix."

12-1-2 The Model

Consider the *regression*-type location for the response and the *progressive* pattern for the variation—as described rather generally in the introduction. Recall the comments at the beginning of Sec. 8-2 that the progressive pattern of variation has some applications but that our interest is primarily to develop methods and techniques for the more general *linear* pattern for the variation. We examine the regression linear case in Secs. 12-3 and 12-4.

Let $f_\lambda(\mathbf{z})$ be the density function for the objective variation, with a possible shape parameter λ taking values in a space Λ. We suppose that f_λ has been suitably standardized as discussed in Sec. 8-2-2.

For the presentation of the response $\mathbf{y}$ we now combine the notations from Chap. 6 and Sec. 8-2:

$$\begin{aligned}
y_1 &= \beta_{11}x_1 + \cdots + \beta_{1r}x_r + \sigma_{(1)}z_1 \\
y_2 &= \beta_{21}x_1 + \cdots + \beta_{2r}x_r + \tau_{21}z_1 + \sigma_{(2)}z_2 \\
&\vdots \\
y_p &= \beta_{p1}x_1 + \cdots + \beta_{pr}x_r + \tau_{p1}z_1 + \cdots + \tau_{p,p-1}z_{p-1} + \sigma_{(p)}z_p
\end{aligned} \tag{12-3}$$

or

$$\mathbf{y} = \mathscr{B}\mathbf{x} + \Upsilon\mathbf{z} = [\mathscr{B}, \Upsilon]\mathbf{z} \tag{12-4}$$

where the regression coefficient matrix is

$$\mathscr{B} = \begin{bmatrix} \beta_{11} & \cdots & \beta_{1r} \\ \cdots & \cdots & \cdots \\ \beta_{p1} & \cdots & \beta_{pr} \end{bmatrix} \tag{12-5}$$

the scaling matrix is

$$\Upsilon = \begin{bmatrix} \sigma_{(1)} & & \cdots & & 0 \\ \tau_{21} & \sigma_{(2)} & & & \\ \vdots & & \ddots & & \\ \tau_{p1} & \cdots & & \tau_{p,p-1} & \sigma_{(p)} \end{bmatrix} \tag{12-6}$$

and the transformation $[\mathscr{B}, \Upsilon]$, with $\mathscr{B}$ as $p \times r$ and Υ as $p \times p$ PLT, is defined implicitly by (12-4) with (12-3).

For n independent performances we then have $\mathbf{y}_i = \mathscr{B}\mathbf{x}_i + \Upsilon\mathbf{z}_i$ for the ith performance and

$$Y = \mathscr{B}X + \Upsilon Z = [\mathscr{B}, \Upsilon]Z = \theta Z \tag{12-7}$$

for the compound performance. The distribution for the variation Z is

$$f_\lambda(Z) = \Pi_1^n f_\lambda(\mathbf{z}_i) \tag{12-8}$$

The transformations $[\mathscr{B}, \Upsilon]$ form a group:

$$\begin{aligned} [B_2, T_2][B_1, T_1] &= [B_2 + T_2 B_1, T_2 T_1] \\ [B, T]^{-1} &= [-T^{-1}B, T^{-1}] \\ [0, I] &= i \end{aligned} \tag{12-9}$$

where as before the key item is that PLT times PLT gives PLT. Thus

$$G = \{[B, T] : B \in \mathbb{R}^{pr}, T \text{ is } p \times p \text{ PLT}\} \tag{12-10}$$

is a group. We will see that G satisfies the exactness Assumption 7-2 on the sample space $\mathbb{R}^{pn}$ provided $n \geq p + r$ and a set of measure zero is excluded.

This gives us the structural model

$$\mathcal{M}_V = (\Omega; \mathbb{R}^{pn}, \mathcal{A}^{pn}, \mathcal{V}, G) \qquad (12\text{-}11)$$

where the parameter space $\Omega = (G \times \Lambda)$, $\mathcal{V}$ is the class of densities f_λ in (12-8), and G is the transformation group (12-10) with action (12-7); note the use of $\mathcal{A}^{pn}$ for the Borel class to avoid confusion with the regression matrix. We can abbreviate the model as

$$\begin{aligned} Y &= \theta Z \text{ with } \theta \text{ in } G \\ Z &\text{ has distribution in the class } \mathcal{V} \end{aligned} \qquad (12\text{-}12)$$

and we assume that $n \geq p + r$.

12-1-3 Preliminary Analysis

Consider an inference base

$$(\mathcal{M}_V, Y^0) \qquad (12\text{-}13)$$

using the model (12-11) with data Y^0. We have the observed orbit

$$GZ = GY^0 \qquad (12\text{-}14)$$

together with the corresponding model, as described by (a) in Sec. 7-1-4, and we have the conditional model for Z given $GZ = GY^0$ together with the equation $[Y] = \theta[Z]$ and the data $[Y^0]$, as described by (b) in Sec. 7-1-4. We now put together some convenient notation using that developed in Chaps. 6 and 8.

We examine the action of the group G on the space $\mathbb{R}^{pn}$. From (12-3) and (12-7) we see that the transformation

$$\tilde{Z} = BX + TZ \qquad (12\text{-}15)$$

can be written as

$$\begin{aligned} \tilde{Z}_1 &= B_1 X + s_{(1)} Z_1 \\ \tilde{Z}_2 &= B_2 X + t_{21} Z_1 + s_{(2)} Z_2 \\ &\vdots \\ \tilde{Z}_p &= B_p X + t_{p1} Z_1 + \cdots + s_{(p)} Z_p \end{aligned} \qquad (12\text{-}16)$$

Thus, under the action of G we see that

$$\begin{aligned} \tilde{Z}_1 &\text{ is in } \mathscr{L}^+(X; Z_1) \\ \tilde{Z}_1 &\text{ is in } \mathscr{L}^+(X; Z_1; Z_2) \\ &\vdots \\ \tilde{Z}_p &\text{ is in } \mathscr{L}^+(X; Z_1, \ldots; Z_p) \end{aligned} \qquad (12\text{-}17)$$

where $\mathscr{L}^+(X; \mathbf{y})$ is defined by (6-14).

We now build up the reference point and transformation sequentially using the methods available from Sec. 6-1-3. Let $B_1(Z)$ be the regression coefficients for Z_1 on the rows of X, $s_{(1)}(Z)$ be the residual length, and $D_1(Z)$ be the unit residual; thus, for example, we have

$$B_1(Z) = Z_1 X'(XX')^{-1} \qquad (12\text{-}18)$$

which is the row vector form of the usual formula (6-16). We obtain

$$Z_1 = B_1(Z)X + s_{(1)}(Z)D_1(Z) \qquad (12\text{-}19)$$

Similarly, let $B_2(Z)$ be the regression coefficients of Z_2 on the rows of X, $t_{21}(Z)$ be the regression coefficient on $D_1(Z)$, $s_{(2)}(Z)$ be the residual length, and $D_2(Z)$ be the unit residual; thus

$$B_2(Z) = Z_2 X'(XX')^{-1} \qquad (12\text{-}20)$$

We then obtain

$$Z_2 = B_2(Z)X + t_{21}(Z)D_1(Z) + s_{(2)}(Z)D_2(Z) \qquad (12\text{-}21)$$

We continue in this way and let $B_p(Z)$ be the regression coefficients of Z_p on the rows of X, $[t_{p1}(Z), \ldots, t_{p,p-1}(Z)]$ be the regression coefficients Z_p on $D_1(Z), \ldots, D_{p-1}(Z)$, $s_{(p)}(Z)$ be the residual length, and $D_p(Z)$ be the unit residual. We then obtain

$$Z_p = B_p(Z)X + \sum_{1}^{p-1} t_{pj}(Z)D_j(Z) + s_{(p)}(Z)D_p(Z) \qquad (12\text{-}22)$$

The preceding can be collected in the following:

$$\begin{bmatrix} Z_1 \\ \vdots \\ Z_p \end{bmatrix} = \begin{bmatrix} B_1(Z) \\ \vdots \\ B_p(Z) \end{bmatrix} X + \begin{bmatrix} s_{(1)}(Z) & \cdots & 0 \\ \vdots & \ddots & \\ t_{p1}(Z) & \cdots & s_{(p)}(Z) \end{bmatrix} \begin{bmatrix} D_1(Z) \\ \vdots \\ D_p(Z) \end{bmatrix} \qquad (12\text{-}23)$$

$$Z = B(Z)X + T(Z)D(Z)$$
$$= [B(Z), T(Z)]D(Z) = [Z]D(Z) \qquad (12\text{-}24)$$

where $[B(Z), T(Z)]$ is a transformation variable that gives the group position of Z relative to the reference point $D(Z)$; note that

$$B(Z) = \begin{bmatrix} B_1(Z) \\ \vdots \\ B_p(Z) \end{bmatrix} = ZX'(XX')^{-1} \qquad (12\text{-}25)$$

records the regression coefficients of the rows of Z on the rows of X and

$$T(Z) = \begin{bmatrix} s_{(1)}(Z) & & \cdots & & 0 \\ \vdots & \ddots & & & \vdots \\ t_{p1}(Z) & \cdots & & t_{p,p-1}(Z) & s_{(p)}(Z) \end{bmatrix} \qquad (12\text{-}26)$$

records the regression coefficients of the rows of Z on the sequentially constructed unit residuals $D_1(Z), \ldots, D_p(Z)$. Note that

$$Z - B(Z)X = T(Z)D(Z) \tag{12-27}$$

is the PLT orthogonal decomposition of the matrix that records the deviation vectors from regression on X.

For exactness the preceding analysis must be feasible; accordingly, we require $n \geq p + r$ and we eliminate the set of measure zero in which $X_1, \ldots, X_r, Z_1, \ldots, Z_p$ are linearly dependent.

As in preceding chapters we have, of course, the separation of the inference base (12-13) by the necessary method RM_3 in Sec. 3-3. Using the preceding notation we have the separation into

(a) The marginal model for $D(Y) = D(Z)$ with observed value $D(Y^0)$.
(b) The conditional distribution for $[B, T] = [B(Z), T(Z)]$ given $D(Z) = D(Y^0) = D^0$ together with the presentation equation

$$[B(Y), T(Y)] = [\mathcal{B}, \Upsilon][B, T] \tag{12-28}$$

or
$$\begin{aligned} B(Y) &= \mathcal{B} + \Upsilon B \\ T(Y) &= \phantom{\mathcal{B} +{}} \Upsilon T \end{aligned} \tag{12-29}$$

and the observed value $[B(Y^0), T(Y^0)]$.

In Sec. 12-1-4, we derive the appropriate distributions.

12-1-4 Density Functions

The Jacobians and measures from Sec. 7-2 can be calculated as in Sec. 8-2.

The transformation $g = [B, T]$ applies column by column to the matrix point Z. Accordingly,

$$\begin{aligned} J_{pn}(g:Z) &= |T|^n \qquad J_{pn}(Z) = |T(Z)|^n \\ dM(Z) &= |T(Z)|^{-n} \, dZ \end{aligned} \tag{12-30}$$

The equation $\tilde{g} = gg^*$ has the form

$$\begin{aligned} \tilde{B} &= B + TB^* \\ \tilde{T} &= \phantom{B +{}} TT^* \end{aligned} \tag{12-31}$$

The left transformation operates column by column on B^*, T^*; for a column of T^* only the relevant part of the matrix T is used (compare with Sec. 8-2-3). Thus

$$\begin{aligned} J(g:g^*) &= (s_{(1)} \ldots s_{(p)})^r (s_{(1)} \ldots s_{(p)}) \ldots (s_{(p)}) = |T|^r |T|_\Delta \\ J(g) &= |T|^r |T|_\Delta \\ d\mu(g) &= \frac{dB\, dT}{|T|^r |T|_\Delta} \end{aligned} \tag{12-32}$$

The right transformation operates row by row on T and, for given T, by location only on B. Thus in the pattern of Sec. 8-2-3 we have

$$J^*(g^*:g) = s_{(1)}^{*p}\ldots s_{(p)}^{*1} = |T^*|_v$$

$$J^*(g) = |T|_v$$

$$dv(g) = \frac{dB\,dT}{|T|_v} \tag{12-33}$$

$$\Delta(g) = \frac{|T|_v}{|T|^r|T|_\Delta}$$

The basic Jacobian $J(D)$ recording volume change on orbit relative to change on group is easily obtained by noting that on the orbit the rows of X are used as basis vectors and X need not be a set of orthonormal vectors. From Sec. 6-1-4 in the pattern of Sec. 8-2-3, we obtain a factor $|XX'|^{1/2}$ for each row of Z and thus have

$$J(D) = |XX'|^{p/2} \tag{12-34}$$

Now consider the analysis of the inference base $(\mathcal{M}_V, Y^0)$. The observed orbit gives the observed value for the variation:

$$D(Z) = D(Y^0) = D^0 \tag{12-35}$$

This gives the inference base (a) from Sec. 7-1-4:

$$(\mathcal{M}_D, D^0) \tag{12-36}$$

The distributions in $\mathcal{M}_D$ are available from (7-46) together with our present calculations:

$$h_\lambda(D) = \int_G f_\lambda(BX + TD)s_{(1)}^{n-r-1}\ldots s_{(p)}^{n-r-p}|XX'|^{p/2}\,dB\,dT \tag{12-37}$$

with integration over $\mathbb{R}$ for each coordinate of B, T except over $\mathbb{R}^+$ for the diagonal elements of T. This, of course, leads to the observed likelihood for λ:

$$L(D^0;\lambda) = ch_\lambda(D^0) \tag{12-38}$$

The unobserved characteristics of the variation are given by $[B,T] = [B(Z), T(Z)]$. For these characteristics we have the inference base (b) from Sec. 7-1-4:

$$(\mathcal{M}_V^{D^0}, [Y^0]) \tag{12-39}$$

The distributions for the structural model $\mathcal{M}_V^{D^0}$ are available from (7-47) together with our present calculations; we have the density

$$h_\lambda^{-1}(D^0)f_\lambda(BX + TD^0)s_{(1)}^{n-r-1}\ldots s_{(p)}^{n-r-p}|XX'|^{p/2} \tag{12-40}$$

for B, T on the group G. This distribution for B, T is used with the transformation

$$B(Y) = \mathcal{B} + \Upsilon B$$
$$T(Y) = \Upsilon T \tag{12-41}$$

The observed values for the response coordinates are given by

$$[Y^0] = [B(Y^0), T(Y^0)] \qquad (12\text{-}42)$$

The response distribution consistent with the identified $D(Y) = D(Y^0) = D^0$ from the actual response is available from (7-49) in the pattern of (8-66):

$$h_\lambda^{-1}(D^0) f_\lambda \left(\Upsilon^{-1}[(B - \mathcal{B})X + TD^0]\right) \begin{bmatrix} s_{(1)} \cdots s_{(p)} \\ \sigma_{(1)} \cdots \sigma_{(p)} \end{bmatrix}^n \frac{|XX'|^{p/2} \, dB \, dT}{|T|^r |T|_\Delta} \qquad (12\text{-}43)$$

where for this particular formula we have written B for $B(Y)$, T for $T(Y)$, and $s_{(i)}$ for $s_{(i)}(Y)$.

12-1-5 Inference for Component Parameters

We have noted in Secs. 7-3 and 8-2-4 that some component parameters may not have the special left coset form and thus not be amenable to the strong inference methods developed in Chap. 7.

There are, of course, various ad hoc, intuitive, and theory-based methods for seeking out tests and confidence regions for component parameters; for some discussion see Fraser and Ng (1977). However, as we move from the one- and two-parameter cases to more complicated models we find these methods to be more elusive and less fruitful.

With left coset parameters we are involved with parameters that are in certain ways natural and fundamental. The corresponding tests and confidence regions have the strong and unequivocal properties that we have discussed in preceding chapters.

However, in discussing these properties we have perhaps not emphasized some important and relevant points. There are a wide range of left coset parameters for multivariate problems, and these parameters do have certain natural and fundamental characteristics. For any particular left coset parameter we have a direct correspondence that gives us unequivocally the appropriate tests and confidence regions. In addition, the appropriate distribution theory is available in a straightforward manner. For details, recall Secs. 7-3-3 and 7-3-5 and Sec. 7-4.

In this section we restrict our attention to the basic location and scale parameters.

For the location parameter $\mathcal{B}$ we have the separation of the equation

$$[B(Y), T(Y)] = [\mathcal{B}, \Upsilon][B, T] \qquad (12\text{-}44)$$

giving the unique $\mathcal{B}$-specific component

$$T^{-1}(Y)[B(Y) - \mathcal{B}] = T^{-1}B = H \qquad (12\text{-}45)$$

We can see that $\mathcal{B}$ indexes left cosets by noting the factorization

$$[\mathcal{B}, \Upsilon] = [\mathcal{B}, I][0, \Upsilon] \qquad (12\text{-}46)$$

Thus $\mathscr{B}$ indexes left cosets of the scale group

$$G_S = \{[0, T]: T_{p \times p} \text{ PLT}\} \tag{12-47}$$

Note that the group coordinates can be separated in the reverse order:

$$[B, T] = [0, T][H, I] \tag{12-48}$$

and thus that H indexes right cosets or orbits on the sample space.
The marginal distribution for H is obtained from (12-40) by using

$$dB = |T|^r \, dH \tag{12-49}$$

for fixed T. The probability differential is

$$h_\lambda^{-1}(D^0) \int_T f_\lambda(T(HX + D^0)) |T|^n \frac{dT}{|T|_\Delta} |XX'|^{p/2} \, dH \tag{12-50}$$

This distribution together with equation (12-45) and the observed values provides tests and confidence regions for the location parameter $\mathscr{B}$.

For the scale parameter Υ we have the separation of the basic equation giving the unique Υ-specific component

$$\Upsilon^{-1} T(Y) = T \tag{12-51}$$

The full parameter can be separated as

$$[\mathscr{B}, \Upsilon] = [0, \Upsilon][\Xi, I] \tag{12-52}$$

where $\Xi = \Upsilon^{-1} \mathscr{B}$ is the coefficient of variation; thus Υ indexes left cosets of the location group

$$G_L = \{[B, I]: B \in \mathbb{R}^{pr}\} \tag{12-53}$$

Note that the group coordinates can be separated in the reverse order:

$$[B, T] = [B, I][0, T] \tag{12-54}$$

and thus that T indexes the right cosets or orbits on the sample space.
The marginal distribution for T is obtained from (12-40) by directly integrating out the location variable B. The probability differential is

$$h_\lambda^{-1}(D^0) \int_B f_\lambda(BX + TD^0) |XX'|^{p/2} \, dB \, |T|^{n-r} \frac{dT}{|T|_\Delta} \tag{12-55}$$

This distribution together with equation (12-51) and the observed value $T(Y^0)$ provides tests and confidence regions for the scale parameter.

The response distribution of $T(Y)$ consistent with the observed $T(Y^0)$ from Y^0 is obtained from (12-55) by direct transformation using the invariant measure $dT/|T|_\Delta$ from (8-83). We obtain

$$h_\lambda^{-1}(D^0) \int_B f_\lambda(BX + \Upsilon^{-1} TD^0) |XX'|^{p/2} \, dB \frac{|T|^{n-r}}{|\Upsilon|^{n-r}} \frac{dT}{|T|_\Delta} \tag{12-56}$$

where for this formula we have used T for $T(Y)$.

12-2 NORMAL MULTIVARIATE REGRESSION MODEL WITH PROGRESSIVE VARIATION

In the preceding section we developed the multivariate regression model using the progressive variation from Sec. 8-2. We now examine this for the special case of a normal pattern for the variation. As noted in Sec. 8-3 many important distinctions vanish when a normal pattern of variation is used. Nevertheless, we are able to illustrate some points from the preceding section, and we do obtain in a simple mechanical way all the basic distribution theory for the multivariate normal regression model.

12-2-1 The Normal Model

Consider a random system with a p-variate response $\mathbf{y}$. From Sec. 12-1 the multivariate regression model $\mathcal{M}_V$ for n performances has the form

$$Y = \mathcal{B}X + \Upsilon Z \tag{12-57}$$

where Y and Z are the $p \times n$ response and variation matrices, $\mathcal{B}$ is the regression coefficient matrix relative to the design matrix X, and Υ is the PLT scale matrix for the variation.

We examine this model using the standard normal model for the variation. From (8-90) we have

$$f(Z)\, dZ = (2\pi)^{-np/2}\, \text{etr}\, (-\tfrac{1}{2} Z'Z)\, dZ \tag{12-58}$$

We can rewrite this in the invariant form

$$(2\pi)^{-np/2}\, \text{etr}\, (-\tfrac{1}{2} Z'Z)\, |T(Z)|^n\, dM(Z) \tag{12-59}$$

where

$$dM(Z) = \frac{dZ}{|T(Z)|^n} \tag{12-60}$$

and $T(Z)$ is the PLT matrix (12-26) appropriate to the multivariate regression model with progressive variation.

The corresponding response distribution can be obtained by direct substitution utilizing the invariant form (12-59). For this we have

$$\begin{aligned}
\text{tr}\, Z'Z &= \text{tr}\, (Y - \mathcal{B}X)'\, \Upsilon'^{-1}\Upsilon^{-1}(Y - \mathcal{B}X) \\
&= \text{tr}\, (Y - \mathcal{B}X)'\, \Sigma^{-1}(Y - \mathcal{B}X)
\end{aligned} \tag{12-61}$$

where $\Sigma = \Upsilon\Upsilon'$, as in Sec. 8-3-1, is the variance matrix for the response $Y = \mathcal{B}X + \Upsilon Z$. We then obtain the response distribution

$$\begin{aligned}
&(2\pi)^{-np/2}\, \text{etr}\, [-\tfrac{1}{2}(Y - \mathcal{B}X)'\, \Sigma^{-1}(Y - \mathcal{B}X)]\, \frac{|T(Y)|^n}{|\Upsilon|^n}\, dM(Y) \\
&= (2\pi)^{-np/2}\, |\Sigma|^{-n/2}\, \text{etr}\, [-\tfrac{1}{2}(Y - \mathcal{B}X)'\, \Sigma^{-1}(Y - \mathcal{B}X)]\, dY
\end{aligned} \tag{12-62}$$

Some general comments are given at the end of Sec. 8-3-1 concerning progressive variation with the normal as the distribution form.

12-2-2 The Analysis

Consider the analysis of the inference base $(\mathcal{M}_V, Y^0)$ using the normal model for variation. In particular, we determine the marginal distribution for $D(Z) = D(Y) = D$ and the conditional model for $[Z]$ given D.

For this we need a simplification of the normal-density exponent expressed in terms of the position variable $[Z] = [B(Z), T(Z)] = [B, T]$:

$$\begin{aligned}
\operatorname{tr} Z'Z &= \operatorname{tr} ZZ' \\
&= \operatorname{tr}(BX + TD)(BX + TD)' \\
&= \operatorname{tr} BXX'B' + \operatorname{tr} TDD'T' \\
&= \operatorname{tr} BXX'B' + \operatorname{tr} TT'
\end{aligned} \qquad (12\text{-}63)$$

where we use the orthogonality $XD' = 0$ and orthonormality $DD' = I$.

First, we obtain the conditional distribution of $[Z] = [B, T]$ given D:

$$h^{-1}(D) f(BX + TD) s_{(1)}^{n-r-1} \cdots s_{(p)}^{n-r-p} |XX'|^{p/2} \, dB \, dT$$

$$= \frac{1}{(2\pi)^{rp/2}} \exp\left(-\tfrac{1}{2}\Sigma B_j XX' B_j'\right) |XX'|^{p/2} \, dB$$

$$\times \frac{A_{n-r}^{(p)}}{(2\pi)^{(n-r)p/2}} \operatorname{etr}\left(-\tfrac{1}{2}TT'\right) s_{(1)}^{n-r-1} \cdots s_{(p)}^{n-r-p} \, dT \qquad (12\text{-}64)$$

$$= \frac{|XX'|^{p/2}}{(2\pi)^{rp/2}} \operatorname{etr}\left(-\tfrac{1}{2}BXX'B'\right) dB \; \frac{A_{n-r}^{(p)}}{(2\pi)^{(n-r)p/2}} \operatorname{etr}\left(-\tfrac{1}{2}TT'\right) \frac{|T|^{n-r}}{|T|_\Delta} dT$$

We have used the normalizing constant $|XX'|^{1/2}(2\pi)^{-r/2}$ for the multivariate normal with inverse variance matrix XX' and we have used the normalizing constants for the standard normal and chi densities as in Sec. 8-3-2. The conditional distributions for B and T can be described simply. The rows B_j and the matrix T are statistically independent and

B_j is normal $[\mathbf{0}'; (XX')^{-1}]$

$$T = \begin{bmatrix} s_{(1)} & \cdots & 0 \\ t_{21} & s_{(2)} & & \vdots \\ \vdots & & \ddots & \\ t_{p1} & \cdots & t_{p,p-1} & s_{(p)} \end{bmatrix} = \begin{bmatrix} \chi_{n-r} & \cdots & & 0 \\ z_{21} & \chi_{n-r-1} & & \vdots \\ \vdots & & \ddots & \\ z_{p1} & \cdots & z_{p,p-1} & \chi_{n-r-p+1} \end{bmatrix} \qquad (12\text{-}65)$$

where the z variables are independent standard normals and the χ variables are independent chi variables with degrees of freedom as subscribed. The distribution of T is the triangular chi $\Delta\chi_p(n-r)$ discussed in Sec. 8-3-2.

The conditional distribution (12-64) gives us the distribution component for the structural model $\mathscr{M}_Y^{D^o}$. Thus as in Sec. 12-1-4 we obtain the component inference base (12-39).

$$(\mathscr{M}_Y^{D^o}, [Y^o]) \qquad (12\text{-}66)$$

where the model $\mathscr{M}_Y^{D^o}$ has the distribution for $[Z]$ given by (12-64) and the transformation

$$B(Y) = \mathscr{B} + \Upsilon B$$
$$T(Y) = \quad \Upsilon T \qquad (12\text{-}67)$$

relating the response $[Y] = [B(Y), T(Y)]$ and the standardized variation $[Z] = [B, T]$; the observed values for the response components are

$$[Y^o] = [B(Y^o), T(Y^o)] \qquad (12\text{-}68)$$

The conditional response distribution for Y given the observed $D(Y) = D(Y^o) = D^o$ is available from (12-43). For this we use (12-67) with the exponent (12-63) and simplify:

$$\operatorname{tr} Z'Z = \operatorname{tr} BXX'B' + \operatorname{tr} TT'$$
$$= \operatorname{tr} \{\Upsilon^{-1}[B(Y) - \mathscr{B}]XX'[B(Y) - \mathscr{B}]'\Upsilon'^{-1}\}$$
$$+ \operatorname{tr} \Upsilon^{-1}T(Y)T'(Y)\Upsilon'^{-1} \qquad (12\text{-}69)$$
$$= \operatorname{tr} [B(Y) - \mathscr{B}]XX'[B(Y) - \mathscr{B}]'\Sigma^{-1} + \operatorname{tr} T(Y)T'(Y)\Sigma^{-1}$$
$$= \operatorname{tr} [B(Y) - \mathscr{B}]XX'[B(Y) - \mathscr{B}]'\Sigma^{-1} + \operatorname{tr} S(Y)\Sigma^{-1}$$

where

$$S(Y) = T(Y)T'(Y) = T(Y)D(Y)D'(Y)T'(Y)$$
$$= [Y - B(Y)X][Y - B(Y)X]' \qquad (12\text{-}70)$$

is the inner product matrix for the deviations from regression on X. The probability element for the response distribution is

$$\frac{A_{n-r}^{(p)}}{(2\pi)^{np/2}} \operatorname{etr} \{-\tfrac{1}{2}[B(Y) - \mathscr{B}]XX'[B(Y) - \mathscr{B}]'\Sigma^{-1}\} \operatorname{etr} [-\tfrac{1}{2}S(Y)\Sigma^{-1}]$$

$$\times \left[\frac{s_{(1)}(Y) \ldots s_{(p)}(Y)}{\sigma_{(1)} \ldots \sigma_{(p)}}\right]^n |XX'|^{p/2} \frac{dB(Y)\, dT(Y)}{s_{(1)}^{r+1}(Y) \ldots s_{(p)}^{r+p}(Y)}$$

$$= \frac{|XX'|^{p/2} |\Sigma|^{-r/2}}{(2\pi)^{rp/2}} \operatorname{etr} \{-\tfrac{1}{2}[B(Y) - \mathscr{B}]XX'[B(Y) - \mathscr{B}]'\Sigma^{-1}\}\, dB(Y) \qquad (12\text{-}71)$$

$$\times \frac{A_{n-r}^{(p)}}{(2\pi)^{(n-r)p/2}} \operatorname{etr} [-\tfrac{1}{2}S(Y)\Sigma^{-1}] \frac{|S(Y)|^{(n-r)/2}}{|\Sigma|^{(n-r)/2}} \frac{dT(Y)}{|T(Y)|_\Delta}$$

We can then read the conditional response distribution from the differential (12-71), or from equation (12-67) in relation to the standardized expressions in (12-64) and (12-65). We have that $B(Y)$ and $T(Y)$ are statistically independent and that

$$B(Y) \text{ is } N[\mathscr{B}; (XX')^{-1} \otimes \Sigma]$$
$$T(Y) \text{ is } \Delta\chi_p(n-r, \Upsilon) \tag{12-72}$$

For this we recall the definition of the scaled triangular chi distribution in Sec. 8-3-2 and we note that $B(Y)$ has a normal distribution located at $\mathscr{B}$ with variance matrix $(XX')^{-1} \otimes \Sigma$ and inverse variance matrix $(XX') \otimes \Sigma^{-1}$. This treats $B(Y)$ as an extended vector formed by taking the columns in succession and it uses the Kronecker product

$$A \otimes B = \begin{bmatrix} a_{11}B & \cdots & a_{1r}B \\ \cdots\cdots\cdots\cdots\cdots \\ a_{r1}B & \cdots & a_{rr}B \end{bmatrix}$$

$$(A \otimes B)^{-1} = A^{-1} \otimes B^{-1} \tag{12-73}$$

Now we obtain the marginal distribution of D. This involves, of course, the integration accomplished in (12-64) which needed only the simple normalizing constants for the normal and chi density expressions. Thus in the pattern of Sec. 8-3-2 we obtain

$$h(D) \, dD = \frac{dD}{A_{n-r}^{(p)}} \tag{12-74}$$

where we continue with our interpretation of dD as volume calculated orthogonal to the orbits of the group G. Thus D has a uniform distribution relative to the volume measure just described.

Note that D consists of p orthonormal vectors in $\mathscr{L}^{\perp}(X)$. The space Q of such matrices forms a manifold in $\mathbb{R}^{pn}$ of dimension $p(n-r) - p(p+1)/2$; compare with Sec. 8-3-2 with $n-1$ replaced by $n-r$. Also from Sec. 8-3-5, note that the surface volume of the manifold is $2^{p(p-1)/4}$ times the orthogonal volume which is $A_{n-r}^{(p)}$.

The marginal distribution (12-74) gives us the distribution component of the model $\mathscr{M}_D$. Thus as in Sec. 12-1-4 we obtain the component inference base

$$(\mathscr{M}_D, D^0) \tag{12-75}$$

where $\mathscr{M}_D$ has the single distribution (12-74) and thus no parameter. Accordingly, the inference base collapses as in Sec. 8-3-2.

12-2-3 The Scale Component

Consider the scale parameter Υ. As noted in Sec. 12-1-5 this indexes left cosets on the parameter space and we have

$$[\mathscr{B}, \Upsilon] = [0, \Upsilon][\Xi, I] \tag{12-76}$$

where $\Xi = \Upsilon^{-1}\mathscr{B}$ is the coefficient of variation. The presentation equation

280 INFERENCE AND LINEAR MODELS

separates, giving the Υ-specific component

$$\Upsilon^{-1}T(Y) = T \tag{12-77}$$

The marginal distribution of T is available from (12-64) and (12-65); we have the probability differential

$$\frac{A_{n-r}^{(p)}}{(2\pi)^{(n-r)p/2}} \text{etr}\left(-\tfrac{1}{2}TT'\right) |T|^{n-r} \frac{dT}{|T|_\Delta} \tag{12-78}$$

which is the triangular chi $\Delta\chi_p(n-r)$ of formula (12-65). This distribution with equation (12-77) gives tests and confidence regions for the parameter Υ exactly as in Sec. 8-3-3.

As noted there the analysis of the multivariate normal is usually in terms of the variance matrix $\Sigma = \Upsilon\Upsilon'$ and the inner product matrix $S(Y) = T(Y)T'(Y)$. The discussion there applies equally here but with the degrees of freedom $n-1$ replaced by $n-r$.

For example, the distribution of $S(Y)$ is Wishart $W_p(n-r; \Sigma)$ with density as indicated by (8-112) and (8-113) but with $n-1$ replaced by $n-r$.

12-2-4 The Location Component

Now consider the location component $\mathscr{B}$. As noted in Sec. 12-1-5, $\mathscr{B}$ indexes left cosets on the parameter space and we have

$$[\mathscr{B}, \Upsilon] = [\mathscr{B}, I][0, \Upsilon] \tag{12-79}$$

The presentation equation separates, giving the $\mathscr{B}$-specific component

$$T^{-1}(Y)[B(Y) - \mathscr{B}] = T^{-1}B = H \tag{12-80}$$

The marginal distribution of this matrix t-type variable can be obtained from (12-64) using (12-50). In the pattern of Sec. 8-3-4 we obtain the density function

$$\int_{G_S} \frac{A_{n-r}^{(p)}}{(2\pi)^{np/2}} \text{etr}\left[-\tfrac{1}{2}T(I + HXX'H')T'\right] \frac{|T|^n}{|T|_\Delta} |XX'|^{p/2} \, dT$$

$$= \int_{G_S} \frac{A_{n-r}^{(p)}}{(2\pi)^{np/2}} \text{etr}\left(-\tfrac{1}{2}TEE'T'\right) |T|^n \frac{dT}{|T|_\Delta} |XX'|^{p/2} \tag{12-81}$$

where $E = (I + HXX'H')^L$ is the PLT square root of the "inner product" matrix $I + HXX'H'$. The integration then duplicates the pattern in Sec. 8-3-4 and we obtain

$$\frac{A_{n-r}^{(p)}}{(2\pi)^{np/2}} \int_{G_S} \text{etr}\left(-\tfrac{1}{2}TEE'T'\right) |TE|^n \frac{dTE}{|TE|_\Delta} \frac{|E|_\Delta |XX'|^{p/2}}{|E|^n |E|_\nabla}$$

$$= \frac{A_{n-r}^{(p)}}{(2\pi)^{np/2}} \frac{(2\pi)^{np/2}}{A_n^{(p)}} \frac{|E|_\Delta}{|E|^n |E|_\nabla} |XX'|^{p/2}$$

$$= \frac{A_{n-r}^{(p)}}{A_n^{(p)}} \frac{1}{|E|^n} \frac{|E|_\Delta |E|_\nabla}{|E|_\nabla^2} |XX'|^{p/2}$$

$$= \frac{A_{n-r}^{(p)}}{A_n^{(p)}} \frac{1}{|I + HXX'H'|^{n/2}} \frac{|I + HXX'H'|^{(p+1)/2}}{|I + HXX'H'|_\nabla} |XX'|^{p/2} \quad (12\text{-}82)$$

Note that the "standardized" variable $H(XX')^L$ has the preceding density function with XX' set equal to the identity. The standardized distribution is called the matrix triangular Student $\Delta t_{p \times r}(n-r)$ with $(n-r)$ degrees of freedom (compare with Sec. 8-3-4). The distribution of the matrix H is called the scaled matrix triangular Student $\Delta t_{p \times r}[n-r, (XX')^L]$ with $(n-r)$ degrees of freedom. See Fraser, Lee, and Streit (1968).

This distribution with equation (12-80) gives tests and confidence regions for the location matrix $\mathscr{B}$.

12-3 MULTIVARIATE REGRESSION MODEL WITH LINEAR VARIATION

In this section we examine the multivariate model with linear variation from Secs. 8-4 and 8-5 as combined with the regression model from Chap. 6. In the next section we specialize this to the normal case.

Consider a random system with a p-variate response and with input variables that can be changed by the investigator. The investigator wants information on how changes in the input variables affect the responses.

As in Sec. 12-1 suppose the background information on the system specifies that the only effect from the input variables is on the location of the response and is linear over the appropriate range for the input variables. Also suppose that the distribution describing the variation has the linear form discussed at the beginning of Sec. 8-4—that the scale of the response variation is unknown to the extent of a positive linear transformation but that otherwise the distribution form is known or known up to a shape parameter λ.

12-3-1 The Model

Let $\mathbf{y} = (y_1, \ldots, y_p)'$ be the response for a single performance and

$$Y = (\mathbf{y}_1, \ldots, \mathbf{y}_n) = \begin{bmatrix} Y_1 \\ \vdots \\ Y_p \end{bmatrix} = \begin{bmatrix} y_{11} & \cdots & y_{1n} \\ \cdots & \cdots & \cdots \\ y_{p1} & \cdots & y_{pn} \end{bmatrix} \quad (12\text{-}83)$$

be the compound response recording $\mathbf{y}_1, \ldots, \mathbf{y}_n$ for n performances of the system. Similarly, let $\mathbf{z}$ and Z be the corresponding variables for the variation.

Now let $x_1, \ldots, x_r$ be the input variables. For a particular response $\mathbf{y}$ we let $\mathbf{x} = (x_1, \ldots, x_r)'$ be the corresponding values or settings for the input variables, and for the compound response Y we let

$$X = (\mathbf{x}_1, \ldots, \mathbf{x}_n) = \begin{bmatrix} X_1 \\ \vdots \\ X_r \end{bmatrix} = \begin{bmatrix} x_{11} & \cdots & x_{1n} \\ \cdots & \cdots & \cdots \\ x_{r1} & \cdots & x_{rn} \end{bmatrix} \quad (12\text{-}84)$$

be the corresponding settings for the input variables. The matrix X is called the design matrix; recall the comments in Sec. 12-1-1 concerning the name design matrix for this X or its transpose.

The basic model for a single performance has the presentation

$$\mathbf{y} = \mathscr{B}\mathbf{x} + \Gamma\mathbf{z} \quad (12\text{-}85)$$

where $\mathscr{B}$ is $p \times r$ and Γ is $p \times p$ with $|\Gamma| > 0$. Also let

$$f_\lambda(\mathbf{z}) = f_\lambda(z_1, \ldots, z_p) \quad (12\text{-}86)$$

be the density function for the objective variation. We suppose that f_λ has been suitably standardized in some appropriate way as indicated in Sec. 8-4-1.

Then for n independent performances we have $\mathbf{y}_i = \mathscr{B}\mathbf{x}_i + \Gamma\mathbf{z}_i$ for the ith performance and

$$Y = \mathscr{B}X + \Gamma Z = [\mathscr{B}, \Gamma]Z = \theta Z \quad (12\text{-}87)$$

for the compound response in terms of the variation Z. The distribution for Z is

$$f_\lambda(Z) = \Pi_1^n f_\lambda(\mathbf{z}_i) \quad (12\text{-}88)$$

The transformations $[\mathscr{B}, \Gamma]$ form a group:

$$[B_2, C_2][B_1, C_1] = [B_2 + C_2 B_1, C_2 C_1]$$
$$[B, C]^{-1} = [-C^{-1}B, C^{-1}] \quad (12\text{-}89)$$
$$[0, I] = i$$

Note that the positive $p \times p$ matrices are closed under multiplication and inversion; compare with (8-127). We then have the regression affine group

$$G = \{[B, C] : B \in \mathbb{R}^{pr}, |C| > 0\} \quad (12\text{-}90)$$

We will see that G is exact on the sample space provided $n \geq p + r$ and a set of measure zero is excluded.

This gives us the structural model

$$\mathscr{M}_V = (\Omega; \mathbb{R}^{pn}, \mathscr{A}^{pn}, \mathscr{V}, G) \quad (12\text{-}91)$$

where $\Omega = (G \times \Lambda)$, $\mathscr{V}$ is the class of distributions f_λ in (12-88), and G is the transformation group (12-90) with action (12-87). We abbreviate this as

$$\begin{aligned} & Y = \theta Z \text{ with } \theta \text{ in } G \\ & Z \text{ has distribution in the class } \mathscr{V} \end{aligned} \quad (12\text{-}92)$$

and assume that $n \geq p + r$.

12-3-2 The Analysis

Consider an inference base

$$(\mathcal{M}_V, Y^0) \tag{12-93}$$

using the model (12-92) with data Y^0. We have the observed orbit

$$GZ = GY^0 \tag{12-94}$$

together with the corresponding model, as described by (*a*) in Sec. 7-1-4. And we have the conditional model for Z given $GZ = GY^0$ together with the presentation $[Y] = \theta[Z]$ and the data $[Y^0]$, as described by (*b*) in Sec. 7-1-4. We first assemble notation from Chaps. 6 and 8, following the pattern in Sec. 12-1.

We examine the action of the group G on the space $\mathbb{R}^{pn}$. From (12-85) and (12-87) we see that the transformation

$$\tilde{Z} = BX + CZ \tag{12-95}$$

can be written

$$\begin{bmatrix} \tilde{Z}_1 \\ \vdots \\ \tilde{Z}_p \end{bmatrix} = B \begin{bmatrix} X_1 \\ \vdots \\ X_r \end{bmatrix} + C \begin{bmatrix} Z_1 \\ \vdots \\ Z_p \end{bmatrix} \tag{12-96}$$

Let $\mathscr{L}^+(X_1, \ldots, X_r, Z_1, \ldots, Z_p)$ be the $(p + r)$-dimensional subspace

$$\mathscr{L}^+(X_1, \ldots, X_r, Z_1, \ldots, Z_p) = \{\Sigma b_u X_u + \Sigma c_j Z_j : b_u, c_j \in \mathbb{R}\} \tag{12-97}$$

together with the orientation of $X_1, \ldots, X_r, Z_1, \ldots, Z_p$ treated as the *positive* orientation for the $(p + r)$-dimensional subspace. A group element $[B, C]$ carries a sequence $Z_1, \ldots, Z_p$ into a sequence $\tilde{Z}_1, \ldots, \tilde{Z}_p$, also in $\mathscr{L}^+(X_1, \ldots, X_r, Z_1, \ldots, Z_p)$ and with the same positive orientation relative to the fixed $X_1, \ldots, X_r$; recall that $|C| > 0$. In fact, from (12-95) we see that we can get any sequence of p vectors in the subspace $\mathscr{L}^+(X_1, \ldots, X_r, Z_1, \ldots, Z_p)$ provided it has the positive orientation.

We wish to choose a basis for the subspace $\mathscr{L}^+(X_1, \ldots, X_r, Z_1, \ldots, Z_p)$. The vectors $X_1, \ldots, X_r$ are the obvious first choice. Let

$$D(Z) = \begin{bmatrix} D_1(Z) \\ \vdots \\ D_p(Z) \end{bmatrix} \tag{12-98}$$

consist of p orthonormal vectors orthogonal to the rows of X and with the positive orientation. We use $X_1, \ldots, X_r, D_1(Z), \ldots, D_p(Z)$ as a basis for the subspace. Note as in Sec. 8-4-2 that the choice of basis must not depend on $Z_1, \ldots, Z_p$ directly, only on the *space* $\mathscr{L}^+(X_1, \ldots, X_r, Z_1, \ldots, Z_p)$. Such a basis can be formed, for example, in the following way, paralleling that in Sec. 8-4-2: take a sequence of p linearly independent vectors, say the first p coordinate axes; project them onto the subspace; and orthonormalize them in sequence as done to obtain $D_1, \ldots, D_p$ in Sec. 12-1-3. Such a procedure gives a basis, except of course for a set of measure zero for which the projections have linear dependence.

Now let $B_j(Z)$, $C_j(Z)$ record the regression coefficients of Z_j on $X_1, \ldots, X_r$, $D_1(Z), \ldots, D_p(Z)$ and write

$$B(Z) = \begin{bmatrix} B_1(Z) \\ \vdots \\ B_p(Z) \end{bmatrix} \quad C(Z) = \begin{bmatrix} C_1(Z) \\ \vdots \\ C_p(Z) \end{bmatrix} \quad (12\text{-}99)$$

We then have

$$\begin{aligned} Z &= B(Z)X + C(Z)D(Z) \\ &= [B(Z), C(Z)]D(Z) = [Z]D(Z) \end{aligned} \quad (12\text{-}100)$$

where $[Z] = [B(Z), C(Z)]$ is a transformation variable that gives the group position of Z relative to the reference point $D(Z)$.

For exactness the preceding analysis must be feasible; accordingly, we require that $n \geq p + r$ and we eliminate the set of measure zero in which $X_1, \ldots, X_r, Z_1, \ldots, Z_p$ are linearly dependent.

As in preceding chapters we have, of course, the separation of the inference base according to the observed

$$D(Z) = D(Y) = D(Y^0) \quad (12\text{-}101)$$

and the unobserved $[Z]$ given $D(Z) = D(Y^0)$. We first derive the appropriate distribution for the component inference bases.

12-3-3 Density Functions

The Jacobians and measures from Sec. 7-2 can be calculated as in Sec. 8-4-2 but with the regression component as in Sec. 12-1-4.

The transformation $g = [B, C]$ applied to the matrix Z operates column by column. Accordingly, we have

$$\begin{aligned} J_{pn}(g:Z) &= |C|^n \quad J_{pn}(Z) = |C(Z)|^n \\ dM(Z) &= |C(Z)|^{-n} dZ \end{aligned} \quad (12\text{-}102)$$

The equation $\tilde{g} = gg^*$ has the form

$$\begin{aligned} \tilde{B} &= B + CB^* \\ \tilde{C} &= CC^* \end{aligned} \quad (12\text{-}103)$$

The left transformation operates column by column on B^*, C^*, and the columns of C^* are fully effective. Thus

$$\begin{aligned} J(g:g^*) &= |C|^{p+r} \quad J([Z]) = |C(Z)|^{p+r} \\ d\mu(g) &= \frac{dB \, dC}{|C|^{p+r}} \end{aligned} \quad (12\text{-}104)$$

The right transformation operates row by row on C and, for given C, by location only on B. Thus in the pattern of Sec. 8-4-2 we have

$$J^*(g^*:g) = |C^*|^p \qquad J^*(g) = |C|^p$$

$$dv(g) = \frac{dB\,dC}{|C|^p} \tag{12-105}$$

$$\Delta(g) = \frac{1}{|C|^r}$$

The basic Jacobian $J(D)$ recording volume change on orbit relative to change on group is easily obtained by the same argument as in Sec. 12-1-4:

$$J(D) = |XX'|^{p/2} \tag{12-106}$$

Now consider the analysis of the inference base $(\mathcal{M}_V, Y^0)$. The observed orbit gives the observed value for the variation

$$D(Z) = D(Y^0) = D^0 \tag{12-107}$$

This gives the inference base (*a*) from Sec. 7-1-4:

$$(\mathcal{M}_D, D^0) \tag{12-108}$$

The distributions in $\mathcal{M}_D$ are available from (7-46) together with our present calculations:

$$h_\lambda(D) = \int_G f_\lambda(BX + CD)|C|^{n-p-r}|XX'|^{p/2}\,dB\,dC \tag{12-109}$$

with integration over $\mathbb{R}$ for each coordinate subject only to $|C| > 0$. This, of course, leads to the observed likelihood for λ:

$$L(D^0; \lambda) = ch_\lambda(D^0) \tag{12-110}$$

The unobserved characteristics of the variation are given by $[B, C] = [B(Z), C(Z)]$. For these characteristics we have the inference base (*b*) from Sec. 7-1-4:

$$(\mathcal{M}_V^{D^0}, [Y^0]) \tag{12-111}$$

The distributions for the structural model $\mathcal{M}_V^{D^0}$ are available from (7-47) together with our present calculations; we have the probability differential

$$h_\lambda^{-1}(D^0)f_\lambda(BX + CD^0)|C|^{n-p-r}|XX'|^{p/2}\,dB\,dC \tag{12-112}$$

for B, C on the group G. This distribution for B, C is used with the presentation

$$B(Y) = \mathcal{B} + \Gamma B$$
$$C(Y) = \phantom{\mathcal{B} + {}}\Gamma C \tag{12-113}$$

The observed values for the response coordinates are given by

$$[Y^0] = [B(Y^0), C(Y^0)] \tag{12-114}$$

The response distribution consistent with the identified $D(Y) = D(Y^0) = D^0$ from the actual response is available from (7-49) in the pattern of (8-149):

$$h_\lambda^{-1}(D^0) f_\lambda(\Gamma^{-1}[(B - \mathscr{B})X + CD^0]) \frac{|C|^{n-p-r}}{|\Gamma|^n} |XX'|^{p/2} \, dB \, dC \quad (12\text{-}115)$$

where for this formula we have written $B(Y) = B$, $C(Y) = C$.

12-3-4 Inference for Component Parameters

Some component parameters may not have left coset form; see the discussion in Sec. 12-1-5.

In this section we restrict our attention to the basic location and scale parameters; for this, note the concluding comments in Sec. 8-4-3 relative to the linear pattern for variation.

For the location parameter $\mathscr{B}$ we have the separation of the presentation

$$[B(Y), C(Y)] = [\mathscr{B}, \Gamma][B, T] \quad (12\text{-}116)$$

giving the unique $\mathscr{B}$-specific component

$$C^{-1}(Y)[B(Y) - \mathscr{B}] = C^{-1}B = H \quad (12\text{-}117)$$

where H is a matrix t-type variable extending the vector $\mathbf{t}$ in Sec. 8-4-3 to a matrix, in the pattern of (12-45). We see that $\mathscr{B}$ indexes left cosets by noting the factorization

$$[\mathscr{B}, \Gamma] = [\mathscr{B}, I][0, \Gamma] \quad (12\text{-}118)$$

Thus $\mathscr{B}$ indexes left cosets of the scale group

$$G_S = \{[0, C] : C \text{ is } p \times p \text{ with } |C| > 0\} \quad (12\text{-}119)$$

as given by (8-153). Note that the group coordinates can be separated in the reverse order:

$$[B, C] = [0, C][H, I] \quad (12\text{-}120)$$

and thus that H indexes the right cosets or orbits on the sample space.

The marginal distribution for it is obtained from (12-112) by using

$$dB = |C|^r \, dH \quad (12\text{-}121)$$

for fixed C. The probability differential is

$$h_\lambda^{-1}(D^0) \int_C f_\lambda(C(HX + D^0)) |C|^n \frac{dC}{|C|^p} |XX'|^{p/2} \, dH \quad (12\text{-}122)$$

This distribution together with equation (12-117) and the observed values provides tests and confidence regions for the location parameter.

For the scale parameter Γ we have the separation of the basic equation giving the unique Γ-specific component

$$\Gamma^{-1}C(Y) = C \quad (12\text{-}123)$$

The full parameter can be separated as

$$[\mathcal{B}, \Gamma] = [0, \Gamma][\Xi, I] \tag{12-124}$$

where $\Xi = \Gamma^{-1}\mathcal{B}$ is the coefficient of variation; thus Γ indexes the left cosets of the location group

$$G_L = \{[B, I] : B \in \mathbb{R}^{pr}\} \tag{12-125}$$

Note that the group coordinates can be separated in the reverse order:

$$[B, C] = [B, I][0, C] \tag{12-126}$$

and thus that C indexes the right cosets or orbits on the sample space.

The marginal distribution of C is obtained from (12-112) by directly integrating out the location variable B. The probability differential is

$$h_\lambda^{-1}(D^0) \int_B f_\lambda(BX + CD^0) |XX'|^{p/2} \, dB \, |C|^{n-p-r} \, dC \tag{12-127}$$

This distribution with the presentation (12-123) and the observed value $C(Y^0)$ provides tests and confidence regions for the scale parameter.

The response distribution for $C(Y)$ consistent with the observed $C(Y^0)$ from Y^0 is obtained from (12-127) by direct transformation using the invariant measure $dC/|C|^p$ for the $p \times p$ positive matrices (available from the case $r = 0$). We obtain

$$h_\lambda^{-1}(D^0) \int_B f_\lambda(BX + \Gamma^{-1}CD^0) |XX'|^{p/2} \, dB \, \frac{|C|^{n-r} \, dC}{|\Gamma|^{n-r}|C|^p} \tag{12-128}$$

where for this formula we have used C for $C(Y)$.

12-4 NORMAL MULTIVARIATE REGRESSION MODEL WITH LINEAR VARIATION

In the preceding section we develop the multivariate regression model using the linear variation from Sec. 8-4. We now examine this for the special case of a normal pattern for the variation.

12-4-1 The Model

Consider a random system with a p-variate response. The multivariate regression model with linear variation has the form

$$Y = \mathcal{B}X + \Gamma Z \tag{12-129}$$

for n performances; Y and Z are the $p \times n$ response and variation matrices, $\mathcal{B}$ is the regression coefficient matrix relative to the design matrix X, and Γ is the positive $p \times p$ scale matrix for the variation.

We examine this model using the standard normal model for the variation. From (12-58) we have

$$f(Z)\,dZ = (2\pi)^{-np/2} \operatorname{etr}(-\tfrac{1}{2}ZZ')\,dZ \tag{12-130}$$

We can rewrite this in the invariant form

$$(2\pi)^{-np/2} \operatorname{etr}(-\tfrac{1}{2}ZZ')|C(Z)|^n\,dM(Z) \tag{12-131}$$

where

$$dM(Z) = \frac{dZ}{|C(Z)|^n} \tag{12-132}$$

is the invariant measure on the sample space and $C(Z)$ is the positive scale matrix defined by (12-99) and (12-100).

The corresponding response distribution can be obtained by direct substitution using the invariant form (12-131). For this we can write

$$\operatorname{tr} ZZ' = \operatorname{tr}(Z'Z)$$
$$= \operatorname{tr}(Y - \mathscr{B}X)'\Gamma'^{-1}\Gamma^{-1}(Y - \mathscr{B}X)$$
$$= \operatorname{tr}(Y - \mathscr{B}X)'\Sigma^{-1}(Y - \mathscr{B}X) \tag{12-133}$$

where $\Sigma = \Gamma\Gamma'$ as in Sec. 8-5-1 is, of course, the variance matrix for $\dot{Y}$; recall the comments there concerning the square roots of a variance matrix. The response distribution is

$$(2\pi)^{-np/2} \operatorname{etr}\left[-\tfrac{1}{2}(Y - \mathscr{B}X)'\Sigma^{-1}(Y - \mathscr{B}X)\right] \frac{|C(Y)|^n}{|\Gamma|^n}\,dM(Y)$$
$$= (2\pi)^{-np/2}|\Sigma|^{-n/2} \operatorname{etr}\left[-\tfrac{1}{2}(Y - \mathscr{B}X)'\Sigma^{-1}(Y - \mathscr{B}X)\right] dY \tag{12-134}$$

which of course agrees with (12-62) from Sec. 12-2 using progressive variation.

In summary we have the following structural model:

$$\mathscr{M}_V = (\Omega; \mathbb{R}^{pn}, \mathscr{A}^{pn}, \mathscr{V}, G) \tag{12-135}$$

where $\Omega = G$ is the parameter space, $\mathscr{V}$ has a single distribution (12-130), and G is the regression affine group (12-90). We abbreviate this as

$$\begin{aligned} &Y = \theta Z \text{ with } \theta \text{ in } G \\ &Z \text{ is standard normal on } \mathbb{R}^{pn} \end{aligned} \tag{12-136}$$

12-4-2 The Analysis

Consider the analysis of the inference base $(\mathscr{M}_V, Y^0)$ using the standard normal model for the variation. We first determine the marginal distribution for D and the conditional distribution for $[Z]$ given D.

The change of variables to get coordinates on the orbit is given by

$$Z = B(Z)X + C(Z)D(Z) = [B, C]D \tag{12-137}$$

where the rows of $D(Z)$ are orthonormal and orthogonal to those of X. The exponent of the normal density involves the following quadratic expression:

MULTIVARIATE REGRESSION MODELS **289**

$$\operatorname{tr} ZZ' = \operatorname{tr}(BX + CD)(BX + CD)'$$
$$= \operatorname{tr} BXX'B' + \operatorname{tr} CDD'C$$
$$= \operatorname{tr} BXX'B' + \operatorname{tr} CC' \qquad (12\text{-}138)$$

where we have used $XD' = 0$ and $DD' = I$.

For the integration on the group we find it convenient to use the PLT orthogonal factorization as used in Sec. 8-5-2:

$$C = TO \qquad (12\text{-}139)$$

where T is PLT and O is a $p \times p$ rotation matrix. We can then further simplify (12-138):

$$\operatorname{tr} ZZ' = \operatorname{tr} BXX'B' + \operatorname{tr} TOO'T'$$
$$= \operatorname{tr} BXX'B' + \operatorname{tr} TT' \qquad (12\text{-}140)$$

Recall then that $\operatorname{tr} TT'$ is just the sum of squares of the elements of the PLT matrix T. The change of variable for the differential is given by (8-181):

$$\frac{dC}{|C|^p} = \frac{dT}{|T|_\Delta} dO \qquad (12\text{-}141)$$

where dO is interpreted as volume orthogonal to the orbits of the PLT scale group in the space $\mathbb{R}^{p^2}$ for $p \times p$ matrices. For the alternative volume measure tangential to the manifold of the rotation matrices see (8-182).

We can now determine the conditional and marginal densities. For the conditional distribution of $[Z] = [B, C] = [B, TO]$ given D we have from (12-112) with (12-140):

$$h^{-1}(D) f(BX + CD) |C|^{n-p-r} |XX'|^{p/2} \, dB \, dC$$

$$= h^{-1}(D)(2\pi)^{-np/2} \operatorname{etr}\left[-\tfrac{1}{2}(BXX'B' + TT')\right] |C|^{n-r} |XX'|^{p/2} \, dB \, \frac{dT}{|T|_\Delta} dO$$

$$= h^{-1}(D)(2\pi)^{-np/2} \operatorname{etr}(-\tfrac{1}{2}BXX'B') \operatorname{etr}(-\tfrac{1}{2}TT') |T|^{n-r} |XX'|^{p/2} \, dB$$

$$\times \frac{dT}{|T|_\Delta} dO$$

$$= \frac{1}{(2\pi)^{rp/2}} \operatorname{etr}(-\tfrac{1}{2}BXX'B') |XX'|^{p/2} \, dB$$

$$\times \frac{A_{n-r}^{(p)}}{(2\pi)^{(n-r)p/2}} \operatorname{etr}(-\tfrac{1}{2}TT') s_{(1)}^{n-r-1} \ldots s_{(p)}^{n-r-p} \, dT \, \frac{dO}{A_p^{(p-1)}}$$

$$= \frac{1}{(2\pi)^{rp/2}} \operatorname{etr}(-\tfrac{1}{2}BXX'B') |XX'|^{p/2} \, dB$$

$$\times \frac{A_{n-r}^{(p)}}{A_p^{(p-1)}(2\pi)^{(n-r)p/2}} \operatorname{etr}(-\tfrac{1}{2}CC') |C|^{n-r-p} \, dC \qquad (12\text{-}142)$$

290 INFERENCE AND LINEAR MODELS

where we have used integration results from (12-64) and (8-183). The conditional for B, T, O can be described simply: the rows of B, the matrix T, and the matrix O are statistically independent and

B_j is normal $[0; (XX')^{-1}]$

$$T = \begin{bmatrix} s_{(1)} & \cdots & 0 \\ t_{21} & s_{(2)} & & \vdots \\ \vdots & & \ddots & \\ t_{p1} & \cdots & t_{p,p-1} & s_{(p)} \end{bmatrix} = \begin{bmatrix} \chi_{n-r} & \cdots & & 0 \\ z_{21} & \chi_{n-r-1} & & \vdots \\ \vdots & & \ddots & \\ z_{p1} & \cdots & z_{pp-1} & \chi_{n-r-p+1} \end{bmatrix}$$

(12-143)

O is uniform on the rotation group (Secs. 8-3-5 and 8-5-2)

where the z variables are independent standard normals and the χ variables are independent chi variables with degrees of freedom as subscripted. The distribution of T is the triangular chi $\Delta\chi_p(n - r)$ distribution discussed in Sec. 8-3-2.

The conditional distribution (12-142) gives us the distribution component for the model $\mathcal{M}_V^{D^0}$. Thus, as in Sec. 12-3-3, we obtain the component inference base (12-111):

$$(\mathcal{M}_V^{D^0}, [Y^0]) \qquad (12\text{-}144)$$

where the model $\mathcal{M}_V^{D^0}$ has the distribution for $[Z]$ given by (12-142) and the presentation

$$\begin{aligned} B(Y) &= \mathcal{B} + \Gamma B \\ C(Y) &= \quad \Gamma C \end{aligned} \qquad (12\text{-}145)$$

relating the response $[Y] = [B(Y), C(Y)]$ and the standardized variation $[Z] = [B, C]$; the observed values for the response components are given by

$$[Y^0] = [B(Y^0), C(Y^0)] \qquad (12\text{-}146)$$

The conditional response distribution for Y given the observed $D(Y) = D(Y^0) = D^0$ is available from (12-115) and (12-142). For this we use (12-145) with the exponent (12-133) and simplify as for (12-69):

tr $Z'Z =$ tr $BXX'B' +$ tr CC'

$$\begin{aligned} &= \operatorname{tr} \{\Gamma^{-1}[B(Y) - \mathcal{B}]XX'[B(Y) - \mathcal{B}]'\Gamma'^{-1}\} \\ &\quad + \operatorname{tr} \Gamma^{-1}C(Y)C'(Y)\Gamma'^{-1} \\ &= \operatorname{tr} \{[B(Y) - \mathcal{B}]XX'[B(Y) - \mathcal{B}]'\Sigma^{-1}\} + \operatorname{tr} C(Y)C'(Y)\Sigma^{-1} \\ &= \operatorname{tr} \{[B(Y) - \mathcal{B}]XX'[B(Y) - \mathcal{B}]'\Sigma^{-1}\} + \operatorname{tr} S(Y)\Sigma^{-1} \end{aligned} \qquad (12\text{-}147)$$

where

$$\begin{aligned} S(Y) &= C(Y)C'(Y) = C(Y)D(Y)D'(Y)C'(Y) \\ &= [Y - B(Y)X][Y - B(Y)X]' \end{aligned} \qquad (12\text{-}148)$$

is the inner product matrix for the deviations from regression. The probability element for

$$[Y] = [B(Y), C(Y)] = [B(Y), T(Y)O(Y)] \qquad (12\text{-}149)$$

is

$$\frac{A_{n-r}^{(p)}}{A_p^{(p-1)}(2\pi)^{np/2}} \text{etr}\left\{-\tfrac{1}{2}[B(Y) - \mathscr{B}]XX'[B(Y) - \mathscr{B}]'\Sigma^{-1} - \tfrac{1}{2}S(Y)\Sigma^{-1}\right\}$$

$$\times \frac{|C(Y)|^n}{|\Gamma|^n}|XX'|^{p/2}\frac{dB(Y)\,dC(Y)}{|C(Y)|^{p+r}}$$

$$= \frac{A_{n-r}^{(p)}}{A_p^{(p-1)}(2\pi)^{np/2}} \text{etr}\left\{-\tfrac{1}{2}[B(Y) - \mathscr{B}]XX'[B(Y) - \mathscr{B}]'\Sigma^{-1} - \tfrac{1}{2}S(Y)\Sigma^{-1}\right\}$$

$$\times \frac{|T(Y)|^{n-r}}{|\Gamma|^n}|XX'|^{p/2}\,dB(Y)\,\frac{dT(Y)\,dO(Y)}{|T(Y)|_\Delta}$$

$$= \frac{|XX'|^{p/2}|\Sigma|^{-r/2}}{(2\pi)^{rp/2}} \text{etr}\left\{-\tfrac{1}{2}[B(Y) - \mathscr{B}]XX'[B(Y) - \mathscr{B}]'\Sigma^{-1}\right\}dB(Y)$$

$$\times \frac{A_{n-r}^{(p)}}{(2\pi)^{(n-r)p/2}} \text{etr}\left[-\tfrac{1}{2}S(Y)\Sigma^{-1}\right]\frac{|S(Y)|^{(n-r)/2}}{|\Sigma|^{(n-r)/2}}\frac{dT(Y)}{|T(Y)|_\Delta}\frac{dO(Y)}{A_p^{(p-1)}} \qquad (12\text{-}150)$$

where we have followed some of the steps in formulas (12-71) and (8-191). We can then read the conditional response distribution from (12-150). We have that $B(Y)$, $T(Y)$, $O(Y)$ are statistically independent and that

$$B(Y) \text{ is } N[\mathscr{B};(XX')^{-1}\otimes\Sigma]$$

$$T(Y) \text{ is } \Delta\chi_p(n - r, \Upsilon) \qquad (12\text{-}151)$$

$$O(Y) \text{ is uniform on the rotation group}$$

where Υ is the PLT root of Σ. This agrees with that for progressive error in (12-72) but has the additional uniform distribution for $O(Y)$.

We can now obtain the marginal distribution of D. This can be obtained by solving for $h(D)$ from the expression (12-142):

$$h(D)\,dD = \frac{A_p^{(p-1)}}{A_{n-r}^{(p)}}\,dD \qquad (12\text{-}152)$$

Note that D has a uniform distribution on the reference point manifold Q using our particular choice of Euclidean volume orthogonal to the orbit GD at D; Q is an $np - rp - p^2 = (n - r - p)p$-dimensional manifold of semiorthogonal $p \times n$ matrices.

The marginal distribution (12-152) gives us the distribution for the model $\mathscr{M}_D$. As in Sec. 12-2-2, the corresponding inference base $(\mathscr{M}_D, D^0)$ has no parameters and thus collapses.

12-4-3 The Scale Component

Consider the scale parameter Γ. As noted in Sec. 12-3-4 we have the separation of the basic equation giving the Γ-specific component

$$\Gamma^{-1}C(Y) = C \tag{12-153}$$

Also, we have that Γ is a left coset parameter relative to the location group (12-125) and that C indexes the right cosets or orbits on the sample space.

The marginal distribution of C is available directly from (12-142):

$$\frac{A_{n-r}^{(p)}}{A_p^{(p-1)}(2\pi)^{(n-r)p/2}} \operatorname{etr}(-\tfrac{1}{2}CC')|C|^{n-r-p} dC \tag{12-154}$$

This is the matrix chi distribution $\chi_p(n-r)$ on $(n-r)$ degrees of freedom; compare with (8-194).

Note that the equation (8-193) in Sec. 8-5-3 is the same as the present equation (12-153) and that the distribution (8-194) is the same as the present distribution (12-154) but with $n-1$ replaced by $n-r$.

In Sec. 8-5 we discussed inference for Γ and for Σ using a linear expression for normal variation. All the results there are applicable here, with just the appropriate replacement of the degrees of freedom $n-1$ there by $n-r$ here.

12-4-4 The Location Component

Now consider the location component $\mathscr{B}$. As noted in Sec. 12-3-4, $\mathscr{B}$ indexes left cosets of the scale group (12-119) on the parameter space. Also, we have the separation of the basic equation giving the $\mathscr{B}$-specific component

$$C^{-1}(Y)[B(Y) - \mathscr{B}] = C^{-1}B = H \tag{12-155}$$

The matrix t-type variable H indexes the corresponding right cosets or orbits on the sample space.

The marginal distribution for H can be obtained from (12-122) together with (12-142) and the integration result (12-154). The probability differential is

$$\int_{G_s} \frac{A_{n-r}^{(p)}}{A_p^{(p-1)}(2\pi)^{np/2}} \operatorname{etr}(-\tfrac{1}{2}CHXX'H'C' - \tfrac{1}{2}CC')|C|^{n-p} dC |XX'|^{p/2} dH$$

$$= \int_{G_s} \frac{A_n^{(p)}}{A_p^{(p-1)}(2\pi)^{np/2}} \operatorname{etr}[-\tfrac{1}{2}C(I + HXX'H')C']$$

$$\times |C|^{n-p}|I + HXX'H'|^{n/2} dC \frac{A_{n-r}^{(p)}}{A_n^{(p)}}|I + HXX'H'|^{-n/2}|XX'|^{p/2} dH$$

$$= \frac{A_{n-r}^{(p)}}{A_n^{(p)}}|I + HXX'H'|^{-n/2}|XX'|^{p/2} dH$$

$$= \frac{A_{n-p}^{(r)}}{A_n^{(r)}}|I + HXX'H'|^{-n/2}|XX'|^{p/2} dH \tag{12-156}$$

The adjustment between the last two expressions is a simple cancellation or insertion of factors.

Consider the preceding distribution in the standardized or canonical case with $XX' = I$. For this canonical case we have the probability element

$$\frac{A_{n-r}^{(p)}}{A_n^{(p)}} |I + HH'|^{-n/2} dH = \frac{A_{n-p}^{(r)}}{A_n^{(r)}} |I + H'H|^{-n/2} dH \qquad (12\text{-}157)$$

where we have used the identity

$$|I + HH'| = |I + H'H| \qquad (12\text{-}158)$$

in which the first I is $p \times p$ and the second I is $r \times r$. This distribution (12-157) for H is called the matrix Student $t_{p \times r}(n - r)$ distribution on $(n - r)$ degrees of freedom; note that H' is the matrix Student $t_{r \times p}(n - p)$ distribution. The distribution (12-156) is a right scaled matrix Student distribution.

Tests and confidence regions for the location $\mathscr{B}$ are available from the distribution (12-156) together with the presentation equation (12-155) and the needed observed values $C(Y^0)$ and $B(Y^0)$.

For some background information on the models and analysis in this chapter see Fraser (1968a), Fraser, Lee, and Streit (1968), and Fraser and Haq (1969, 1970).

REFERENCES AND BIBLIOGRAPHY

Dickey, J. M.: Matricvariate Generalizations of the Multivariate t Distribution and the Inverted Multivariate t Distribution, *Ann. Math. Stat.*, vol. 38, pp. 511–518, 1967.

Fraser, D. A. S.: "The Structure of Inference," Huntington, Krieger, New York, 1968a.

——: The Conditional Wishart: Normal and Nonnormal, *Ann. Math. Stat.*, vol. 39, pp. 593–605, 1968b.

—— and M. S. Haq: Structural Probability and Prediction for the Multivariate Model, *Jour. Roy. Stat. Soc.*, ser. B, vol. 31, pp. 317–331, 1969.

—— and ——: Inference and Prediction for the Multilinear Model, *Jour. Stat. Res.*, vol. 4, pp. 93–109, 1970.

—— Y. S. Lee, and T. F. Streit: "Structure of Inference: Solutions," Department of Mathematics, University of Toronto, 1968.

—— and Kai-W. Ng: Inference for the Multivariate Regression Model, in P. R. Krishnaiah (ed.), "Multivariate Analysis," vol. IV, North-Holland Publishing Company, pp. 35–53, 1977.

Mitra, S. K.: A Density-Free Approach to the Matrix Variate Beta Distribution, *Sankhyā*, ser. A, vol. 32, pp. 81–88, 1970.

Olkin, I., and H. Rubin: Multivariate Beta Distributions and Independence Properties of the Wishart Distribution, *Ann. Math. Stat.*, vol. 35, pp. 261–269, 1964.

INDEX

Acceptance region, 92
Adaptive inference, 46
Alternative set, 136
Ancillarity, 75
Ancillarity principle, 78
Ancillary reduction, 75
    weak, 84
Ancillary statistic, 76
Antecedent set, 136

Bayes, 5, 104
    empirical, 5
Bioassay, 233

Chi
    see distribution
Coherence, 135, 137
Conditioning, 54
Confidence distribution, 92, 157
Confidence interval
    location, 24
    scale, 25
Confidence region, 91, 151
    component parameter, 156
Confidence regions
    in sequence, 162
Correlation coefficient
    distribution of, 173
Cross-section, 139

Darwin data, 26
Data, 1
Decisions, 103
Density allocation, 69
    ancillarity, 75
    likelihood, 75
    sufficiency, 70
Descriptive property, 3
Diffeomorphism, 62
Dilution series, 233

Disguised Student
    see distribution
Distribution
    disguised Student, 189, 281
    extreme value, 32, 239
    logistic, 235
    matrix chi, 202, 292
    matrix Student, 204, 293
    matrix triangular Student, 189, 281
    multinomial, 78, 83
    normal, 64, 67, 73, 86, 101, 103
    normal on circle, 207
    normal on sphere, 212
    Poisson, 83
    projected normal, 220
    subgroup, 155
    triangular chi, 185, 187, 277
    uniform, 50, 93
    von Mises, 207
    Wishart, 188, 280
Distribution form, 8
Dosage, 234, 238

ED50, 234
Events,
    identification, 136
Exactness, 138
Exhaustive property, 3
Experiment, 2
Experimental design, 56
Exponential model, 73
Extreme value
    see distribution

Gould data, 226

Haar measure
    see invariant measure

Index set, 50
Inference base, 4

Information
   no direct, 85
Imput variable
   random choice of, 59
Invariant group, 62, 66
Invariant measure
   left, 143
   on sample space, 142
   right, 144
Investigation, 1

Jacobians
   group, 145
   group element to inverse, 148
   location-diagonal scale, 169
   location scale, 142, 144, 148
   location-square scale, 193
   location-triangular scale, 177
   regression-square scale, 284
   regression-triangular scale, 272
   triangular to inner product, 188

Kronecker product, 279

Lieblein and Zelen data, 31
Life testing, 30
Likelihood, 98
   conditional, 101
   consistency principle, 102
   extended, 253
   integrated, 253
   invariant, 262
   marginal, 23, 101, 117, 150
      extended, 258
   profile, 100, 252
   section, 100
   transit, 263
Likelihood function, 70, 71
Likelihood map, 71
Likelihood modulation, 173
Likelihood principle
   strong, 99
   weak, 74
Location distribution
   location-scale, 21
   multivariate, 180, 195
   multivariate regression 275, 286
   on the circle, 211, 226
   on the sphere, 218
   regression, 116
Location-scale model, 16
Logistic
   see distribution
Logit function, 237

Measuring instrument, 56, 93
Mixed population
   sample from, 58

Model, 2
   definition of, 3, 4
   for the investigation, 3
   for the system, 2
   pivotal type, 12
   response based, 6, 7
   variation based, 11, 12
Modular function, 147
Multinomial
   see distribution

Necessary reduction, 5, 49
   by reexpression, 61
   factorization, 59
   on the parameter space, 49
   on the sample space, 54
   parameter component, 65
Normal
   see distribution

Options set, 50

Personal preferences, 5
Pivotal function, 13
Poisson
   see distribution
Position, 139
Presentation, 11
Probabilistic property, 3
Probit function, 237

Randomization, 2, 56
Reduction, 137
Reexpression transformations, 7
Reference point, 138
Regression model, 109
   nonnormal variation, 125
   normal, 119
   restricted variation, 51
   transformed, 263
Regression-scale transformations, 111
Reid data, 236
Resistance, 37
Robustness, 37
Rope spores data, 247

Sample size,
   random, 57
Sample survey, 56
Scale distribution
   location-scale, 22
   multivariate, 180, 196
   multivariate regression, 275, 287
   regression, 116
Serial correlation, 120
Specification, 1
Spurious observation, 42
Standardization, 9
Stiefel manifold, 186, 191

Stimulus-response curve, 236
Strength, 234, 238
Structural model, 133
   generalized, 255
Student
   see distribution
Sufficiency principle, 74
Sufficiency reduction, 70
   weak, 84
Sufficiency (ψ), 81
Sufficiency-ancillarity reduction, 82
Sufficient statistic, 71
   minimal, 72
System, 1
   random, 2

Tests of significance, 89, 151
   in sequence, 160
Tolerance distribution, 234
Transformations,
   location-scale, 9
   power, 261

Transformations (*continued*)
   regression-scale, 111, 269, 282
   vector location-scale, 168, 175, 192
True value, 4
Turtle data, 226

Uniform
   see distribution
Universal family, 85
Unknowns, 1

Variable
   controllable, 2
Variation, 11, 135
von Mises
   see distribution

Weibull analysis, 30
Wheat consumption data, 123
Wishart
   see distribution

RAYMOND H. FOGLER LIBRARY
DATE DUE